STATISTICAL ANALYSIS
OF GEOLOGICAL DATA

STATISTICAL ANALYSIS
OF GEOLOGICAL DATA

GEORGE S. KOCH, JR.
RICHARD F. LINK

John Wiley & Sons, Inc.
New York • London • Sydney • Toronto

Library of Congress Catalogue Card Number: 70–104190

ISBN 0–471–49690–1

Printed in the United States of America

10   9   8   7   6   5   4   3   2   1

Preface

All geologists who work with numbers use statistical methods, whether or not formally, and can profit from a knowledge of applied statistics. The purpose of this book is to explain some effective statistical procedures for the analysis of geological data and to discuss methods to obtain reliable data that are worth analyzing.

We write for the person who has numerical data from which he wants to draw conclusions or who has a problem whose solution will require obtaining and interpreting numerical data. We stress basic statistical methods and emphasize that thoughtful application of these methods will yield valid results. No statistical methods are introduced for their own sake but only because they have proved useful for data analysis. Because geology is a complicated and diverse science, we have purposely included some involved geological arguments. The mathematics is relatively simple, however, and requires for its understanding only elementary algebra and geometry.

This book is for readers with some geological training, who may or may not have professional experience. Although most readers may be geologists, the book will also interest mining and petroleum engineers, geochemists and geophysicists, mineral economists, and others.

The scope and arrangement of the book, outlined in Chapter 1, reflect our selection of the statistical methods that are most useful now, given the present state of geology, statistics, and electronic computing. The book is divided into two volumes. The first chapters, constituting most of Volume 1, place statistics first and geology second because the same statistical methods serve for varied geological problems. Most of the example analyses in this volume are chosen for numerical properties that illustrate specific statistical methods and are fictitious. Volume 2 places geology first and statistics second, and most of the examples are real ones for which the statistical principles introduced in Volume 1 are put to work, extended, and refined if necessary. Nowhere in the book is a large body of geological data analyzed completely because to do so would require a monographic presentation, too long for a textbook.

Although data for examples come from many fields of geology, most are from economic geology, and many are data that we ourselves have analyzed. We think that the reader will profit more from our experience with real data than from analyses of fictitious data or the second-hand data of others. This opportunity to learn from real analyses should offset the two disadvantages

v

that readers relatively unfamiliar with economic geology may need to invest a small amount of thought in order to make a transfer to the fields of geology that interest them and that our own work is mentioned more often than we would like.

We intend this book for the practicing geologist as well as for the student in a formal course. We have closely tied it to J. C. R. Li's book on applied statistics, which is excellent for self-instruction, and have adopted Li's notation. A page of notation follows the table of contents; we hope that readers familiar with other mathematical symbols will not be seriously inconvenienced.

Our collaboration on the statistical analysis of geological data began at Oregon State University in 1956 and has been continued since 1962 in the Mine Systems Engineering Group of the U.S. Bureau of Mines. This book was outlined and partly written in the spring of 1966, when we were visiting research statisticians at Princeton University.

We are grateful for financial support to Oregon State and Princeton universities, the American Philosophical Society, and the National Science Foundation.

Several mining companies, cited in the text, have generously supplied data. We thank in particular two Mexican companies which have, at some expense and inconvenience, furnished data. Henry B. Hanson, General Manager, Minera Frisco, S.A., has permitted us access to assay records, and his staff has been most cooperative in assembling the data. J. B. Stone, formerly General Manager, Compañia Fresnillo, S.A., and his successor Ing. Luis Villaseñor, also allowed us to copy assay records; G. K. Lowther, Chief Geologist, discussed Fresnillo geology with Koch on many occasions.

Many of the statistical analyses in the book were first made for the U.S. Bureau of Mines and published in the Bureau reports cited in the text. S. W. Hazen, Jr., formerly chief of the Bureau's Mine Systems Engineering Group, encouraged and aided us in this work. Many past and present members of this group have also helped us, particularly J. H. Schuenemeyer, G. W. Gladfelter, and Velma Sturgis in computer programing. This book is not an official statement or representation of the U.S. Bureau of Mines, however.

Many of the geologists and statisticians cited in the text have helped us by discussions as well as by their published works. We have also profited from the criticisms of the following men who read all or part of Volume 1: J. C. Davis, R. C. Flemal, David Hoaglin, D. F. Merriam, A. T. Miesch, D. B. Morris, C. W. Ondrick, G. S. Watson, and Alfred Weiss. The errors and misjudgments that remain are ours alone.

We have also been helped by librarians of the U.S. Geological Survey, Denver, Colorado; the Colorado School of Mines, the Denver Public Library, and especially by Samuel Shephard, librarian of the U.S. Bureau of Mines library in Denver. G. W. Johnson and Sally Konnak improved the English expression and organization of the book by careful editing. The book was typed by Vicky Yen Contreras and Verna Bertrand.

Finally, GSK would like to acknowledge a personal debt of gratitude to L. C. Graton, who taught him the importance of careful observation and verification of scientific data, to Ernst Cloos, who taught him to draw conclusions from numerical measurements, and to the late H. E. McKinstry, who taught him that information of fundamental geologic interest resides in the numerical data of the mining industry and urged him, in 1952, to begin the investigations that eventually led to this book.

Similarly RFL would like to acknowledge the help and inspiration of W. J. Dixon, who introduced him to statistics and computing, and to J. W. Tukey, who contributed materially to his development in statistics and data analysis.

George S. Koch, Jr.
Richard F. Link

Denver, Colorado
New York, New York
February 1970

Contents

STATISTICAL ANALYSIS
OF GEOLOGICAL DATA

PART I

INTRODUCTION

Chapter 1

Introduction

Every geologist obtains numerical data—the number of hand specimens collected from a formation; strike and dip of a bed or fault; chemical rock analyses; porosity measurements; assays of ore, coal, or oil; etc. And every geologist summarizes these numerical data when he prepares the report that, as an essential part of the scientific process, communicates his findings to others. He summarizes because it is usual in the discipline of geology to obtain many more data than can be reported verbatim; for instance, a geologist may omit strike and dip measurements on a quadrangle map because they are too crowded, or he may average chemical or modal rock analyses. Exactly what he summarizes, and how, will determine much of the value of the report.

The geologist who has substantial amounts of data must use statistics—whether formally or informally. Our thesis is that he can sharpen his thinking and improve the reliability of his conclusions through purposefully devised statistical methods that yield incisive results. He may thus avoid the end effect too often produced today—tables of uninterpreted or misinterpreted numerical data, attached only as ornaments to his report.

Although an increasingly large number of geologists study statistics, a gap remains between academic study and useful practical application. Most text-books on statistics stress such problems as drawing black and white marbles from urns, measuring the life of light bulbs, and counting beer bottles that leak. Problems of time and space relations, problems that are at the heart of geological thinking, are seldom found. Therefore it is difficult for geologists to make the translation from statistical theory to their immediate problems.

In this book we emphasize data analysis rather than the application of formal statistics. Although the geologist interested in applying statistics to geology must learn some formal techniques, it is even more important for him to develop taste and judgment. We would rather explain the effective allocation of effort to real problems than develop complicated analyses that are likely to be mathematically unstable and computationally unreliable. Our point of view owes much to J. W. Tukey (1962, p. 2), who has written the following:

> I have come to feel that my central interest is in *data analysis*, which I take to include, among other things: procedures for analyzing data, techniques for interpreting the results of such procedures, ways of planning the gathering of data to make its analysis easier, more precise or more accurate, and all the machinery and results of (mathematical) statistics which apply to analyzing data. Large parts of data analysis are inferential . . . but these are only parts, not the whole. Large parts of data analysis are incisive, laying bare indications which we could not perceive by simple and direct examination of the raw data, but these too are only parts, not the whole. Some parts of data analysis, as the term is here stretched beyond its philology, are allocation, in the sense that they guide us in the distribution of effort and other valuable considerations in observation, experimentation, or analysis. Data analysis is a larger and more varied field than inference, or incisive procedures, or allocation.

1.1 SCOPE AND ARRANGEMENT OF THE BOOK

Because we write for those who want to apply statistics to geology, we assume that our readers know some geology, but not necessarily any statistics. The book starts from elementary statistical principles that are sufficiently developed to be put to work on geological data. No attempt is made to survey applied statistics or to review the literature on statistics applied to geology.

Although the discussion of statistical methods starts with first principles, it sometimes ends in specialized procedures not treated in elementary statistics books. Only those statistical methods are introduced that advance interpretation of geological data and are needed to explain a geological problem. Many numerical examples, most of them based on real geological data, are calculated. Simple algebraic arguments are given but complicated proofs requiring calculus are omitted. Even the reader who knows statistics will find it helpful to scan the first, elementary chapters to become acquainted with our viewpoint and notation.

All of the above is within the scope of the book. In the next paragraphs, some related subjects that are not treated in the book are mentioned, and works that explain them are suggested.

Because we are concerned with selected statistical subjects, a broad exposition of statistics, even on an elementary level, is not attempted. Of the many excellent textbooks on statistics, two of the best for geologists are by J. C. R. Li (the two volumes are referenced as 1964, I; and 1964, II) and Dixon and Massey (1969). Li's book (especially vol. I) is well suited to self-instruction. It is written for scientists and engineers and explains statistics in great detail for readers whose mathematics is limited or rusty. Anyone who reads one or both of these books and works the problems will obtain a good grounding in the basic statistics useful to a geologist.

We do not discuss mathematical statistics, a subject explained in many books, for example, the introductory books by Hoel (1962 and 1966) and an advanced book by Wilks (1962). Probability, a central subject in theoretical statistics, is discussed here only briefly; the reader interested in this subject is referred to an elementary text by Mosteller and others (1961) and an advanced book by Feller (vol. I, 1957 and vol. II, 1966).

Five excellent books on statistics in geology cover some additional topics not touched upon in this book. Miller and Kahn (1962) provide an excellent summary of the literature up to 1962, after which so many papers have appeared that a full review volume could contain little else. Besides reviewing the literature, Miller and Kahn discuss such subjects as probability, probability density functions, and paleobiometrics, to all of which we devote little attention. Krumbein and Graybill (1965) stress model formulation and draw most of their examples from sedimentation and oil geology. In *U.S. Bureau of Mines Bulletin 621*, Hazen (1967) covers mining technology, special kinds of sampling, and size distributions of particles in ores and rocks. Griffiths (1967) discusses statistical methods of studying sediments, but his book has wider application, contains statistical methods not mentioned in our book, and offers perceptive comments on the science of geology. Finally, Harbaugh and Merriam (1968) emphasize statistical methods implemented by electronic computers for studying stratigraphy.

This book is divided into two volumes and six parts, three in each volume. Part I is this first introductory chapter. In Part II, which covers Chapters 2 to 6, univariate statistical methods for the analysis of single variables are explained and illustrated through examples; only enough geological data are introduced to help explain the statistics. Part III, which is Chapters 7 and 8, takes up geological sampling and variability. In Part III the formidable problems of data interpretation as it pertains to geology are stressed, and attention is given, therefore, to geological as well as statistical problems of data analysis. Part IV, comprising Chapters 9 to 11 in Volume 2, develops multivariate statistical methods for the analysis of two or more variables. Part V, which includes Chapters 12 to 16, is about the statistical analysis of data from applied geology, mainly mining geology; the chapters pertain to exploration for natural resources and their valuation, to decision making through operations research, and to some specialized methods of sampling and data analysis. Finally, Part VI, or Chapter 17, is about the use of electronic computers to implement the statistical methods.

1.2 PROBLEM SOLVING

The geological investigation that statistics is likely to benefit is that which focuses on solving a specific problem. Good advice, simply summarized, is as follows: have a goal, plan ahead, and use any valid geological and statistical methods, but only those methods that will lead to that goal.

First and foremost, the investigator should think carefully about the problem he is about to pursue: the scope, limitations, ramifications, and objectives. Then through appropriate methodology he may construct that type of hypothesis known as a model (sec. 1.5), purposefully collect data to test the hypothesis, and solve the problem within stated limits of reliability—perhaps in the process revising his model in the light of the preliminary results and then, if necessary, collecting additional data to refine and to verify his conclusions. Thus, he will neither become overwhelmed with enormous amounts of data, nor fail to collect that which is pertinent. He will be able to reach an effective solution to a specific problem while keeping within limits of cost—whether reckoned in money or in time, or in both.

An example of the success to be achieved by striving for one objective is the exploration program that resulted in the discovery in the United States after World War II of large new uranium deposits by government and industry geologists. Much of the new uranium was found by systematic search in environments deemed unfavorable by theoretical geology. It is interesting to note that later, when the data collected primarily for finding ore were used for secondary purposes, such as stratigraphic and geochemical studies, they were found to be not wholly suitable. Thinking in retrospect, some of the investigators wished that, for the new studies directed at different problems, they had collected new data.

In problem solving, rather than being tied only to traditional geology, one should accept the imaginative, integrated use of any and all sciences and engineering fields that have anything to contribute. It is in this spirit that this book includes approaches from a number of disciplines. Cloud (1964) writes in a similar vein:

What is most characteristic philosophically, and most gratifying to me personally, about the earth sciences today is their blending of the useful parts of classical science with the most exciting aspects of advancing science. . . . All of the systematic sciences, and I use *systematic* in the broad sense of classifying *and* explaining, are in a state of ferment as new equipment, new measurements, and improved computer facilities provide different and in some instances more fundamental bases for classification and rapid quantitative methods of evaluation—this is true, not only of mineralogy but also of paleontology and petrology. . . .

If we were to tabulate the things that most generally characterize the earth sciences in the modern world we might include:

1. A growing restiveness with traditional methods of investigation.
2. An increasing tendency to express observations and conclusions quantitatively wherever it is possible to do so.
3. An increasing degree of interaction with other sciences.
4. The assumption or requirement of an increasing degree of familiarity with mathematics as a form of communication, and with physics, chemistry, and biology. . . .
5. A high degree of sophistication of instrumentation that is increasing our resolving power in all fields and permitting us to make new observations and discoveries, both in new fields and in fields that once appeared on the verge of foreclosure.

William Hambleton (1966), associate director of the Kansas Geological Survey, takes a similar view toward the work of state geological surveys in particular. He writes as follows:

As to the program of a modern survey, one might say that it is characterized by change, urgency, and involvement in the social, economic, and political problems of the state and region which it influences. . . . One hundred years ago, the purpose of our organization was to survey the mineral resources of the state. Today, it is an instrument of economic development; its research serves to catalyze activity in the mineral industries; and its plans, purposes, and programs are characterized by innovation. For a number of years we have engaged in projects involving computer techniques for fundamental geologic problems. . . . For the past several years we have involved people from geophysics, statistics, petroleum engineering, mining engineering, geology, and econometrics in studies to develop methodologies for regional economic analysis.

In other words, we are dealing increasingly with the whole field of systems analysis and operations research in the economics of the mineral industries. We are launching urban development and environmental geology studies that relate to the problems of environmental health, transportation, land use, and urban and regional planning. . . .

Most of the studies that I have mentioned emphasize change and new directions. The methods used involve transference of ideas from one discipline to another. The systems approach of the engineer is evident; we have drawn heavily upon management science, biometrics, the correlation techniques of psychology, mathematics and statistics, as well as chemistry and physics.

1.3   THE NATURE OF GEOLOGICAL DATA

Geology differs from the experimental sciences in that most geological data are fragmentary and are derived from the surface manifestations of natural processes that are uncontrolled by the investigator. When a geologist inspects a particular place on the earth, he finds a unique situation developed over a long time through more processes than he can take into account. He cannot erase the natural processes that produced this environment and do it over with a simpler, controlled laboratory experiment. Nor can he observe the natural processes; most of them are finished, and the results are fixed. Furthermore, he finds that the natural processes worked to destroy or remove part of the evidence. And most of the remaining evidence is buried inaccessibly deep in the ground, while the surface outcrops are contaminated by water, weather, and the works of mankind.

So the geologist must make do with the data available, which are seldom those with which he would prefer to work. It is here that statistics may enable him to plan data collection and deduce inferences that are not readily discernible from the raw fragmentary observations that he collects.

Despite these difficulties the investigator will find before him ten thousand times more potential data than he can collect, most of it useless to his purpose. He must, first, be aware that this is the situation and then be selective in choosing the data he will collect. Sampling guided by statistical design will serve him well in choosing these data. For this reason, designed sampling and deliberate selection of data are the major concerns of this book. The available material and the purpose govern the design, the design governs the sampling, and sampling provides the data from which valid conclusions may be inferred.

The sampling results—raw data, derived observations, and conclusions—should be verified before being accepted as valid information on which to base an analysis. Verification, an essential requirement of the scientific method, is too often done informally or not at all in geology. The reader has undoubtedly visited areas whose geology does not correspond to the published maps and has searched for fossil or mineral localities whose positions are inadequately described. Even "quantitative" data such as rock analyses are often suspect, as the well-known silicate rock studies of the U.S. Geological Survey showed ("B-1" and "G-1" rocks, R. E. Stevens and others, 1960). Yet with reasonable care verification is possible. Duplicate or replicate samples or specimens can be sent to two or more analysts, petrographers, or paleontologists. If ore is sold, the mine's assay is verified by the smelter or by an umpire if buyer and seller disagree.

Classes of Geological Data

Of the many ways to classify geological data, one of the most suitable for data analysis is according to the method of collection. Four classes of data can be distinguished by the *operations* used to collect them. They are: *measurement, counting, identification*, and *ranking*.

One class is primarily derived from *measurement*, involving such operations as measuring the deflection of a needle on a dial, the width of an x-ray line, or the thickness of a sedimentary bed. Examples are compass measurements of the strike and dip of a plane or of the bearing and inclination of a line; distance measurements by scale, tape, or alidade; and microscopic measurements, including optic angle, extinction cleavage angle, and index of refraction.

A second class of data is derived primarily from *counting*. Examples are the number of zircon grains recognized in a microscope traverse, the number of right- and left-handed clam shells found in a study of life-and-death fossil assemblages, and the number of oil fields in a state.

In a third class of geological data the interest is in *identification*. For instance, in the discovery of a fossil, such as the first brachiopods found in metamorphic rocks in New Hampshire (Billings, 1935), the fact of discovery and the veracity of identification are important rather than the number of specimens or other numerical data. Many data about rocks and minerals are recognized intuitively much as one recognizes the face of a friend, rather than by a formal procedure. In this stage, common in field work, counting and measurement may be unnecessary. Just as the physician will diagnose at the bedside and later confirm his diagnosis through laboratory tests, so will the geologist confirm field identifications through tests, such as laboratory measurements on thin sections or on powders.

To these three classes of data may be added a fourth: data that can be *ranked*, where a scale of measurement is difficult if not impossible to assign. Examples of data in this category are color descriptions, "favorability" of rocks for oil or ore, and a ranking of oil fields or mineral prospects in the order of which to drill first.

Thus, as Philip Frank (1957, pp. 311 ff.) explains in his excellent book, *Philosophy of Science*, when he discusses operational definitions, scientific data are derived from many kinds of human activity and cannot be divorced from the methods of collection. Regardless of the philosophical concept of the geologist and the elegance of his scientific theory, the data to support a study are obtained by some person reading a needle on a compass dial, counting pebbles, or performing some other physical operation. The choice of operations and the details of how they are performed may be of as much or more consequence than the philosophy and theory of the geologist.

Sources of Data

Many of the difficulties in statistical analysis of geological data stem from the data sources. In science, data are derived ideally from controlled experiments rather than from uncontrolled natural processes. The monitoring of seismic quakes resulting from nuclear explosions at the Nevada test site is an example of a new and controlled experiment. In geology most data are obtained either from uncontrolled natural processes such as volcanic gases and earthquakes, or, more often, from the results of events concluded in the geologic past, for instance from basalt flows and granitized rocks. The data collected may therefore be regarded as coming from already completed as well as uncontrolled experiments. Furthermore, the data that must be used may have been collected years ago under circumstances that today are often unclear. It is a tribute to the geologist that he can make any sense at all from such inferior data. Yet such an investigation may yield valuable results, as when Chayes (1963) studied analyses of basalt made in the past from which he was able to draw valid conclusions.

Because the scientific method demands that data be verified and because the inferences of the geologist working in construction, or in oil or mineral extraction will become apparent during implementation of his decisions, the reliability of data must always be considered. This is a familiar exercise for geologists, who customarily distinguish observed, inferred, and indicated contacts in geologic mapping; differentiate proved, inferred, and indicated ore; and predict volcanic eruptions and earthquakes.

Although the geologist who uses old existing data is restricted in the breadth of his statistical analysis and is often doubtful of the validity of his conclusions, the geologist who plans to collect new data is in a better situation. He can choose those kinds of data that will illuminate a problem. He can also seek kinds of data that are easy to obtain and as useful as data that are difficult to collect. This subject is further developed in section 5.11 of this book. Whitten (1964) has also written about it.

Still another distinction about geological data has been made by Krumbein (1962, p. 1088), who contrasts *observational* data of geology, comprised of the qualitative or quantitative data of natural objects obtained in the field or laboratory, with *experimental* data obtained only in a laboratory. In the laboratory, with data obtained under controlled experimental conditions, a geological problem must be highly simplified. Indeed one experimentalist has told us, in a pessimistic moment, that the only chemical systems about which he feels confident are those of no geological interest! Krumbein (1962, p. 1088) provides an instructive table contrasting observational and experimental data.

Data Analysis

We are often approached by a geologist with a thick pile of paper representing the results of a large drilling campaign and are asked to interpret them statistically. Often we can offer very little help because the data are diffuse and the problem ill-defined. With the advent of new methods for gathering data rapidly—by electronic well logging, rapid rock analyses, and aerial geophysics—a geologist can collect data very quickly. In a recent year, one laboratory alone (the Exploration Research Laboratory of the U.S. Geological Survey in Denver, Colorado) made one and a half million chemical analyses of geological samples. Unless a geologist takes the time and has the training to plan data collection and interpretation, he is soon submerged in a mass of information, and cataloging alone will take most of his time.

For data analysis to be effective the geologist must be willing to manage the data gathering and analysis from start to finish. He must be alert for the odd findings that signal the unusual; that is, he must be serendipitous. Nothing is worse than a too-systematized data collection, such as providing printed forms with no space for extra comments, so that valuable data are neglected only because they do not fit a preconceived classification. This point is well made by Stephan and McCarthy (1958), who, in discussing the social sciences, stress that the principal investigator or project leader should do some of the data collecting himself.

Many geologists are making excellent analyses of data, with and without formal statistics. Any one of several books will give good examples of proficient analysis of geological data. Besides the books primarily on statistics in geology (sec. 1.1), three recent books are outstanding in their analysis of geological data: Irving (1964) on paleomagnetism, Ager (1963) on paleoecology, and Leopold and others (1964) on geomorphology. Among the best of the older books is one on mining geology by McKinstry (1948).

1.4   SOURCES OF DATA FOR EXAMPLES OF ANALYSES

In principle, numerical data for examples of analyses are available from most if not all fields of geology. In practice, suitable published data are hard to find, although more appear every year. Numerical data are common in hydrology and in special studies in paleontology and geomorphology. From sedimentology come data on size and shape characteristics of particles. There are many data in petroleum geology, although those most interesting are locked up in company files. Petrology also furnishes many data, although those potentially most interesting are rock analyses difficult to compare because of marked inconsistencies from one laboratory to another.

Although some example data in this book come from these fields, most are from mining geology, for two reasons. First, we have been analyzing mining data for long enough to deal with them with some assurance. Second, numerical data are abundant in mining geology. Mining geologists are concerned with sampling in order to determine whether an ore deposit is rich enough to mine and, if so, how the mine should be designed. Because of the large variability within a set of mine data and the substantial amounts of money that will be spent as a result of their conclusions, mining geologists require more data than other geologists and usually more data than investigators in other fields of the natural and social sciences.

Although many of our example data come from mining geology, the reader can readily interpret the statistical analyses in terms of any field of geology in which he is interested; for instance, our studies of distribution of metals in ore deposits are, in principle, like those in quantitative petrology. To make the transition easier, data are introduced from other fields of geology. Made-up data are designated "fictitious illustrative data" to separate them from real data.

Several examples are drawn from investigations based on assay data from the Frisco and Fresnillo mines in Mexico. Because these studies are repeatedly referred to, the geology of the ore deposits and some of their problems are summarized for convenience in this section. Other sources of data for examples are characterized briefly the first time they are mentioned.

The Frisco Mine

The Frisco mine, located in San Francisco del Oro, Chihuahua, Mexico (fig. 1.1), is 550 kilometers south of El Paso, Texas. The mine yields monthly some 70,000 metric tons of ore with a grade of 0.4 gram (0.01 ounce) gold per metric ton, 180 grams (5 ounces) silver, 5 percent lead, 0.6 percent copper, and 8 percent zinc. The primary ore minerals visible without a microscope are galena, chalcopyrite, sphalerite, and fluorite. The ore is produced from about 70 typical fissure veins that cut calcareous argillite of Mesozoic age. The mine occupies a block of ground about 3 kilometers long, 1 kilometer across, and 800 meters deep. Details are given in a general account of the geology (Koch, 1956).

The data that we have studied are assay results from 19,050 chip samples cut at 2-meter intervals in all drifts on mine levels 7, 10, 13, 14, and 15 and from 7676 mine samples cut at the same intervals in all drifts, raises, and

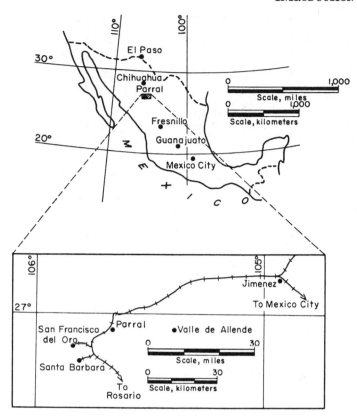

Fig. 1.1. Location map. Frisco and Fresnillo mines.

stopes on the Don Tomás and Brown veins. These levels were selected to provide information on changes in mineralization in depth, from level 7, one of the upper levels, to level 15, the lowest developed level. The Don Tomás and Brown veins were chosen as typical of large veins in this mine.

The general pattern of veins in the Frisco mine is shown in composite plan in figure 1.2 and in cross section in figure 1.3. A longitudinal section of the Don Tomás vein appears as figure 4.11.

The mining company assayed each sample for gold, silver, lead, copper, and zinc; the variables that we calculated and analyzed statistically are vein widths, metal contents (calculated by multiplying grades by vein widths), logarithms of metal contents, and logarithms of metal ratios (except those formed with gold, which are mostly indeterminate because most gold assay results were zero). Metal contents are measured in meter-grams per metric ton for precious metals and in meter-percent for base metals. The reasons

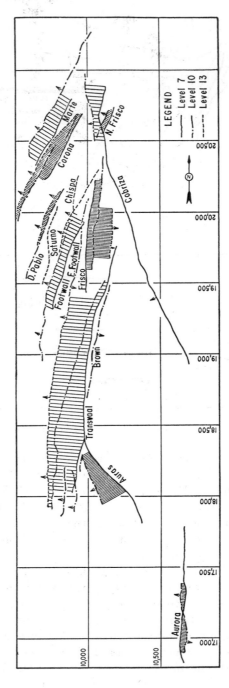

Fig. 1.2. Frisco mine. Composite plan of drifts following veins on levels 7, 10, and 13. (For clarity most short drifts and a few long ones are omitted.)

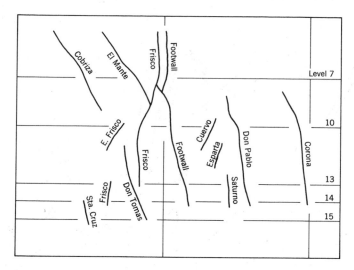

Fig. 1.3. Frisco mine. Vertical, east-west cross section at N-20,000. Observer looking south.

for working with metal contents rather than with unweighted assays are reviewed in section 13.2.

The Fresnillo Mine

The Fresnillo mine is at the city of Fresnillo (fig. 1.1), near the center of the state of Zacatecas, Mexico. Fresnillo is on the Pan-American Highway, 766 kilometers (476 miles) north of Mexico City, and 1316 kilometers (818 miles) south of El Paso, Texas.

The Fresnillo mine has been developed below vein outcrops on Proaño hill, which rises 100 meters above the general level of the plain. For the most part, the Fresnillo veins strike northwestward and dip northeast or southwest at angles ranging from 50 degrees to nearly vertical. There are 6 major and more than 50 minor veins. The veins cut graywackes, carbonaceous and calcareous shales of Mesozoic age, and sediments of Tertiary age. The primary ore minerals are galena, chalcopyrite, sphalerite, pyrargyrite, and other silver sulfides. The general geology of the mine has been described by Stone and McCarthy (1948).

The Fresnillo data that we have studied are assay results from 16,400 chip samples cut at 2-meter intervals in all drifts on the 2137, 2137 Footwall, 2200, 2630, and Esperanza veins. We treat the data in the same manner as the Frisco data described in the preceding subsection. The general pattern of these five veins is shown in composite plan in figure 1.4, in cross section in

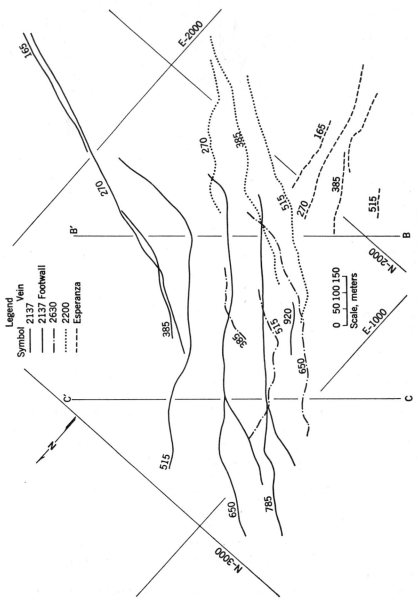

Fig. 1.4. Fresnillo mine. Composite plan of five veins.

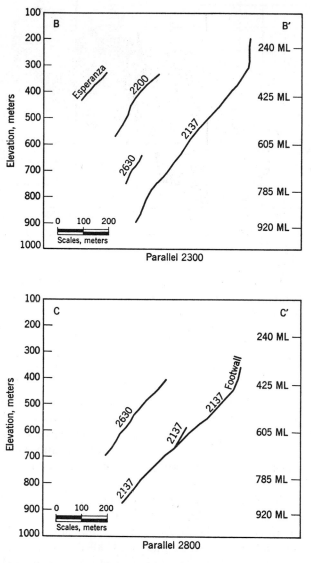

Fig. 1.5. Fresnillo mine. Vertical cross sections at Parallels 2300 and 2800. Observer looking northwest.

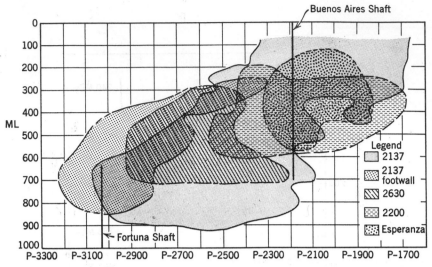

Fig. 1.6. Fresnillo mine. Composite longitudinal section. Observer looking northeast.

figure 1.5, and in composite longitudinal section in figure 1.6. The relation of these five veins to the rest of the mine is shown by illustrations in the paper by Stone and McCarthy (1948).

1.5   MODELS

A *model* is a representation of a natural phenomenon or process. Because the concepts and usefulness of models are not familiar to all geologists, elementary ideas are introduced in this section and expanded and refined in subsequent chapters. Although models may be defined and classified in many ways, we distinguish three main kinds while also explaining some principles of model formation.

A *physical model* is a tangible object representing a natural phenomenon or process. Some examples are the model of the Mississippi River system in the U.S. Corps of Engineers laboratory in Vicksburg, Mississippi; clay models simulating rock deformation; wax models of salt domes; and atomic models made of wood and wire. The advantage of a physical model is that a part of nature whose elements can be manipulated is isolated for study. Providing that the proper part of nature is selected, the principal difficulty in using a physical model is that the model is usually only a scaled representation of nature. For physical models in geology, the most troublesome scale changes are liable to be in size and time. Other scale changes, such as in temperature and pressure, may also be large but are usually easier to control.

A *geologic model* is an abstract formulation of a geologic concept that may be tested by collecting geologic data. For example, the model has been postulated (Gross and Nelson, 1966) that the change in radioactivity of present-day sediments with increasing distance from the mouth of the Columbia River is due solely to radioactive decay and that there is no mixing of material from the Hanford, Washington, nuclear reactor with previously deposited sediments. Another geologic model is the concept that every oil field consists of three essential elements: a trap, a seal, and a reservoir rock. Still another is the hypothesis that metals in some hydrothermal ore deposits, such as that at Butte, Montana, were deposited under a temperature gradient decreasing outward from a hot center.

Physical and geological models are the bases for devising mathematical models. A *mathematical model* is a set of formal rules defining exact relationships among variables in order to describe the essential elements of a physical process for a specified range of conditions. Often the relationship can be expressed in equations with mathematical symbols. An example of a mathematical model is Boyle's gas law that pressure multiplied by volume is equal to a constant, for a certain pressure range. Another example is Gross and Nelson's (1966) assertion that the relationship between radioactive concentration in sediments and the distance from the Hanford reactor is defined by the linear equation

$$y = a + bx,$$

where y is the activity ratio of Zn^{65} and Co^{60}, x is distance, and a and b are experimentally determined constants.

From the several subtypes of mathematical models, it is convenient to define two. One is the *deterministic model*, in which the relationship among variables is completely predictable so that, if one or more are known, the other or others can be exactly calculated. An example is the previous equation for radioactive concentration in sediments. The other is the *statistical* or *stochastic model*, in which the relationship among variables is not entirely predictable because a *random* or *chance element* is added to an otherwise deterministic model. An example of a random element is the "experimental error" introduced in measuring distances in surveying or in making weighings on an analytical balance. The formal definition of a random element and its relation to a mathematical model requires complicated reasoning that is explained as the subject is developed.

The purpose of a model, whether physical, geological, or mathematical, is to abstract, simplify, and organize reality in order to focus attention on one or a few factors in a geological situation. Thus the geologist, rather than trying to describe every small detail, can use a model to define a specific problem for which a reliable solution can be found.

Models, particularly geological and physical ones, have been used in geology since earliest times. However, they have been used less than in the simpler physical sciences, because to be of much value, they must deal with conceptually complex natural phenomena. Nonetheless, modern geology is making the first steps toward wide application of models, particularly mathematical ones.

To clarify the definitions and to illustrate the sequence of models that may be used in a geologic problem we discuss a simple example which utilizes three types of models, one after another. The example comes from a study of shearing stress in soil. The aim of this investigation was to prevent the failure of engineering structures caused by soil movement. The example uses both a physical apparatus and a mathematical formula of the sort developed by Terzaghi (1948; the book includes references to the original works) and other pioneer students of soil mechanics.

The first step is to devise a geological model. A list of variables and some tentative relationships among them is drawn up. Some of the variables that may affect shearing stress in soils and cause soil failure are to be found in a list of questions such as these:

1. Is the type of soil important?
2. Is the pressure on the soil a factor?
3. What is the role of water and other natural fluids?
4. Must the force applied to the soil reach a definite value before failure occurs?
5. At what displacement of the soil does rupture take place?
6. Is time a factor?
7. Does the temperature make a difference?

From these questions one may develop a geological model. Terzaghi's model was the hypothesis that shearing stress in soils is determined mainly by pressure, water content of the soil, displacement of the soil, and time. These factors can be regulated with Terzaghi's (1948, p. 79) direct-shear apparatus, one of the many devices, or physical models, used to study soil failure.

In the direct-shear device (fig. 1.7) the soil sample is placed in an apparatus consisting of a movable upper frame and a fixed lower frame. The sample is contained on the sides by the two frames and on the top and bottom by two porous, grooved stones which function to prevent slippage of the soil and to allow water to be drained from moist soils. The shearing force is applied by pulling the upper frame of the shear box. As the displacement of the upper

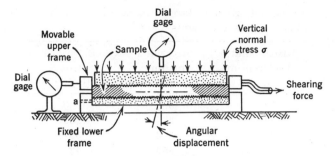

Fig. 1.7. Direct-shear apparatus (after Terzaghi, 1948, p. 79).

frame increases the force required to increase the displacement increases and approaches a maximum, as shown by the graph in figure 1.8.

By varying the experimental conditions one may use the physical model to collect data like that graphed in figure 1.8. Different types of soils with various moisture contents can be used. Different pressures can be applied by varying the load on the upper stone. Shearing force can be applied in various ways, either by a steady pull or by increments, and the time taken to apply it can be fast or slow. Temperature can be changed.

Consider the results of one series of tests (Akroyd, 1957). Horizontal pull was applied in increments, each corresponding to a horizontal displacement

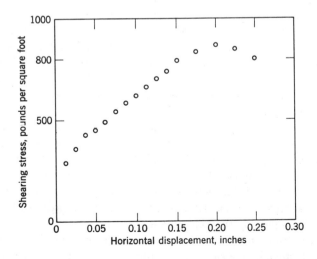

Fig. 1.8. Graph of shearing stress versus horizontal displacement in direct-shear apparatus (data from Akroyd, 1959).

TABLE 1.1. EXPERIMENTAL DATA OBTAINED IN A
DIRECT-SHEAR APPARATUS *

Horizontal displacement (in.)	Stress dial reading (1/10,000 in.)	Shearing stress (lb/ft^2)
0.0	300	0
0.0125	334	289
0.0250	342	357
0.0375	350	425
0.0500	353	451
0.0625	358	493
0.0750	364	544
0.0875	369	587
0.1000	373	621
0.1125	378	663
0.1250	383	706
0.1375	387	740
0.1500	393	791
0.1750	398	833
0.2000	402	867
0.2250	400	850
0.2500	394	799

* Data from Akroyd, 1957.

of 0.05 inch per minute. A stress dial on an instrument called a proving ring
performed like a spring balance to measure the horizontal stress in the system
at each displacement interval, as recorded in column 2 of table 1.1. After the
initial reading of 300 was subtracted, the stress dial reading was multiplied
by a constant to obtain the shear stress in column 3 of the table. The constant
was found by dividing the proving-ring constant, 0.33/0.00001 inch, specific
to the instrument, by the area of the sample in square feet, 0.3882, to yield a
result in pounds per square foot.

When the experimental data in table 1.1 were plotted in figure 1.8, the
maximum shearing stress was found to occur at a horizontal displacement of
0.2000 inch. The purpose of plotting figure 1.8 was to obtain this maximum
point, which could also have been found, at least in principle, by means of a
maximum reading dial like that on a maximum reading thermometer. The
shear stresses listed in table 1.2 and plotted against pressure in figure 1.9
were found by repeating the experiment four times.

These data from the physical model may be interpreted in terms of a
mathematical model. One such model, developed by the early investigators
in soil mechanics, may be written simply as

$$w = (\tan \theta)x,$$

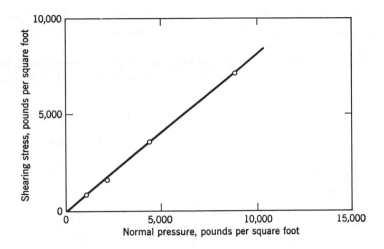

Fig. 1.9. Shear stress versus pressure (data from Akroyd, 1957).

where w is shear strength, x is pressure, and θ, named the angle of internal friction, is the angle between a straight line through the data points and the x axis (fig. 1.9). From similar experiments on cohesive rather than non-cohesive soils, a more general model can be formulated as

$$w = K_1 + (\tan \theta)x,$$

where K_1, a constant for a particular soil, is named the coefficient of cohesion.

The formulation of the mathematical model in the preceding paragraph is rather arbitrary. An alternative formulation is

$$w = K_1 + K_2 x,$$

TABLE 1.2. SHEAR STRESSES OBTAINED BY REPEATING THE EXPERIMENT OF TABLE 1.1 FOUR TIMES [*]

Experiment number	Normal pressure (lb/ft^2)	Shear stress (lb/ft^2)	Bulk density (lb/ft^2)	Moisture content $(\%)$
1	1100	869	115	29.0
2	2200	1603	116	26.8
3	4400	3580	118	29.1
4	8800	7090	118	27.8

[*] Data from Akroyd, 1957.

where $K_2 = \tan \theta$. However, the first formulation is more meaningful because $\tan \theta$ is the angle of repose of dry, loose sand and therefore affords a tie to nature.

The mathematical model that has been described is deterministic. It can be rewritten as a statistical (stochastic) model by adding a random fluctuation to form the new equation

$$w = K_1 + (\tan \theta)x + e,$$

where e is the random fluctuation or "experimental error," including factors such as temperature and type of soil packing not explicitly accounted for in the model. In the example data, random fluctuation was reflected by the failure of the graphed points (fig. 1.9) to lie exactly on a straight line, a discrepancy discussed in detail in section 9.2.

The simple shear device is only one of the many models used in geology. Other models are described in the following chapters and the reader will be able to think of many more, even though they may not have been formally or explicitly defined; for instance, the method of multiple working hypotheses of T. C. Chamberlain (1897) is essentially a geologic model.

In choosing between a deterministic or a statistical model for a mathematical study of a geological problem, one relies on taste and judgment, influenced above all by one's view of the real nature of the world. An example of a deterministic model is that for the accumulation of sand in dunes, the geometry of which may sometimes be rather well specified if wind direction, velocity, and distribution of sand-grain sizes are measured. Another example is the formation of coral reefs, which grow at a fairly constant rate if water temperature, salinity, etc., are constant. Examples of statistical models are those that describe the distribution of gold in conglomerate at Witwatersrand, South Africa, the dispersal of fossil populations, and the occurrence of earthquakes in California. These examples are each in a sensible category, but any one of the deterministic models may be regarded as statistical, and vice versa, to suit the purpose of the investigator.

Once a model is formulated, its correspondence to the real world is tested by gathering and analyzing data. Usually, the model turns out to be less than ideal, and the process of data collection itself suggests a better model. The sequence is like that followed by the field geologist who gathers data during the field season, interprets it the following winter, and in the next field season collects new data to test a revised hypothesis.

Models are discussed throughout this book. The first extensive and explicit treatment is in section 5.11 on linear models, and the next is in section 8.6 on experimental designs.

An outstanding proponent of using models in geology is W. C. Krumbein, whose book, *An Introduction to Statistical Models in Geology* (1965), coauthored by F. A. Graybill, summarizes thinking developed through many years. In Chapter 2 of their book models in geology are introduced with a somewhat different set of definitions and with many references to the literature.

REFERENCES

Ager, D. V., 1963, Principles of paleoecology: New York, McGraw-Hill, 371 p.

Akroyd, T. N. W., 1957, Laboratory testing in soil engineering: London, G. T. Foulis & Co., 233 p.

Billings, M. P., and Cleaves, A. B., 1935, Brachiopods from mica schist, Mt. Clough, New Hampshire: Am. Jour. Sci., 5th ser., v. 30, no. 180, p. 530–536.

Chamberlin, T. C., 1897, Studies for students: the method of multiple working hypotheses: Jour. Geology, v. 5, p. 837–848.

Chayes, Felix, 1963, Relative abundance of intermediate members of the oceanic basalt-trachyte association: Jour. Geophys. Research, v. 68, no. 5, p. 1519–1534.

Cloud, P. E., Jr., 1964, Earth science today: Science, v. 144, p. 1428–1431.

Dixon, W. J., and Massey, F. J., Jr., 1969, Introduction to statistical analysis: New York, McGraw-Hill, 638 p.

Feller, William, 1968, An introduction to probability theory and its applications, v. 1: New York, John Wiley & Sons, 509 p.

————, 1966, An introduction to probability theory and its applications, v. 2: New York, John Wiley & Sons, 626 p.

Frank, Philipp, 1957, Philosophy of science: Englewood Cliffs, N.J., Prentice-Hall, 360 p.

Griffiths, J. C., 1967, Scientific method in analysis of sediments: New York, McGraw-Hill, 508 p.

Gross, M. G., and Nelson, J. L., 1966, Sediment movement on the continental shelf near Washington and Oregon: Science, v. 154, p. 879–885.

Hambleton, W. W., 1966, Education of geologists for geological surveys: Jour. Geol. Education, v. 14, no. 3, p. 83–86.

Harbaugh, J. W., and Merriam, D. F., 1968, Computer applications in stratigraphic analysis: New York, John Wiley & Sons, 259 p.

Hazen, S. W., Jr., 1967, Some statistical techniques for analyzing mine and mineral-deposit sample and assay data: U.S. Bur. Mines Bull. 621, 223 p.

Hoel, P. G., 1962, Introduction to mathematical statistics: New York, John Wiley & Sons, 427 p.

————, 1966, Elementary statistics: New York, John Wiley & Sons, 351 p.

Irving, E., 1964, Paleomagnetism: New York, John Wiley & Sons, 399 p.

Koch, G. S., Jr., 1956, The Frisco mine, Chihuahua, Mexico: Econ. Geology, v. 51, p. 1–40.

Krumbein, W. C., 1962, The computer in geology: Science, v. 136, no. 3522, p. 1087–1092.

Krumbein, W. C., and Graybill, F. A., 1965, An introduction to statistical models in geology: New York, McGraw-Hill, 475 p.

Leopold, L. B., Wolman, M. G., and Miller, J. P., 1964, Fluvial processes in geomorphology: San Francisco, W. H. Freeman, 503 p.

Li, J. C. R., 1964, Statistical inference: Ann Arbor, Mich., Edwards Bros., v. 1, 658 p.; v. 2, 575 p.

McKinstry, H. E., 1948, Mining geology: Englewood Cliffs, N.J., Prentice-Hall, 680 p.

Miller, R. L., and Kahn, J. S., 1962, Statistical analysis in the geological sciences: New York, John Wiley & Sons, 483 p.

Mosteller, Frederick, Rourke, E. K., and Thomas, G. B., Jr., 1961, Probability with statistical applications: Reading, Mass., Addison-Wesley, 478 p.

Stephan, F. F., and McCarthy, P. J., 1958, Sampling opinions, an analysis of survey procedure: New York, John Wiley & Sons, 451 p.

Stevens, R. E., and others, 1960, Second report on a cooperative investigation of the composition of two silicate rocks: U.S. Geol. Survey Bull. 1113, 126 p.

Stone, J. B., and McCarthy, J. C., 1948, Mineral and metal variations in the veins of Fresnillo, Zacatecas, Mexico: Am. Inst. Mining Engineers Trans., v. 178, p. 91–106.

Terzaghi, Karl, 1948, Soil mechanics in engineering practice: New York, John Wiley & Sons, 566 p.

Tukey, J. W., 1962, The future of data analysis: Annals Math. Statistics, v. 33, p. 1–67.

Whitten, E. H. T., 1964, Process-response models in geology: Geol. Soc. Amer. Bull., v. 75, p. 455–464.

Wilks, S. S., 1962, Mathematical statistics: New York, John Wiley & Sons, 644 p.

UNIVARIATE STATISTICAL METHODS

Because the same statistical methods are useful for different geological problems, Part II, which contains Chapters 2 to 6, is organized statistically rather than geologically. The methods given are for the analysis of single variables.

In Chapters 2 to 5 is introduced the minimum number of statistical methods essential for conducting a serious study. The reader prepared in statistics may decide to skim over these chapters. Chapter 2 explains frequency distributions of observations, for instance, grain-size measurements or a count of oil wells per square mile, and relates empirical distributions to theoretical distributions and to probability. Chapter 3 is about statistical principles of sampling, and Chapter 4 is about statistical inferences that can be drawn from samples. Chapter 5, on the analysis of variance and the sources of variability, concludes the introduction to the statistics that must be understood before real geological data can be effectively studied.

Distributions of observations are examined in more detail in Chapter 6, and transformations are introduced, which are devices to change original observations into distributions that can be more easily handled with mathematics.

Chapter 2

===

Distributions

Chapter 2 introduces some methods to organize numerical information about geological phenomena and relates the resulting empirical arrangements of data to some theoretical principles of statistics and probability. Additional methods to organize geological data are given later in the book, particularly in Chapters 6, 9, 10, and 11.

2.1 OBSERVATIONS

In principle almost any geological phenomenon or item can be represented by a numerical expression. The numerical expression may be obtained by various means, including measuring, counting, or ranking. The expression may be a single number defining copper grade in a stope, the number of zircon grains in a microscope slide, or a series of numbers designating, as a vector, the plunge of a fold axis. Whatever its origin, the numerical expression is named an *observation*.

In this chapter only the technical details for manipulating observations are considered. At this point, it is not necessary to consider the important matter that the assignment of a number may be of no interest, as, for example, in the discovery of the first trilobite in Oregon, where the fact of discovery rather than the number of fossils is the point of interest, nor is it necessary to consider that the assignment of a numerical expression may be very difficult, as in describing earthquake intensity. To simplify we restrict the discussion to *univariate observations*, that is, observations that can be represented by a single number. In Chapter 9 (Vol. 2) *multivariate observations*, those represented by more than one number, are taken up for the first time. However, essentially all of the discussion of univariate observations is also required for the analysis of multivariate data.

2.2   EMPIRICAL FREQUENCY DISTRIBUTIONS

A person cannot comprehend a large number of observations, say 10,000, by reading the list of individual observations. To make sense of the numbers, one must organize them by sorting, grouping, and averaging. In this section, the use of *frequency distributions* to organize observations is explained. The object is to summarize numerical information—no profound reasoning is involved, nor are there any optimum methods to employ.

Table 2.1 is a frequency distribution for 224 furnace-shale phosphate assays from the Phosphoria formation (Permian) near Fort Hall, Idaho (Hazen, 1964, p. 6). In column 1, assay intervals of 2 percent P_2O_5 are tabulated to include the range of assays from lowest to highest. In column 2 the midpoints of the assay intervals are tabulated, and in column 3 the number of assays in each interval, named the *frequency*, is recorded. To prepare the entries for column 3 one reads the list of assays and keeps a count for each interval. In column 4 the frequency is converted to a percentage of the 224 total assays. In column 5 the cumulative frequencies obtained by summing

TABLE 2.1. EXAMPLE FREQUENCY DISTRIBUTION OF 224 PHOSPHATE ASSAYS *

Assay interval ($\%$ P_2O_5)	Interval midpoint, w	Frequency, f	Relative frequency ($\%$)	Cumulative frequency, c.f.	Relative cumulative frequency, r.c.f. ($\%$)
14–16	15	1	0.45	1	0.45
16–18	17	1	0.45	2	0.90
18–20	19	8	3.57	10	4.47
20–22	21	21	9.37	31	13.84
22–24	23	44	19.64	75	33.48
24–26	25	54	24.12	129	57.60
26–28	27	56	25.00	185	82.60
28–30	29	30	13.39	215	95.99
30–32	31	7	3.12	222	99.11
32–34	33	2	0.89	224	100.00

* After Hazen, 1964, p. 6.

the successive lines in column 3 up to the final total of 224 are recorded, and in column 6 the values in column 5 are converted to percents.

In order to prepare the frequency distribution in table 2.1, the choice had to be made of interval width and the related number of groups. By use of an interval width of 2 percent, 10 groups were obtained. In general, both practical experience and theory (Tukey, 1948) have shown that, if the observations in any set are grouped into 10 to 50 intervals, all of the essential information is preserved. Because the interval width is generally taken to be the same for all groups, the more groups chosen, the narrower the interval width. However, if a set of data contains extreme values, it is sometimes convenient to widen the widths for the intervals with small or large midpoints.

2.3 HISTOGRAMS

Because it is even easier to look at a picture than to read a table, it is generally useful to draw a *histogram* to represent the frequency table pictorially. Figure 2.1 is a histogram of the phosphate data represented by the frequency distribution of table 2.1. The diagram is made by plotting a series of contiguous rectangles whose base length corresponds to the interval width from the frequency table (all are equal if the interval widths are equal) and whose height corresponds to the number of observations in each interval.

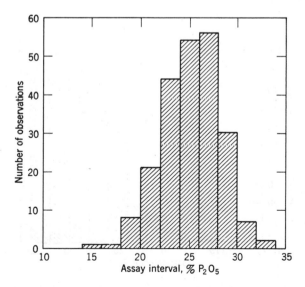

Fig. 2.1. Histogram, showing frequency distribution of 224 phosphate assays.

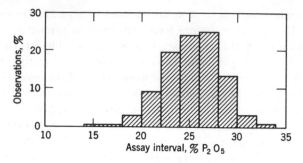

Fig. 2.2. Histogram, showing frequency distribution of 224 phosphate assays, recalculated to 100 percent area.

In figure 2.1, the height corresponds directly to the number of observations. In figure 2.2, for the same data, the height corresponds to the percentage of observations, so that the total area is 100 percent, or unity. The purpose of a unit area plot is to allow ready comparison of histograms from two sets of observations that differ in number. Because the total area is 100 percent, it is also possible to see at once from the histogram the percentage area of observations in a given range; for example, figure 2.2 shows that about 25 percent of the observations, corresponding to one-fourth the area of the histogram, are between 26 and 28 percent P_2O_5.

Figure 2.3 is a graph of the relative frequency curve from the phosphate data in table 2.1. The relative frequency curve can be used to determine

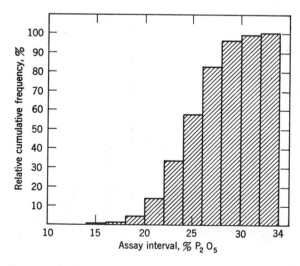

Fig. 2.3. Percent relative cumulative frequency plot for 224 phosphate assays.

directly the proportion of observations that are larger or smaller than an observation of given size. For example, figure 2.3 shows that only 4.5 percent of the observations are smaller than 20 percent P_2O_5.

2.4 THEORETICAL FREQUENCY DISTRIBUTIONS

In order to progress in interpreting observations beyond the summary afforded by the frequency table and histogram, the abstraction of the *theoretical frequency distribution* must be introduced. The starting point is the histogram of the phosphate data (fig. 2.1) which is for only 224 observations grouped into only 10 classes. If more phosphate observations were made and if more class intervals were taken at narrower widths, the rectangle heights in the histogram would take on more and more the appearance of a smooth curve. The smooth curve would resemble the dashed line on figure 2.4. Because the transition from a histogram to a smooth curve implies an infinite number of observations, it is essential to recognize that a smooth curve represents all *potential* observations of a specified kind, rather than actual observations.

Such a smooth curve is a theoretical representation, for which a mathematical model can be found, of potential observations. Thus a theoretical frequency distribution is a mathematical model which represents potential

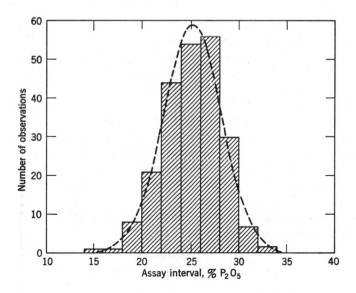

Fig. 2.4. Smooth curve fitted to histogram in figure 2.1.

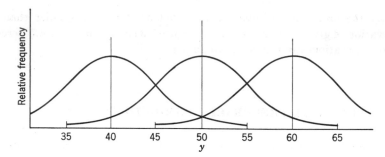

Fig. 2.5. Three normal curves with different means but the same variance (after Li).

observations. The value of a theoretical frequency distribution, in contrast to an empirical frequency distribution, is that the theoretical frequency distribution can be manipulated mathematically to develop statistical methods. Therefore a theoretical frequency distribution is essential, whether one's interest is in theoretical statistics or in applied statistics.

By way of illustration the *normal distribution*, the most important theoretical frequency distribution, is introduced. The equation for the normal distribution is†

$$f(w) = \frac{1}{\sigma\sqrt{2\pi}} \exp\left[-\frac{1}{2}\left(\frac{w-\mu}{\sigma}\right)^2\right],$$

where w is an observation, and μ and σ are variable constants, which must be evaluated to specify the distribution uniquely. Constants like μ and σ, whose values are different for different normal distributions, are called *parameters*, to distinguish them from fixed constants, which are 2, e, and π in the above formula. Regardless of the parameter values, the normal curve is always symmetrical and bell-shaped, but its form changes as the parameters change, as illustrated by figures 2.5 and 2.6 for normal curves with different parameter values. The normal distribution is also known by other names, for example, *Gaussian* and *error*. In present usage, the name is arbitrary and does not mean that other theoretical frequency distributions are abnormal. The normal curve is fully described in textbooks on mathematical statistics, for example, that by Hoel (1962, p. 98).

Not all symmetric bell-shaped distributions are normal; for example, the Cauchy distribution is bell-shaped but its equation is

$$f(w) = \frac{k}{\pi[k^2 + (w-\eta)^2]},$$

† The notation exp (x) is another form for the exponent e^x.

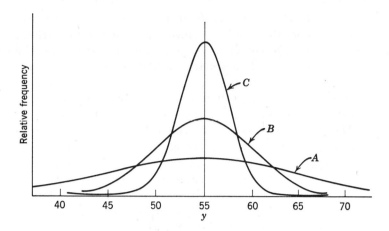

Fig. 2.6. Three normal curves with different variances but the same mean (after Li).

where k and η are parameters. Figure 2.7, a plot of a Cauchy distribution and a normal distribution, is presented to show that two distributions with different mathematical functions can look much alike.

There is, in principle, an infinite number of theoretical frequency distributions, many of which may under some conditions have frequency curves that look alike, just as in figure 2.7, which compares the normal and Cauchy distributions. From an empirical frequency distribution, there is no way to find a theoretical frequency distribution that is a *unique* representation of a set of actual observations. Nevertheless, because the detailed behavior of

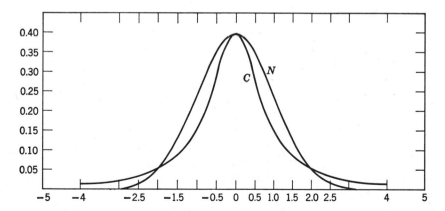

Fig. 2.7. Comparison of curves from normal and Cauchy distributions.

observations, as developed by statistical theory, depends very much on the assumed distribution, the choice of a theoretical frequency distribution has always been a matter of concern. Theoretical considerations may lead to a choice; for example, one may postulate a model for radioactive decay which yields a specific theoretical distribution, the exponential distribution, for the time of particle emission. Most often, however, there is no choice, much less a clear choice, for the theoretical distribution.

Mainly because it is mathematically convenient to work with, the normal distribution is by far the most used theoretical distribution. Fortunately, the normal distribution is one that can be applied in practice. Many observations are normally, or at least approximately normally, distributed. Even for observations that are far from normally distributed, the normal distribution can be applied if a problem can be answered by considering the behavior of a variable formed by constructing statistics from groups of observations rather than by considering the behavior of the original observations. Or, we may use a mathematical function to transform each observation into a new observation that follows the normal distribution. Moreover, many theoretical frequency distributions that are not followed by observations are themselves derived from the normal distribution, and these derived distributions may be useful even if the original observations are not normal.

Chapters 3 to 5 take up properties of the normal distribution and other distributions derived from it. Only that material necessary to supply essential tools for use on geological data is presented. Then, in Chapter 6, statistical methods are presented for observations that do not follow or even approximately follow the normal distribution. Also in Chapter 6, other distributions that have been useful in the analysis of geological data are presented, as well as mathematical transformations than can be used to change observations from other distributions into observations that follow the normal distribution.

2.5   PROBABILITY

The concepts of probability lie at the heart of statistical theory, and at least a minimal understanding of them is necessary to be able to interpret the results of statistical analyses. Probability is a large subject of study. This book gives only the briefest outline in order to introduce the subject to those who have never studied it and to refresh the memory of those who have. Only the most elementary concepts and calculations are introduced and some fictitious problems are solved.

Probability theory represents, in the abstract, the results of various physical occurrences, called *experiments*. A mathematical *point* is defined for each possible outcome, or *event*, of an experiment. Each mathematical point is therefore associated with an event. The collection or *set* of points associated with all possible events of the experiment is named the *sample space* for the experiment.

If there are two sets, named A and B, their *logical* sum, designated $A + B$, is the collection of the points in either set A or set B. Their *logical product*, designated AB, is the collection of the common points in both set A and set B. If sets A and B have no points in common, the logical product is an *empty set*, and sets A and B are said to be *mutually exclusive*.

The meaning of these terms will become clearer as the explanation of probability progresses.

Some Probability Calculations

These definitions are illustrated in the two idealized experiments of tossing two coins and tossing two dice. Two formulas for calculating probabilities, the *addition formula* and the *multiplication formula*, are introduced and illustrated. These experiments are then related to three fictitious geological illustrations.

Suppose that a coin is tossed. It may fall heads or tails. If these two outcomes are represented conceptually by two mathematical points, the points describe all possible results of the experiment and define a sample space. The phrase "all possible results" is an oversimplification, of course, since the points do not reflect such real happenings as a coin landing on edge or being lost under the table. The simplifications will help the reader to understand probability theory, which can then be applied to practical statistical problems from real-life situations.

If two coins are tossed, the sample space contains four points: HH, HT, TH, and TT (H means heads and T means tails). The event HH means that both coins landed heads, the event HT means that the first landed heads but the second landed tails, the event TH means that the first landed tails but the second landed heads, and the event TT means that both landed tails. If five coins are tossed, the sample space contains 32 points.

Suppose that two ordinary dice, each a cube whose sides are numbered 1 to 6, are tossed. The sample space has 36 points, as shown in figure 2.8, corresponding to the 36 possible combinations of points on the two dice. Thus (1, 1) means that both dice are one's, (1, 2) that the first die is 1 and the second die is 2, (2, 1) that the first die is 2 and the second die is 1, etc. As before, the experiment has been idealized in that one die could have fallen under the table or leaned up against the wall so that the experimenter could not decide which side was up.

In these experiments with coins and dice, and in real experiments, such as drilling oil wells or sending prospecting teams to the field, numbers named

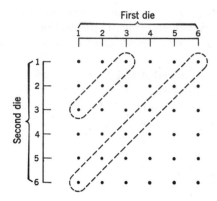

Fig. 2.8. Sample space for tossing two dice.

probabilities may be associated with each of the points of the sample space. Probabilities must be positive or zero numbers and their sum taken over the whole sample space must be 1. Since each event is represented by one or more points in the sample space, the probability of any one event is the sum of the probabilities of the points for that event. If only one point is associated with an event, the probability associated with the point is the probability of the event.

The *addition formula* allows the probabilities for some events to be calculated when the events are represented by two or more points. Consider the sample space associated with the two coins. Let A designate the event that coin 1 lands heads and B designate the event that coin 2 lands heads. What is the probability of the event, designated $(A + B)$, that *at least* one of the two coins lands heads? This situation is represented by figure 2.9, in which

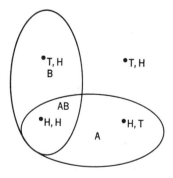

Fig. 2.9. Sample space for two coins, A and B.

events A and B are circled. The event $(A + B)$ is the logical sum of the events A and B. The addition formula, which relates these probabilities, can be written

$$P(A + B) = P(A) + P(B) - P(AB),$$

where $P(A + B)$ is the probability of the event $(A + B)$, $P(A)$ is the probability of event A, $P(B)$ is the probability of event B, and $P(AB)$ is the probability of both events. The event $(A + B)$ is that at least one coin lands heads, the event A is that the first coin lands heads, the event B is that the second coin lands heads, and the event AB is that both coins land heads.

If the events A and B are *mutually exclusive*, $P(AB)$ is zero, because no points are associated with AB, and the addition formula becomes

$$P(A + B) = P(A) + P(B).$$

The probability of the sum of any number of mutually exclusive events is simply the sum of the probabilities associated with the events; that is, by applying the addition formula,

$$P\left(\sum C_i\right) = \sum P(C_i),$$

where C_i denotes the events from C_1 to C_n.

The *multiplication formula* is used to calculate the probabilities of two events which occur simultaneously. Thus AB is the probability that both coins, tossed simultaneously, will land heads. The multiplication formula introduces the concept of *conditional probability*, which is the probability of an event, *given* the fact that another event has already occurred. The probability that coin 1 lands heads $[P(A|B)]$, *given* that coin 2 has already landed heads, is

$$P(A|B) = \frac{P(AB)}{P(B)},$$

which, rewritten, becomes

$$P(AB) = P(A|B)\, P(B),$$

the *multiplication formula*, and, by symmetry,

$$P(AB) = P(B|A)\, P(A).$$

If $P(A|B) = P(A)$, the events A and B are said to be *independent* in a probability sense, and

$$P(AB) = P(A)\, P(B).$$

Both the addition and multiplication formulas are illustrated by the following example. When two dice are tossed, it is reasonable from symmetry to believe that each of the 36 events (fig. 2.8) is equally probable, so that the probability associated with each of the 36 points is 1/36. Through the addition formula, the probability of the sum of the two dice being 7 can readily be calculated; the sample space is encircled in figure 2.9. A 7 corresponds to one of the six mutually exclusive events (1, 6), (2, 5), (3, 4), (4, 3), (5, 2), and (6, 1). Thus the probability of a sum of 7 is simply

$$P(7) = 6(1/36) = 1/6.$$

Next consider the calculation of the probability of the event (1, 6) through the multiplication formula, by considering separately the two events of getting a 1 on the first die and a 6 on the second die. For each die the probability associated with the event of getting a 1 or a 6 is 1/6 because there are six faces of each die and therefore six points in the sample space. Into the multiplication formula

$$P(AB) = P(A) \, P(B),$$

are substituted numerical values to yield

$$P(1, 6) = P(1) \, P(6) = (1/6)(1/6) = 1/36.$$

The concept of independence is intuitively apparent in this result since the one die is obviously not controlled by the other.

In geological applications of probability one of the most important concepts is that of *odds*, for instance, the odds that a fossil will be found in a certain outcrop or that an exploration program will discover a profitable oil field. The word odds in probability has the same meaning as in everyday life and may always be applied to the probability associated with two events, one of which is certain to happen. In the previous illustration, the probability of getting a 1 or a 6 was 1/6; therefore the odds of getting a 1 and a 6 are

$$P(1, 6) + P(6, 1) = 1/36 + 1/36 = 1/18,$$

or 5.56 percent. In general, if the odds of event A to event B are x to y, the probability of event A is $x/(x + y)$ and that of event B is $y/(x + y)$. If the probabilities of two events are known, the corresponding odds are simply the ratio of the two probabilities.

The experiments with coins and dice may readily be related to a fictitious geological illustration. Suppose that instead of a coin being tossed, a wildcat oil well is drilled. It can be classified as a discovery well or a dry hole, just as the coin can be classified as heads or tails. As in the coin example, it may not be possible in practice to classify the oil well since the records may have been lost or because only traces of oil were found. Similarly, instead of two coins being tossed, two wildcat oil wells might be drilled, or, instead of two dice being tossed, two states might be explored for iron ore, the value of which could be referred to six classes, each with an equal chance of occurring. The calculations made for the coins and dice apply equally well to these fictitious geological situations. The reader will find it instructive to rework the calculations in terms of these or other geological situations of interest to him.

Finally, *Bayes' theorem*, a formula for the calculation of conditional probabilities, is introduced. (An application for it is described in sec. 4.8.) The theorem may be stated as follows: If A_n is a series of mutually exclusive events such that $\sum P(A_n) = 1$, and if B is any other event,

$$P(A_n|B) = \frac{P(B|A_n)\,P(A_n)}{\sum P(B|A_i)\,P(A_i)}.$$

Bayes' theorem, fully explained by Feller (1957, p. 114) and Lindley (1965, p. 114) may be illustrated by a simple fictitious illustration.

Table 2.2 represents the situation that a geologist collected 60 fossils that are divided between two species, B and C, from three outcrops, designated A_1, A_2, and A_3. There were 10 fossils at A_1, 20 at A_2, and 30 at A_3. If one of the 60 fossils is picked at random, the probability $P(A_1)$ that it came from outcrop A_1 is $10/60 = 1/6$. Similarly, $P(A_2) = 1/3$, and $P(A_3) = 1/2$. Moreover, given that a fossil is from outcrop A_1, the conditional probability $P(B|A_1)$ that it is of species B is $7/10$. Similarly, $P(B|A_i)$ can be calculated for outcrops A_2 and A_3. The probability statements are listed in table 2.2. Now, by using Bayes' theorem, the probability that a given fossil of species B came from outcrop A_1 can be calculated. The arithmetic is shown in the table and yields the result $7/27$. This development was given to illustrate Bayes'

TABLE 2.2. FICTITIOUS ILLUSTRATION OF COLLECTING FOSSILS, TO ILLUSTRATE BAYES' THEOREM

Outcrop names	Number of observations	Number of fossils	
		Species B	Species C
A_1	10	7	3
A_2	20	10	10
A_3	30	10	20

Probability statements:

$P(A_1) = 1/6$	$P(A_2) = 1/3$	$P(A_3) = 1/2$
$P(B|A_1) = 7/10$	$P(B|A_2) = 1/2$	$P(B|A_3) = 1/3$

Calculation of probability $P(A_1|B)$:

$$P(A_1|B) = \frac{7/10 \times 1/6}{7/10 \times 1/6 + 1/3 \times 1/2 + 1/2 \times 1/3} = 7/27 = 0.259$$

theorem; for this fictitious illustration, the probability 7/27 can be obtained directly, because there are 27 fossils of species B altogether, and 7 of these were found at outcrop A_1.

REFERENCES

Feller, William, 1968, An introduction to probability theory and its applications, v. 1: New York, John Wiley & Sons, 509 p.

Hazen, S. W., Jr., 1964, Statistical analysis of some sample and assay data from the bedded deposits of the Phosphoria Formation in Idaho: U.S. Bur. Mines Rept. Inv. 6401, 29 p.

Hoel, P. G., 1962, Introduction to mathematical statistics: New York, John Wiley & Sons, 427 p.

Lindley, D. V., 1965, Introduction to probability and statistics from a Bayesian viewpoint, pt. 2, inference: Cambridge Univ. Press, 286 p.

Tukey, J. W., 1948, Approximate weights: Annals Math. Statistics, v. 19, p. 91–92.

Chapter 3

Sampling

Chapter 3 is about statistical principles of sampling. First some statistical principles that can help the geologist devise a sampling plan are explained in sections 3.1 and 3.2. In the remainder of the chapter organization and preliminary interpretation of the data obtained from sampling are discussed.

3.1 POPULATIONS AND SAMPLES

In Chapter 2 some simple methods to organize and summarize a set of numerical geological data are explained. Generally, a set of data represents only a small fraction of the total number of observations that might in principle be obtained about the subject of interest. The total set of potential observations is named a *population*, and the set of actual observations in hand is named a *sample* from this population.

As explained in detail in section 3.2 and throughout the book, a population is just what the investigator chooses to make it through a definition that is arbitrary, although it should be thoughtful and purposeful. If analyses of quartz percentage are made for several hand specimens from a batholith, the total population of observations may be the set of hand specimens that could be taken from the rock body; the sample of quartz percentage values consists of the small proportion of values actually obtained. If microfossil counts are made for several specimens of sediment, the total population may be defined to be the set of specimens that could conceivably be collected; the sample consists of the small number of counts actually obtained.

43

Unfortunately the statistical definition of *a sample* as a group of observations does not agree with the geological usage of sample as a single hand specimen or bag of rock chips yielding one observation on chemical analysis, assay, or petrographic study. Therefore, whenever there is a possibility of confusion, the clumsy terminology *statistical sample* or *geological sample* is adopted.

Now, several questions arise. If the set of observations at hand is only a sample of the total information of interest, what conceptual framework encompasses all of the information? Moreover, if the set of observations is only a sample, what can be inferred from the sample about the population? When the behavior of all possible samples from a population is discovered, then it is possible to learn what information about a population is conveyed by one particular statistical sample.

This chapter is mainly statistical and the mathematics is purely deductive. Geology is almost always concerned with the reverse process of making inductive inferences about populations from a single statistical sample. Many or all possible statistical samples in this chapter are considered in order to learn how they behave, not to devise a direct guide for action. In later chapters the geological consequences of the material in this chapter are discussed. If, as rarely if ever happens, an entire population is available for study, population *parameters* can be calculated directly from the available observations. Moreover, if this population is described by a theoretical frequency distribution, the values of the population parameters are known.

The normal distribution has two parameters, the mean and the variance, which are defined in section 2.4. Almost always, however, the set of observations under study constitutes only a sample of the entire population that interests the investigator, and it is therefore necessary to estimate the parameters by corresponding numbers named *statistics* that are derived from the sample. In this book the common convention of designating parameters by Greek letters and statistics by corresponding English letters is followed; for example, s^2 designates a statistic derived from a sample that estimates the parameter σ^2. The notation is consistent with that of Li (1964).

3.2   TARGET AND SAMPLED POPULATIONS

In section 3.1 the concept of a population comprising the total number of potential observations is introduced and contrasted with the statistical sample of observations actually taken. In this section, two kinds of populations are distinguished, the one from which the sample is actually drawn, named the *sampled population*, and the generally larger one of geological interest, named the *target population*. These two types of populations were distinguished and named by Cochran et al. (1954, p. 18), who writes as follows:

> We have found it helpful in our thinking to make a clear distinction between two population concepts. The *target* population is the population of interest, about which we wish to make inferences or draw conclusions. It is the population which we are trying to study. The *sampled* population requires a more careful definition but, speaking popularly, it is the population which we actually succeed in sampling.
>
> The sampled population is an important concept because by statistical theory we can make quantitative inferential statements, with known chances of error, from sample to sampled population. It must be carefully distinguished from the target population, the population of interest, about which we are tempted to make similar inferential statements.
>
> Insofar as we make statistical inferences beyond the sample to a larger body of individuals, we make them to the sampled population. The step from sampled population to target population is based on subject-matter knowledge and skill, general information, and intuition—but not on statistical methodology.

The terms target and sampled populations were introduced into geology by Rosenfeld (1954), and the concept has been extensively used, particularly by Krumbein (1960; 1965, pp. 147–169) and his students. Krumbein's discussions, which are particularly clear and thoughtful, merit close attention.

We compare target and sampled populations here by presenting examples. Because of the importance of the subject and the complicated nature of the argument, the discussion is purposely made detailed. As an example of a target population, we choose the Jurassic Morrison Formation, which contains the source beds for many uranium deposits in the western United States. Assume that the problem is to sample the formation for uranium. To make the problem more definite, assume further that the formation is to be drilled to obtain NX core that will be cut transversely to yield geological samples consisting of rock cylinders each 3 inches long by about 2 inches in diameter. However the population is defined, the Morrison Formation clearly contains an essentially infinite number of such cylinders.

The first task is to define precisely the target population. Although the definition might appear easy because the entire formation is of interest, it at once becomes necessary to choose among alternatives, including:

1. The entire Morrison Formation as originally deposited in Jurassic time.

2. The entire Morrison Formation excluding that part removed by erosion or stoped out by magma since Jurassic time.

3. The Morrison Formation excluding the part covered by younger rocks and colluvium.

4. The Morrison Formation excluding the part covered by younger rocks.

5. The Morrison Formation available for mining under certain specifications, e.g., covered by at most 500 feet of younger rocks, in areas not withdrawn from mining by law, or in areas not owned by competing mining companies.

The list is not intended to be complete, and it is at once clear that there are additional alternative definitions of the target population, as well as combinations among the listed alternatives. Decisions among some alternative definitions may require geological judgment; for example, has a certain rock formation been dissolved by magma or is the rock found today a granitized sedimentary rock formation? Other decisions may require information that can be obtained only in the course of field sampling, for example, the depth below the surface of a formation. These complications are purposely introduced to emphasize that the definition of the target population may be difficult. Yet in the appropriate definition may lie the best way of posing a problem to obtain an interesting or useful answer.

By and large the definition of a target population relies on geological judgment and knowledge, which often must be applied with considerable skill. In specifying an appropriate target population, one can often formulate in geological terms a good question, the answer to which will illuminate the science. The purpose of the sampling must be kept in mind. In order to indicate that the problem is a real one, a real example may be mentioned. We once wrote (Link and Koch, 1962) that, in sampling a granite body, "if there were extraneous material at a grid point, for example a dike, there is a real question as to whether the dike should in fact be sampled. . . ." Another worker (Exley, 1963, p. 650) commented that "the question as to whether or not a dike should be sampled in a granite survey simply does not arise; the dike is not the granite and is excluded," a comment that indicates how misunderstanding can stem from different concepts of target population. If a granite is to be sampled as a potential uranium source and if the dikes would be mined along with the granite, the dikes certainly should be included, although perhaps the samples should be kept separate. If a granite is to be sampled to determine its origin and if the decision was made before sampling that the dikes were post-granite, the dikes certainly should be excluded. If a granite is to be sampled to determine its origin and if the relative age of granite and dike were unknown (possible post-dike granitization), probably the dikes should be included, although the samples should be kept separate.

Thus, after some travail, a target population can be defined precisely. The definition should never be made in a routine way. If a target population is not defined explicitly and precisely, it will be defined implicitly and imprecisely, both by the worker and by those concerned with the result; and if the definitions are hazy, they will lead to conflict. Therefore the geologist should think first—rather than just set off, Brunton compass and topographic map in hand, to collect hand specimens.

Once a target population is defined the next step is to define the sampled population. Some of the target population may clearly be unavailable for sampling, for example, because of time. It is too late to sample the part of the Morrison Formation that has been eroded away, and it is too early to sample the gas from next year's volcanic eruption in Guatemala. Some of the target population may be expensive or inconvenient to sample. For the example of the Morrison Formation the sampled population might be defined in one of the following ways:

1. The entire formation exposed in outcrop, in outcrop below colluvium, and in suboutcrop to a depth of 500 feet.
2. The entire formation exposed in outcrop and in outcrop below colluvium.
3. All outcrops in the state of Utah.
4. All outcrops in roadcuts.
5. All outcrops within 1 mile of a road.
6. All outcrops below 10,000 feet elevation.

The connection between the sampled and the target population must be made for geological reasons. Particularly because of the time element, in geology the sampled population can seldom if ever correspond to the target population. The connection must be made on geological rather than on statistical grounds, although the statistician may be able to offer some help from his general experience. However, geological problems tend to be more complicated and the populations less accessible for sampling than in such simpler situations as sampling cars of coal delivered to a steam generating station, sampling light bulbs, or quality control in a brewery.

Thus, in defining a sampled population for the Morrison Formation, such questions as the following would have to be faced:

1. Is the Morrison Formation different at higher topographic elevations, because it is more resistant to erosion, or for another reason?
2. Even though it may be almost impossible to sample in river valleys, is the formation different there because of zones of weakness, or fractures that may carry mineralization?
3. How deep does weathering extend? It will certainly be absent if the formation is sampled deep enough, but will such sampling be prohibitively expensive?
4. If the sampled population is restricted to outcrops in Utah, perhaps because the work is done by the Utah Geological Survey, how applicable will

the results be to Colorado? Can a larger scale subsampling scheme be set up to extend the validity of results into Colorado?

All these questions are basically geological, whether posed by a geologist or by some other person such as a mining engineer or geophysicist.

Definitions of target and sampled populations and distinctions between them are taken up repeatedly in this book. Exploration of these subjects will sharpen one's geological thinking and improve his proficiency as a geologist. In Chapter 7 on geological sampling the two kinds of populations are treated further in some detail.

3.3   MEAN, VARIANCE, STANDARD DEVIATION, AND COEFFICIENT OF VARIATION

In Chapter 2 ways to organize and summarize data by a frequency table and a histogram are explained. Here, methods of formulating numerical expressions that summarize information about observations in mathematical terms, which can be manipulated, are introduced. For many types of observations, particularly those that follow the normal distribution, the important summary expressions are the *mean* (a measure of central tendency), the *variance* (a measure of spread of a distribution), the *standard deviation* (the square root of the variance), and the *coefficient of variation* (the ratio of standard deviation to mean).

These summary expressions are only a few of the many that have been devised by statisticians; however, they constitute those that are generally of most use for geological data. One additional expression, the *correlation coefficient*, is introduced in section 9.1. Information on other more specialized expressions, such as measures of skewness, are given by Hoel (1962) and in other books on mathematical statistics.

If all the observations of a population are available, the population mean is simply the arithmetic average calculated by dividing the sum of the values of all observations by the number of observations; that is,

$$\mu = \frac{\sum w}{n},$$

where μ is the population mean, w is an observation, and n is the number of observations. If the population is represented by a theoretical frequency distribution, n is indeterminate, because the population is conceptually infinite in size, and therefore the simple formula above cannot be applied. In this case the population mean may be obtained by the formula

$$\mu = \int_{-\infty}^{\infty} w f(w) \, dw,$$

where $f(w)$ is the expression for the particular theoretical distribution. For example, for the normal distribution introduced in section 2.4, the formula obtained is

$$\mu = \int_{-\infty}^{\infty} \frac{w}{\sigma \sqrt{2\pi}} \exp \left[\frac{1}{2} \left(\frac{w - \mu_0}{\sigma} \right)^2 \right] dw.$$

When this integral is evaluated, the result obtained is

$$\mu = \mu_0.$$

Thus the parameter μ in the normal distribution is in fact the population mean.

If, as is almost always the case, the set of observations is a sample rather than a population, the mean is computed by the parallel formula

$$\overline{w} = \frac{\sum w}{n}$$

where the mean is designated by the English letter $\overline{w}$ rather than by the Greek letter μ. The relation of $\overline{w}$ to μ depends on many things, and investigation of this relation is an important subject that reappears repeatedly in this book. In table 3.1, the calculation of the sample mean $\overline{w}$ is illustrated for a sample with $n = 5$, $\sum w = 25$, and $\overline{w} = 5$.

TABLE 3.1. CALCULATION OF MEAN AND VARIANCE FROM A FICTITIOUS SAMPLE OF SIZE 5

Observations, w	Observations squared, w^2	Deviation from mean, $(w - \overline{w})$	Squared deviation from mean, $(w - \overline{w})^2$
2	4	-3	9
4	16	-1	1
6	36	1	1
6	36	1	1
7	49	2	4
Sum 25	141	0	16

Calculations:

$$\overline{w} = 25/5 \qquad s^2 = 16/4$$
$$= 5 \qquad\qquad = 4$$
$$s = \sqrt{4}$$
$$= 2$$

Although the calculation of a mean is commonplace, the calculation of a variance may be less so, although the concept of variation is familiar. The idea that observations vary is familiar, for no one expects all men to be of the same height, all eggs to be identical in size, or all molybdenum assays to be of the same value. In order to measure the spread of observations, some calculation is required of how close on the average the observations are to the mean. One measure of variation might be obtained by subtracting the mean from each observation and summing according to the expression

$$\sum (w - \mu).$$

However, this formula is useless because the sum is equal to 0 for all populations. In order to overcome this difficulty, each term $(w - \mu)$ is squared to make its value positive and to yield the expression

$$\sum (w - \mu)^2.$$

The sum of squared deviations from the mean is used rather than the sum of absolute values, $\sum |w - \mu|$, because it is a more convenient quantity to manipulate mathematically. To obtain an average measure of variation, one divides the expression $\sum (w - \mu)^2$ by the sample size n according to the formula

$$\sigma^2 = \frac{\sum (w - \mu)^2}{n},$$

where σ^2 is named the population variance. Thus, the population variance σ^2 is the average squared deviation of observations from the population mean. The square root of the variance σ^2 is the standard deviation σ.

If the population is represented by a theoretical frequency distribution, n is indeterminate, and the formula for the variance becomes

$$\sigma^2 = \int (w - \mu)^2 f(w) \, dw,$$

where, again, $f(w)$ is the expression for the particular theoretical frequency distribution, and the integration extends from $-\infty$ to $+\infty$. For the normal distribution, the formula obtained is

$$\sigma^2 = \int_{-\infty}^{\infty} \frac{(w - \mu)^2}{\sigma_0 \sqrt{2\pi}} \exp \left[-\frac{1}{2} \left(\frac{w - \mu}{\sigma_0} \right)^2 \right] dw,$$

and when the integral is evaluated, the result obtained is

$$\sigma^2 = \sigma_0^2.$$

Thus the parameter σ^2 in the normal distribution is in fact the population variance.

If the observations constitute only a sample rather than a population and if, as almost always happens, the population mean μ is unknown, the variance is computed as

$$s^2 = \frac{\sum (w - \overline{w})^2}{n - 1},$$

where s^2 is the sample variance, and its square root s is the sample standard deviation. The term $(n - 1)$ rather than n is used as the divisor, as explained in section 3.7. In table 3.1 the sample variance is calculated for the sample of five observations. The mean deviation, tabulated in column 3, is 0 and is therefore useless for measuring variation. Dividing the squared mean deviation, which is 16, by $(n - 1)$ which is 4, yields an s^2 value of 4. The numerator $\sum (w - \overline{w})^2$ in the formula for sample variance is called the *sum of squares*, which is a short form for "sum of squared deviations from the sample mean" and is designated SS. The method of calculating sample variance in table 3.1, given to illustrate the principle, is inconvenient to use in practice, because the sample mean must be calculated before the sum of squares can be calculated. In table 3.2 a short-cut procedure is given that is well suited for use with a

TABLE 3.2. CALCULATION OF VARIANCE FROM THE FICTITIOUS SAMPLE OF TABLE 3.1 BY SHORT-CUT COMPUTING FORMULA

$$
\begin{aligned}
\text{SS} &= \sum (w - \overline{w})^2 \\
&= \sum w^2 - (\sum w)^2/n \\
&= 141 - (25)^2/5 \\
&= 141 - 125 \\
&= 16 \\
s^2 &= 16/4 \\
&= 4
\end{aligned}
$$

desk calculator (but not with a digital computer, Chap. 17). The short-cut procedure depends on the algebraic identity

$$\text{SS} = \sum (w - \overline{w})^2 = \sum w^2 - \frac{(\sum w)^2}{n},$$

which is proved in statistical textbooks (Li, 1964, I, p. 89). In table 3.2, this formula, which is applied to the data of table 3.1, yields an SS value of 16 and the same value of 4 for s^2.

The mean, variance, and standard deviation of a sample are calculated from the observations. If the observations are changed to new ones by adding a constant to each observation or by multiplying each observation by a constant, the mean, variance, and standard deviation, in general, also change.

If all of the original observations w are changed to new observations u by adding a constant a,

$$u = w + a,$$

the new mean is equal to the original mean plus the constant,

$$\overline{u} = \overline{w} + a,$$

and the variance and standard deviation are unchanged,

$$s_u^2 = s_w^2,$$
$$s_u = s_w.$$

If all of the original observations w are changed to new observations u by multiplying by a constant b,

$$u = bw,$$

the new mean is equal to the original mean multiplied by the constant,

$$\overline{u} = b\overline{w},$$

the new variance is equal to the original variance multiplied by the constant squared,

$$s_u^2 = b^2 s_w^2,$$

and the new standard deviation is equal to the original standard deviation multiplied by the constant,

$$s_u = b s_w.$$

The change in an original observation by introducing one or more constants by addition or multiplication is an example of a *linear transformation* (Chap. 6).

An illustration of linear transformations applied to the data of table 3.1 is given by table 3.3. In column 2, the constant 5 is added to each observation; in column 3, each observation is multiplied by the constant 4; and in column 4, each observation is multiplied by the constant 4, and the constant 5 is also added. The calculations verify the remarks in the preceding paragraph.

An especially important linear transformation is to subtract the population mean μ, which is a constant, from the observations w, and divide the result by the population standard deviation σ, which is also a constant, according to the formula

$$u = \frac{w - \mu}{\sigma}.$$

This transformation converts a set of w values into u values whose mean is 0 and whose variance is 1. In the first part of this transformation,

$$u' = w - \mu,$$

TABLE 3.3. EFFECT ON MEAN AND VARIANCE CAUSED BY CHANGING THE OBSERVA-
TIONS IN THE FICTITIOUS SAMPLE OF TABLE 3.1 BY ADDING OR MULTIPLYING BY A
CONSTANT

	Original observations, w		Observations $+ 5$, $u = w + 5$	Observations $\times 4$, $u = 4w$	Observations $\times 4$ and $+ 5$, $u = 5 + 4w$
	2		7	8	13
	4		9	16	21
	6		11	24	29
	6		11	24	29
	7		12	28	33
$\sum w$	25	$\sum u$	50	100	125
n	5		5	5	5
$\overline{w}$	5	$\overline{u}$	10	20	25
$(\sum w)^2$	625	$(\sum u)^2$	2,500	10,000	15,625
$(\sum w)^2/n$	125	$(\sum u)^2/n$	500	2,000	3,125
$\sum w^2$	141	$\sum u^2$	516	2,256	3,381
SS	16		16	256	256
d.f.	4		4	4	4
s^2	4		4	64	64
s	2		2	8	8

where u' is the intermediate u value. The mean of u', designated $\mu_{u'}$, is 0
because

$$\mu_w = \mu,$$

and the variance of u', designated $\sigma_{u'}^2$, is still σ^2. In the second part of the
transformation, u' is multiplied by $1/\sigma$, to yield the final result, that is,

$$u = u'\frac{1}{\sigma}.$$

Then, because the new variance is equal to the original variance multiplied
by the constant squared, the variance of the final result u is

$$\sigma_u^2 = \frac{1}{\sigma^2}\sigma_{u'}^2,$$

but, because $\sigma_{u'}^2$ is equal to σ^2,

$$\sigma_u^2 = \frac{1}{\sigma^2}\sigma^2 = 1.$$

This important transformation is applied in section 3.6.

The *coefficient of variation*, the ratio of standard deviation to mean, is a useful measure of relative variability of observations, provided that they are all either positive or negative. For a population, the coefficient of variation is found by the formula

$$\gamma = \frac{\sigma}{\mu},$$

and, for a sample, the coefficient of variation is found by the formula

$$C = \frac{s}{\overline{w}}.$$

For geological data, most of which are positive, and whose variability tends to be larger than that of many other kinds of data, the coefficient of variation is an important measure that is used repeatedly in this book, particularly in Chapter 6, as a guide to whether to transform observations nonlinearly, and in Chapter 8, to evaluate variability in geological data.

If a frequency distribution of observations has been made, an alternative calculation of the elementary statistics discussed in this section may be made. As an example, phosphate assays from section 2.2 have been used. When the calculations are performed on the original data, the results obtained are

$$\overline{w} = 25.13$$
$$s = 2.95$$
$$C = 0.117.$$

The calculations can also be performed on the classed data, as shown in table 3.4, which is part of table 2.1 with two additional columns to give the results of multiplying the frequency in each class by the midpoint of the interval and

TABLE 3.4. CALCULATION OF MEAN AND VARIANCE FOR CLASSED DATA FROM FIGURE 2.1

Assay interval	Interval midpoint, w	Frequency, f	fw	fw^2
14–16	15	1	15	225
16–18	17	1	17	289
18–20	19	8	152	2,888
20–22	21	21	441	9,261
22–24	23	44	1,012	23,276
24–26	25	54	1,350	33,750
26–28	27	56	1,512	40,824
28–30	29	30	870	25,230
30–32	31	7	217	6,727
32–34	33	2	66	2,178

Calculations:

$$w = \sum fw / \sum f$$
$$\sum fw = 5652$$
$$\sum f = 224$$
$$\bar{w} = 25.23$$
$$s^2 = \frac{SS}{\sum f - 1} = \frac{\sum fw^2 - (\sum fw)^2 / \sum f}{\sum f - 1}$$
$$\sum fw^2 = 144{,}648$$
$$(\sum fw)^2 = 31{,}945{,}104$$
$$(\sum fw)^2 / n = 142{,}612$$
$$SS = 2{,}036$$
$$s^2 = 9.13$$
$$s = 3.02$$
$$C = 0.119$$

by the squared midpoint of the interval. When the calculations are performed as indicated in the table, the new results obtained are

$$\bar{w} = 25.23$$
$$s = 3.02$$
$$C = 0.119.$$

It can be seen that the two methods of calculation yield essentially the same results. The classed method is useful for desk-calculator computations or if only classed data are available; otherwise, with a digital computer it is generally easier to work with the individual observations.

This section will now be summarized. For any set of observations, the concepts of average value and variation of individual observations from average value are familiar ones. Based on these concepts are two specific measures that are particularly useful, the mean and the variance. For a population, the mean is μ and the variance is σ^2, corresponding to the same parameters in the theoretical normal distribution. For a sample, the mean is $\bar{w}$ and the variance is s^2. The standard deviation, the square root of the variance, is σ or s. The coefficient of variation, the ratio of standard deviation to mean, is γ for a population and C for a sample.

3.4 RANDOM SAMPLES

Before considering the behavior of samples from a population, in this section we first explain the method used to obtain a *random sample*, sometimes named a *probability sample*. The sampled population is a collection of potential observations, generally defined to be infinite in number through appropriate definition of the sampled population. If a scheme is devised to take a fraction of these observations so that each potential observation has an equal chance of being selected, the sample obtained is named a *random sample*. All statistical theory depends essentially on the behavior of random samples. When samples have been selected by nonrandom processes, the validity of the conclusions is always indeterminate, although the consequences, as discussed later (especially in secs. 7.2 and 8.6), may or may not be serious.

In order to take a random sample from a population, unambiguous rules must be thoughtfully made and carefully followed. One useful device is to assign to each item in the population a unique serial number in addition to the number or numbers representing the observation; for example, everyone eligible to collect a social security payment in the United States has a social security number in addition to the monthly payment, if any, that he receives. The social security numbers are serial numbers, and the monthly payments, including the zero ones, are observations. Although many individuals may be eligible to receive the same payment each month, so that the observations are the same, the individuals may be differentiated by the serial numbers. In order to take a random sample from this population, it is only necessary to find a method of specifying the serial numbers for the individuals to be selected.

The serial numbers may be specified and the random samples then selected by methods of chance familiar to everyone. For example, a coin may be tossed, a card may be drawn, a number may be chosen in a lottery, or dice may be thrown. In practice, it is more convenient to use a table of random numbers, because the serial numbers can be selected more rapidly and because any simple mechanical process is liable to be biased. Random numbers are generated by random processes, which are devised so that each of the digits from 0 to 9 has an equal chance of occurring. Because there are ten digits in the range 0 to 9, the chance or probability of one occurring each time is 10 percent, or 0.1. The mathematical process is such that there is no memory of which digit was generated last time, just as the toss of an unbiased coin does not depend on the result of the previous toss. In technical terms, the digits are generated with statistical independence.

Tables of random numbers are available, for example, the Rand table (Rand Corporation, 1955) of one million digits. Table A-1, in the appendix to this volume, is a small table of random digits from the Rand table. The table is arranged with serial numbers in the far-left column and with 10 five-digit columns of random numbers to the page. To use the table, choose any starting point, for example, by picking a page, column, and row at random. Read in any consistent direction, for example, down columns or across rows, and record numbers of the desired size. For example, in order to obtain two-digit numbers, the first or last two digits of a group of five could be read.

Because random numbers may be assembled to form numbers of any required size, tables of random numbers may be used to obtain random samples from populations with serially numbered observations. If serial numbers are not explicitly attached, it is sometimes possible to attach them implicitly. If serial numbers cannot be attached, the sampling becomes more or less obscure, and it may be difficult to determine whether a sample is indeed random. For example, pulverized rock might be passed through a Jones splitter three times to obtain eight fractions. If the eight fractions are serially numbered, and if one is selected at random for an analysis, then this phase at least of the sampling is indeed random. On the other hand, if the eight fractions are not serially numbered and preserved for a final random selection, there is always the question whether the sample splitting was indeed random, or whether some bias was introduced, perhaps by the splitter not doing an unbiased job of dividing a powdered sample.

In order to illustrate a specific method of taking a random sample, an example of sampling diamond-drill core from the Homestake mine, Lead, South Dakota, is described here in detail. Because our study (Koch and Link, 1967) supplies other examples for this book, pertinent details of the mine geology are reviewed here. In the Homestake mine, the gold is localized in the Precambrian Homestake Formation. The Homestake Formation, originally a low-grade iron formation, has been metamorphosed and intensely and intricately folded. Its present composition is mainly chlorite, cummingtonite, and quartz—with subordinate ankerite, garnet, iron sulfides, and other minerals. Typically, the ore contains 0.33 troy ounce of gold per ton, 8 to 10 percent of sulfide minerals (mainly pyrrhotite), and a little silver. The Homestake Formation has been folded into steeply dipping ore bodies of banded and foliated rock. The ore is developed on each level by crosscuts and between levels by diamond-drill holes from sublevel stations.

The purpose of our study was to investigate the distribution of gold in gold ore of the world. Such distribution is of use both for theoretical and applied geology and for mining. Thus our initial target population was gold distribution in gold ore throughout the world. The gold-bearing Homestake Formation was the second target population. Because we wanted to compare variability in gold in this formation at short range as well as at longer range, we put the further requirement on our target population that we would sample in two locations some tens of feet apart, with more than one borehole at each location.

Having specified the target population, we could next obtain a potential sampled population corresponding to the specifications. The actual sampled population obtained was ore at two locations (fig. 3.1) selected as typical. Thus, a long chain of decisions and events extended from our initial vague concept through to the actual sampling, connected by geological reasoning within a framework of geological assumptions. To sample the ore at the two

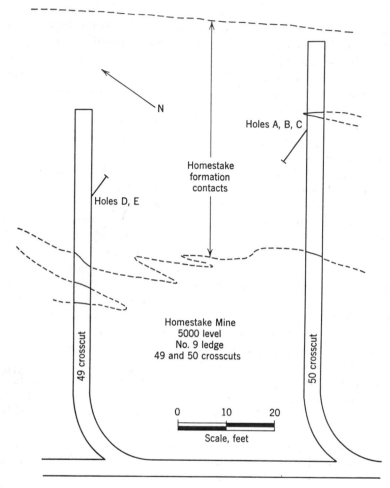

Fig. 3.1. Locations of diamond drill holes for experimental sampling in the Homestake Mine.

chosen locations, we bored five drill holes, three designated A, B, C, in the 50 crosscut; and two, designated D and E, in the 49 crosscut. Because the total combined hole length was 56 feet and the core was sawed transversely into 1-inch-long cylinders, our original sample consisted of 672 (56×12) cylinders. We prefer to think of the sample as composed of the actual rock cylinders rather than as a collection of 672 potential gold assay values, because other observations could have been made on the cylinders, such as mineralogical composition or silver assay. This usage puts our samples into a more geological framework.

The purpose of choosing 1 inch for the length of the rock cylinders was to obtain geological samples for assaying, each weighing about 1 assay ton (29.1667 grams), so that no variation would be introduced by splitting pulverized material or by handling coarse gold that could not be ground. To save time and expense, we decided to sample one-third of the original sample of 672 cylinders, under the restriction that two pairs of adjacent cylinders, or a total of four cylinders would be taken from each foot. This restriction is a *stratification* of the sampling, a method, discussed in section 7.2, that is not considered here.

In order to select randomly the four rock cylinders per foot required by our scheme, the cylinders were serially numbered as shown in table 3.5 and

TABLE 3.5. RANDOM SAMPLING SCHEME FOR DIAMOND-DRILL CORE FROM THE HOMESTAKE MINE

Foot interval	Core pair numbers	Random number from table	Constant	Random number plus constant
1	1			
	2			
	3	3	0	3
	4			
	5	5	0	5
	6			
2	7	1	6	7
	8			
	9	3	6	9
	10			
	11			
	12			
3	13	1	12	13
	14			
	15			
	16			
	17	5	12	17
	18			
4	19			
	20	2	18	20
	21			
	22			
	23	5	18	23
	24			

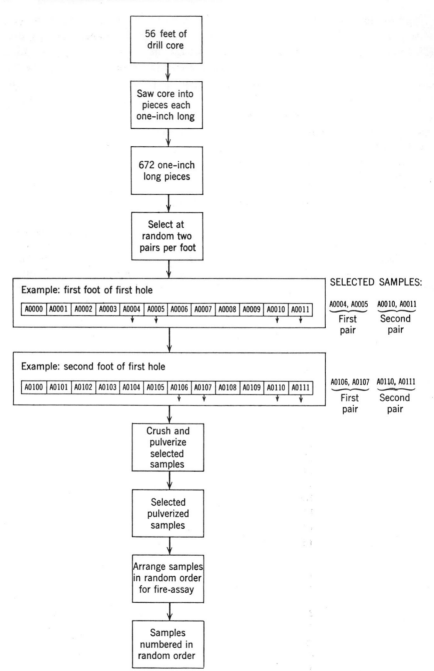

Fig. 3.2. Flow chart for sampling drill core from the Homestake mine.

figure 3.2. Because paired samples were required, for the first foot it was necessary to choose from a random number table the first two numbers between 1 and 6. For the second foot, the next two numbers between 1 and 6 were chosen, and 6 was added to each. Through this process, 224, or one-third of the total 672 cylinders, were selected. The rock from one selected inch was lost in drilling; 223 cylinders and 111 pairs were assayed.

3.5 SAMPLING DISTRIBUTIONS

In this chapter the relation of sample statistics to population parameters has been introduced, and the calculation of some parameters and statistics has been explained. From the observations in a sample, various statistics can be calculated, for instance, the sample mean and the sample variance. Consider, in particular, one statistic and a particular sample size. For repeated samples from the population the calculated value of the statistic fluctuates from sample to sample because the observations in the various samples are different. The calculated values can be summarized in a frequency distribution. The *sampling distribution* of a statistic is simply the theoretical frequency distribution of the values of a statistic calculated for all possible samples of a particular size from a population. The values used to form the sampling distribution are new observations that together compose a new population.

A sampling distribution is valuable because it relates a statistic to an underlying population parameter. In practice only one statistical sample of a population is taken at a time, as we took only one random sample at the Homestake mine. But one must know the behavior of statistics from all possible samples in order to relate the statistic from the single sample to the population parameter that we wish to estimate.

For a small population it is easy to illustrate the sampling distribution by enumerating all possible samples of a given size. Two types of sampling must be distinguished: sampling with replacement and sampling without replacement. Although in practice it is almost never necessary to distinguish between these two methods of sampling, both should be understood.

Consider a population with only three observations—2, 4, and 6—from which a sample of two observations, that is a sample of size 2, is to be drawn. In table 3.6, the population is listed, and the mean μ and variance σ^2 are calculated. The population may be thought of as three playing cards of the appropriate values. When the cards are shuffled and one is drawn at random, the first observation must be either 2, 4, or 6. In *sampling with replacement*, the playing card first selected is put back in the deck before the second is

Table 3.6. Calculation of mean and variance of fictitious population of size 3

	Observation, w	Deviation, $w - \mu$	Squared deviation, $(w - \mu)^2$
	2	-2	4
	4	0	0
	6	2	4
Sum	12		8
Calculations:			
	$\mu = \frac{12}{3} = 4$		$\sigma^2 = \frac{8}{3} = 2\frac{2}{3} = 2.667$

drawn; accordingly, the second observation may again be 2, 4, or 6. The total number of possible samples is 3^2 or 9, according to the scheme of table 3.7 (Li, 1964, I). The order in which observations are selected counts—the sample with observation 2 followed by 6 is different from that with observation 6 followed by 2, even though a particular statistic such as the mean is the same for both samples, because other statistics, such as one that gives more weight to the first observation than to the second, are different. In general, the sample size in sampling with replacement is equal to N^n, where N is the population size and n is the sample size. Thus for the example population of size 3 there are three possible samples of size 1 ($3^1 = 3$), namely, 2, 4, and 6; and there are 27 of size 3 ($3^3 = 27$).

Table 3.7. Scheme for sampling with replacement from population of size 3*

First observation	Second observation	Sample
	2	2, 2
2	4	2, 4
	6	2, 6
	2	4, 2
4	4	4, 4
	6	4, 6
	2	6, 2
6	4	6, 4
	6	6, 6

* After Li, 1964, I, p. 37.

Table 3.8. Scheme for sampling without replacement
from population of size 3

First observation	Second observation	Sample
2	4	2, 4
	6	2, 6
4	2	4, 2
	6	4, 6
6	2	6, 2
	4	6, 4

From the same population of size 3, next consider *sampling without replacement*. The first observation is obtained as before; the cards are shuffled and one is drawn at random. Again, the first observation may be 2, 4, or 6. But the first card is not replaced in the deck as before; therefore the second observation cannot be the same as the first. Thus, as table 3.8 shows, there are fewer possible samples, 6 instead of 9. In general the sample size is equal to $N!/(N - n)!$. Thus for the example population of size 3, in sampling without replacement, there are 3 possible samples of size 1, 6 of size 2 $[3!/(3 - 2)!]$, and 6 again of size 3 $[3!/(3 - 3)!]$.

For each sample of a given size from a population, the value of a particular statistic can be calculated. When the means are calculated for each sample of size 2 from the population of size 3, the results obtained are those listed in table 3.9 for sampling with replacement and in table 3.10 for sampling

Table 3.9. Calculation of sample means from samples of size 2
selected without replacement *

First observation	Second observation	Sample	$\sum w$	$\overline{w}$
2	2	2, 2	4	2
	4	2, 4	6	3
	6	2, 6	8	4
4	2	4, 2	6	3
	4	4, 4	8	4
	6	4, 6	10	5
6	2	6, 2	8	4
	4	6, 4	10	5
	6	6, 6	12	6

* From Li, 1964, I, p. 37.

Table 3.10. Calculation of sample means from samples of size 2 selected with replacement

First observation	Second observation	Sample	$\sum w$	$\overline{w}$
2	4	2, 4	6	3
	6	2, 6	8	4
4	2	4, 2	6	3
	6	4, 6	10	5
6	2	6, 2	8	4
	4	6, 4	10	5

without replacement. The frequency distributions of the sample means are graphed as histograms in figures 3.3 and 3.4 to compare with the original distribution of observations in the population.

In sampling with replacement, the sampling is said to be *statistically independent*, because no observation is influenced by the previous one. Among many familiar examples of sampling with statistical independence are tossing a fair coin, where the observations are 1 or 0, depending on whether the coin comes up heads or tails; throwing fair dice, where the sum of the numbers is the observation; and making repeated potassium analyses on a specimen by a nondestructive method such as x-ray fluorescence.

In contrast, in sampling without replacement, the sampling is said to be *statistically dependent*, because each observation is influenced by the preceding one. A familiar example of sampling with statistical dependence is drawing cards from a deck, when the cards are not replaced after each draw.

In general, in sampling, the probabilities change with each observation taken, so that all stages of sampling from complete dependence to complete

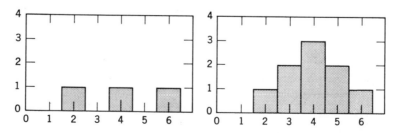

Fig. 3.3. Histogram of original population of size 3 and samples of size 2; sampling with replacement.

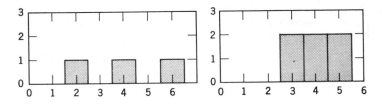

Fig. 3.4. Histogram of original population of size 3 and samples of size 2; sampling without replacement.

independence are possible. For example, if successive cards are drawn from a 52-card deck, the value of the card being the observation, the probability of the first card being a 10 is 4/52 (7.7 percent). The probability of the second card being a 10 is not 4/52, because it depends on the value of the first card: it is 4/51 (7.8 percent) if the first card is not a 10, and 3/51 (5.9 percent) if the first card is a 10. Such probabilities are known, intuitively if not formally, to all poker and blackjack players.

Appropriate mathematical theory has been devised both for sampling with replacement (corresponding to statistical independence) and sampling without replacement (corresponding to statistical dependence). However, as suggested by the changing probabilities in the card-drawing example, the mathematical theory of sampling with replacement is simpler. Fortunately, most if not all geological sampling may be regarded as sampling with replacement, by making an appropriate definition of the sampled population. If the sampled population is regarded as being infinite in size, the theoretical considerations are identical, whether the sampling is done with or without replacement, because drawing 1 or even 100,000 observations from a conceptually infinite number of observations does not change the probabilities.

Although the population of tables 3.6 to 3.8 contains only three observations, the total number of samples of size 2 is 9 in sampling with replacement and is 6 in sampling without replacement. For populations and samples of sizes ordinarily of interest, enumeration of the samples is impractical. Table 3.11 gives the number of all possible samples of sizes 2, 5, and 10, drawn from populations of sizes 10 and 100; even for these small sizes, the number of samples is formidable. Fortunately, it is unnecessary to enumerate the samples in the manner of tables 3.7 and 3.8 because the same results can be obtained through mathematical theory. The details of the mathematics, which are given in books on mathematical statistics and probability, for example, those by Hoel (1962) and Cochran (1963), are not reviewed here. In sections 3.6 and 3.7, the results for the sample mean and the sample variance

Table 3.11. Number of all possible samples of sizes 2, 5, and 10 from populations of sizes 10 and 100, sampling with and without replacement

Population size	Sample size	Number of samples	
		With replacement	Without replacement
	2	100	90
10	5	100,000	30,240
	10	10^{10}	$10! = 3.6288 \times 10^6$
	2	10^4	9900
100	5	10^{10}	$100!/95! = 9.0345 \times 10^9$
	10	10^{20}	$100!/90! = 6.2815 \times 10^{19}$

Note: "$n!$" denotes factorial n; for example, "$10!$" denotes the product of integers from 1 to 10.

are presented. Detailed discussions based on the population 2, 4, and 6 are given to illustrate the similarities and differences of sampling with and without replacement.

3.6   DISTRIBUTION OF THE SAMPLE MEAN

From sampling distributions in general attention is now turned to the sampling distribution of a specific statistic, the sample mean $\bar{w}$. The *sampling distribution of the sample mean* is the theoretical frequency distribution of all sample means calculated from all possible samples of the same size. The purpose of this discussion is to relate the mean and variance of the distribution of the sample mean to the mean and variance of the population. The discussion is introduced with the same simple example of table 3.7 for sampling with replacement from the population of three observations 2, 4, and 6, which defines the theoretical frequency distribution plotted in figure 3.3. From calculations in table 3.6, the mean μ of this population is 4, and the variance σ^2 is $2\frac{2}{3}$. The nine sample means define a new secondary population, whose observations are listed in table 3.9, and graphed in figure 3.3 as a secondary theoretical frequency distribution called the *distribution of the sample mean*. From the nine sample means a mean of sample means $\mu_{\bar{w}}$ and a variance of sample means $\sigma_{\bar{w}}^2$ can be calculated (see table 3.12). Regardless of the initial theoretical frequency distribution of w, comparison of the two kinds of means and variances illustrates two important results:

Table 3.12. Distribution of sample means from samples of size 2 selected with replacement

First observation	Second observation	Sample	$\sum w$	$\overline{w}$	$(\overline{w} - \mu)$	$(\overline{w} - \mu)^2$
	2	2, 2	4	2	−2	4
2	4	2, 4	6	3	−1	1
	6	2, 6	8	4	0	0
	2	4, 2	6	3	−1	1
4	4	4, 4	8	4	0	0
	6	4, 6	10	5	1	1
	2	6, 2	8	4	0	0
6	4	6, 4	10	5	1	1
	6	6, 6	12	6	2	4
Sum				36		12

Calculations:

$$\mu_{\overline{w}} = \tfrac{36}{9} = 4 \qquad \sigma_{\overline{w}}^2 = \tfrac{12}{9} = 1.333$$

1. The mean of sample means is equal to the population mean; that is,

$$\mu_{\overline{w}} = \mu.$$

For the numerical example, the two quantities are equal to 4.

2. The variance of sample means is equal to the population variance divided by the sample size; that is,

$$\sigma_{\overline{w}}^2 = \frac{\sigma^2}{n}.$$

For the numerical example, $\sigma_{\overline{w}}^2$ is equal to $1\tfrac{1}{3}$, and σ^2 is equal to $2\tfrac{2}{3}$.

In table 3.13 the calculations of table 3.12 are repeated for sampling without replacement to illustrate the following results:

1. As before, the mean of sample means is equal to the population mean; that is,

$$\mu_{\overline{w}} = \mu.$$

2. A new formula,

$$\sigma_{\overline{w}}^2 = \frac{\sigma^2}{n} \left(1 - \frac{n}{N}\right)\left(\frac{N}{N - 1}\right),$$

relates the sample variance to the population variance. Because sampling without replacement is seldom if ever important in geological situations, the difference between the two variance formulas is not discussed.

Table 3.13. Distribution of sample means from samples of size 2 selected without replacement

First observation	Second observation	Sample	$\sum w$	$\overline{w}$	$(\overline{w} - \mu)$	$(\overline{w} - \mu)^2$
2 {	4	2, 4	6	3	−1	1
	6	2, 6	8	4	0	0
4 {	2	4, 2	6	3	−1	1
	6	4, 6	10	5	1	1
6 {	2	6, 2	8	4	0	0
	4	6, 4	10	5	1	1
Sum				24		4

Calculations:

$$\mu_{\overline{w}} = \tfrac{24}{6} = 4 \qquad \sigma_{\overline{w}}^2 = \frac{4}{6} = 0.6667 = \frac{\sigma^2}{n}\left(1 - \frac{n}{N}\right)\left(\frac{N}{N-1}\right)$$

The method of calculating the parameters of the distribution of sample means from the population parameters has been stated. If the theoretical frequency distribution followed by the observations is known, the distribution of the sample means can be calculated. It can be shown mathematically (Wilks, 1962, p. 206) that, if the theoretical frequency distribution of the observations is normal, the theoretical frequency distribution of the sample mean is also normal for all sample sizes. It can also be shown mathematically that, even if the theoretical distribution of the observations is not normal, the distribution of the sample mean tends to be nearly normal for moderate sample sizes and becomes more nearly normal as the sample size increases. Even for a sample of size 2 from the far-from-normal population of figure 3.3, the distribution of the sample mean is symmetrical and somewhat bell-shaped. For a sample of size 4 from the same population, the distribution of the sample mean is almost normal.

The preceding facts are based on a remarkable theorem named the *central limit theorem*, which may be stated as follows: *If random samples of fixed size are drawn from a population whose theoretical distribution is of arbitrary shape, but with a finite mean and variance, the distribution of the sample mean tends more and more toward a normal frequency distribution as the size of the sample increases.* The theorem provides no guidance on how large the sample size must be for the distribution of the sample mean to be nearly normal in shape. The size depends on the shape of the original frequency distribution of the observations. For most geological distributions a sample size of 50 to 100 is adequate for the distribution of the sample mean to be nearly normal. The outstanding exceptions to this rule are some distributions of gold assays and other trace elements.

Because of the central limit theorem, whenever the observations are normally distributed, or whenever there is a moderate number of observations even though they are not normally distributed, a discussion of the distribution of the sample mean reduces to a discussion of the normal distribution. Of course, many distributions depend on the values of the two parameters: the mean, which is the population mean, equal to the mean of sample means; and the variance, which is the variance of sample means, equal to σ^2/n. The different distributions may be expressed in terms of the single normal distribution obtained by the linear transformation in section 3.3 because, when the statistic u is defined as

$$u = \frac{\overline{w} - \mu}{\sigma/\sqrt{n}},$$

it is clear from section 4.3 that u has a mean of 0 and a variance of 1. When this equation is solved for $\overline{w}$, the result is

$$\overline{w} = u \frac{\sigma}{\sqrt{n}} + \mu.$$

In table A-2 the normal distribution with μ equal to 0, and σ equal to 1, is tabulated. The first column gives percentage points, and the second gives the standardized normal distribution. The entries in the first line mean that 99.9 percent of calculated values of the statistic

$$\frac{\overline{w} - \mu}{\sigma}$$

will be larger than -3.0902.

The importance of random sampling now becomes apparent. The standardized normal distribution gives the distribution of all possible sample means. If a sample is random, any particular sample has the same chance of being drawn as any other; hence, the chance of its lying between any two percentage points, for example, the 0 and 95 percent points, can be specified. If the sample is not random, no such probability calculation can be made. Although the sample need not be a simple random sample, because restricted or extended sampling methods can be used, the element of randomness must somehow be introduced.

Finally, the variability of the sample mean depends both upon the variability of the original population and upon the sample size; the variability decreases as the sample size increases. Specifically, the standard deviation of the sample mean, $\sigma/\sqrt{n}$, also named the *standard error of the mean*, decreases

TABLE 3.14. FREQUENCY DISTRIBUTIONS OF MEANS OF SAMPLES
OF VARIOUS SIZES DRAWN AT RANDOM FROM A POPULATION OF
900 GOLD ASSAYS FROM THE HOMESTAKE MINE

Class interval (ppm Au)	Sample size			
	1	5	25	100
0–1	439	175	0	0
1–2	120	121	3	0
2–3	67	105	36	1
3–4	44	88	82	5
4–5	35	69	124	38
5–6	26	66	131	149
6–7	23	57	146	222
7–8	14	40	114	201
8–9	23	45	83	198
9–10	14	35	73	98
10–11	16	31	56	49
11–12	13	16	38	22
12–13	13	12	25	10
13–14	13	23	34	5
14–15	9	15	15	2
15–16	10	19	15	
16–17	8	11	7	
17–18	7	5	5	
18–19	9	4	7	
19–20	1	1	1	
20–21	6	5	0	
21–22	7	7	4	
22–23	1	5	2	
23–24	6	7	0	
24–25	1	1	0	
25–26	3	4	0	
26–27	3	5	0	
27–28	4	1	0	
28–29	2	3	0	
29–30	6	6	1	
30–31	2	4		
31–32	6	1		
32–33	7	0		
33–34	1	1		
34–35	0	1		
35–49	14	7		
50–99	21	3		
100–	8			

as the square root of the sample size increases. Furthermore, unless a significantly large fraction of the population is sampled, as seldom if ever happens in geological sampling, the fraction taken has no effect on the reliability of the sample mean, except as reflected in the sample size. Thus, even though the fraction of the population sampled is minute, precision can be attained by taking sufficiently large statistical samples.

The central limit theorem may be illustrated by randomly drawing samples of various sizes from 900 gold values obtained by assaying 1-foot-long EX diamond-drill cores from the Homestake mine. The original frequency distribution, defined for the present purpose as a population of size 900 and normalized in column 1 of table 3.14 to a basis of 1000 observations, is extremely skewed to the right because there are a few very high observations. From the population of size 900, we drew 1000 independent random samples with replacement of sizes 5, 25, and 100; the frequency distributions of the means of these samples are listed in table 3.14. With increasing sample size, the distributions become less and less skewed; the distribution for sample size 100 is skewed slightly and looks nearly normal.

3.7 DISTRIBUTION OF THE SAMPLE VARIANCE AND THE CHI-SQUARE DISTRIBUTION

The discussion of sampling distributions is continued with that of the sample variance s^2. In section 3.6, the distribution of the sample mean is shown to be symmetrical and to approach the normal distribution rapidly as the sample size increases. However, the distribution of the sample variance is more complicated and, for small sample sizes, is noticeably skewed. This discussion of sampling distributions is somewhat different from that in the previous sections of this chapter in that sampling with replacement is discussed first in detail, then sampling without replacement is discussed, at the end of the section.

The distribution of the sample variance is of interest for two reasons: first, the sample variance can be used to estimate the population variance; second, and usually more important for geological problems, the sample variance is required to estimate the reliability of a sample mean, a reliability that is nearly always of great interest.

The distribution of the sample variance is first considered for the samples of size 2 drawn with replacement from the population 2, 4, 6. In table 3.15 the sample variances and standard deviations are calculated for each sample. Figure 3.5, a histogram of distribution of sample variances which can be compared with figure 3.3 for the distribution of population and sample

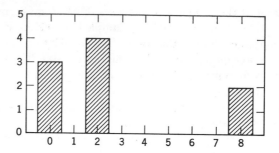

Fig. 3.5. Histogram of sample variances from samples of size 2; sampling with replacement.

means, shows that the distribution of sample variances is skewed in comparison with that of sample means. However, the mean of the sample variances,

$$\frac{\sum s^2}{n} = \frac{24}{9} = 2\tfrac{2}{3} = \sigma^2,$$

is equal to the population variance. The calculation explains why s^2 is calculated by using $(n - 1)$ rather than n as the divisor; the purpose is to make the sample variance equal on the average to the population variance. Thus table 3.15 illustrates the relation, always true, that the mean of sample

TABLE 3.15. DISTRIBUTION OF SAMPLE VARIANCES FROM SAMPLES OF SIZE 2 SELECTED WITH REPLACEMENT *

	Sample	Sample variance, s^2	Sample standard deviation, s
	2, 2	0	0.000
	2, 4	2	1.414
	2, 6	8	2,828
	4, 2	2	1.414
	4, 4	0	0.000
	4, 6	2	1.414
	6, 2	8	2.828
	6, 4	2	1.414
	6, 6	0	0.000
Total		24	11.312
Mean		$\frac{24}{9} = 2\tfrac{2}{3}$	1.257
Corresponding Parameter		$\sigma^2 = 2\tfrac{2}{3}$	$\sigma = 1.633$

* From Li, 1964, I, p. 70.

variances calculated from all possible samples of a given size drawn with replacement from any population is equal to the population variance. However, the mean of sample standard deviations is not equal to the population standard deviation, although it becomes more nearly equal as the sample size increases. Table 3.15 shows the considerable discrepancy for a sample size of 2 for these observations.

Although the relation of sample variance to population variance could be conveniently introduced by the simple example of table 3.15, the most useful means of relating these two kinds of variance is by a new statistic, *chi-square* defined by the formula

$$\chi^2 = \frac{(n-1)s^2}{\sigma^2}.$$

Because, as illustrated in table 3.15, the mean value of the ratio s^2/σ^2 is 1, the mean value of the statistic chi-square must be $(n-1)$; that is, the mean value depends only on the sample size. The quantity $(n-1)$ is named the *degrees of freedom*, abbreviated d.f.; the meaning of the name is revealed as the subject develops. Because the chi-square distribution has a single parameter, the mean, equal to the number of degrees of freedom, there is a separate curve with tabled percentage points for each different number of degrees of freedom. Relating the sample and population variances by the chi-square formula rather than by the simple ratio s^2/σ^2 may seem unnecessarily complicated, but it is done just as the standardized normal deviate u is defined as

$$u = \frac{w-\mu}{\sigma},$$

because table construction is facilitated.

If the original observations come from a normal distribution, the chi-square statistic follows a distribution named the *chi-square distribution*. For the chi-square distribution, as the degrees of freedom increase, the basic shape of the distribution changes. Figure 3.6 gives frequency curves for the chi-square distribution for degrees of freedom from 1 to 6. Although all of the curves are skewed to the right, the skewness is less for the larger values. Although s^2 can never be smaller than 0, the skewness occurs because there is no upper limit to its size, and inclusion of a few extreme values of w in a small sample can lead to a very large value of s^2.

Table A-3 gives percentage points for the chi-square distribution. Each row of the table is for a specific number of degrees of freedom, and each column is for a certain percentage point. Any entry in the table gives the value of the statistic $(n-1)s^2/\sigma^2$ that will be exceeded by a certain percentage of values calculated from a random sample with a certain sample size of number of degrees of freedom. For example, the entry 10.8508 in the column for 95

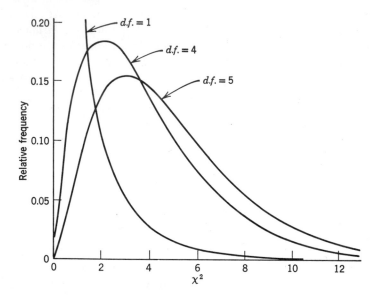

Fig. 3.6. Chi-square distributions for 1, 4, and 5 degrees of freedom.

percent and 20 d.f. means that 95 percent of the values calculated from random samples of size 21 will be *larger* than the tabulated value.

The chi-square table is used in applied statistics, in situations when the variance of a sample needs to be controlled. An example is in the manufacture of lamp bulbs, where the variation in life needs to be controlled within limits so that the bulbs will burn long enough on the average to meet specifications. In this book several applications to geology are given in later chapters for the construction of interval estimates for the variance in section 4.6, and for the curve fitting in section 6.1.

Figure 3.6 indicates that, as the degrees of freedom increase, the skewness of the chi-square distribution decreases. Because the sample variance s^2 is a kind of average, the central limit theorem, that averages tend to follow the normal distribution with increasing sample size, should apply. Table 3.16 presents results and calculations for 50 degrees of freedom to illustrate the good correspondence of percentage points calculated by using the normal distribution to the exact values in the chi-square table. As shown, the central limit theorem indeed applies even to this originally highly skewed chi-square distribution. The calculations are based on the relation we have presented— that the mean of the chi-square distribution is equal to the number of degrees of freedom—and on the relation, not illustrated in this book, that the standard deviation of the chi-square distribution with $n - 1$ degrees of freedom is

TABLE 3.16. APPLICATION OF THE CENTRAL LIMIT
THEOREM TO THE CHI-SQUARE DISTRIBUTION

Source of value	95% point	5% point
Exact values from table A-3	34.8	67.5
Values from normal approximation, calculations below	33.5	66.5

Calculations:

$$\text{d.f.} = 50$$
$$n - 1 = 50$$
$$\mu = n - 1$$
$$= 50$$
$$\sigma = \sqrt{2(n - 1)}$$
$$= \sqrt{2(50)}$$
$$= \sqrt{100}$$
$$= 10$$
$$\text{Percentage point} = 1.645$$
$$95\% \text{ point} = 50 - 1.645 \times 10$$
$$5\% \text{ point} = 50 + 1.645 \times 10$$

$\sqrt{2(n - 1)}$. According to normal theory (see table A-2) the value for the 95-percent point is 1.645.

The name *degrees of freedom* is somewhat troublesome because its usage is partly descriptive. It is related in a sense to the usage in the phase rule of physical chemistry or the mineralogical phase rule, but it is not exactly comparable. As this book progresses, the usage will become clearer. In general, the number of degrees of freedom is simply the number used to divide the sum of squared deviations from the mean to get a sample variance s^2 that is an unbiased estimate of the population variance σ^2. Because

$$s^2 = \frac{\text{SS}}{n - 1},$$

the chi-square statistic,

$$\chi^2 = \frac{(n - 1)s^2}{\sigma^2},$$

can be rewritten, substituting SS for $(n - 1)s^2$, to obtain

$$\chi^2 = \frac{\text{SS}}{\sigma^2}.$$

Indeed, the chi-square statistic is often introduced in the form SS/σ^2. This formulation has been postponed to this point to emphasize the key importance of the degrees of freedom. Although thus far in this book the sample variance has been obtained by dividing the SS by $(n-1)$, some more complicated calculations to be presented later require division of SS by some different quantity to get the sample variance. Then the mean value of the chi-square statistic is not $(n-1)$ but equal to whatever divisor is used.

In many older books, s^2 is defined as exactly analogous to the sample mean by using a divisor of n, as used for the sample mean calculation, rather than $(n-1)$. In most modern books, including this one, only $(n-1)$ is used for several reasons, perhaps the most important being that, because all statistical tables involving the variance are constructed on the basis of degrees of freedom, they cannot be applied exactly if s^2 has been calculated by using n. Also, although the numerical difference between n and $(n-1)$ is small for large sample sizes, numerically large discrepancies may arise even for large sample sizes if complicated estimates of s^2 are required.

The preceding discussion entirely concerns variances of random samples drawn from a normal population. But even for samples drawn from a non-normal population, the mean of sample variances for all possible samples is equal to the population variance, as is illustrated by variances calculated for the samples previously drawn for the 900 Homestake gold values (sec. 3.6). In table 3.17 are recorded the population mean μ equal to 7.59 ppm Au and the variance σ^2 equal to 327.21 for the 900 observations. For the samples of sizes 5, 25, and 100, the calculated mean of sample means and the mean of sample variances are tabulated. As expected, the mean of sample means is nearly equal to the population mean, that is,

$$\mu_{\bar{w}} \approx \mu,$$

TABLE 3.17. DISTRIBUTION OF SAMPLE VARIANCES FOR GOLD DATA FROM THE HOMESTAKE MINE

	$\mu = 7.59$ ppm Au, $\sigma^2 = 327.21$				
Sample size	Mean		Variance, $s^2_{\bar{w}}$	Ratio of population variance to sample variance, $\sigma^2/s^2_{\bar{w}}$	
	$\bar{w}$	s^2		Predicted	Observed
5	6.67	300	57.40	5	6
25	7.59	313	12.60	25	26
100	7.56	324	3.08	100	106

and the mean of sample variances is nearly equal to the population variance, that is,

$$\mu_s^2 \approx \sigma^2.$$

The fourth column lists the variances of the sample means, which should be equal to the population variance divided by the sample size, according to the formula

$$\sigma_{\bar{w}}^2 = \frac{\sigma^2}{n}.$$

The ratios of population variance to predicted and observed variances of sample means are given in columns 5 and 6. If all possible samples had been taken, this ratio would be exactly equal to the sample size; because only 1000 were taken, it is only approximately equal to the sample size, but close enough to verify the relationship. However, if the observations are not normally distributed, the distribution of the quantity SS/σ^2 is not necessarily close to the chi-square distribution because the distribution of this quantity is quite sensitive to the normal assumption being met. Fortunately, because the sample variance seldom need be estimated for itself, this problem is not serious.

Finally, the distribution of the sample variance in sampling without replacement is illustrated through table 3.18, which is constructed like table

TABLE 3.18. DISTRIBUTION OF SAMPLE VARIANCES FROM SAMPLES OF SIZE 2 SELECTED WITHOUT REPLACEMENT

Sample	Sample variance, s^2	Sample standard deviation, s
2, 4	2	1.414
2, 6	8	2.828
4, 2	2	1.414
4, 6	2	1.414
6, 2	8	2.828
6, 4	2	1.414
Sum	24	11.312

Mean
$$s^2 = \tfrac{24}{6} = 4$$
$$s = 1.885$$

Corresponding parameter
$$\sigma^{*2} = \left(\frac{N}{N-1}\right)\sigma^2$$
$$= \tfrac{3}{2} \times \tfrac{8}{3} = 4$$
$$\sigma^* = 2$$

3.15. Table 3.18 shows that for sampling without replacement, the mean of sample variances is 4 rather than the $2\frac{2}{3}$ calculated for sampling with replacement. Therefore, when sampling without replacement, either the sample variance or the population variance must be redefined to make the sample variance an unbiased estimate of the population variance. It is conventional to redefine the population variance σ^2 as a new variance σ^{*2} according to the formula,

$$\sigma^{*2} = \frac{N}{N-1}\,\sigma^2,$$

where N, as before, is the population size. When this is done, the table illustrates that the sample variance, $s^2 = 4$, is an unbiased estimate of the redefined population variance $\sigma^{*2} = 4$. As in sampling with replacement, the sample standard deviation is not an unbiased estimate of the population standard deviation.

REFERENCES

Cochran, W. G., Mosteller, Frederick, and Tukey, J. W., 1954, A report on sexual behavior in the human male: Am. Statistical Assoc., p. 18.

Cochran, W. G., 1963, Sampling techniques: New York, John Wiley & Sons, 413 p.

Exley, C. S., 1963, Quantitative areal modal analysis of granitic complexes: a further contribution: Geol. Soc. Am. Bull., v. 74, p. 649–654.

Hoel, P. G., 1962, Introduction to mathematical statistics: New York, John Wiley & Sons, 427 p.

Koch, G. S., Jr., and Link, R. F., 1967, Gold distribution in diamond-drill core from the Homestake mine, Lead, South Dakota: U.S. Bur. Mines Rept. Inv. 6897, 27 p.

Krumbein, W. C., 1960, Some problems in applying statistics to geology: Applied Statistics, v. 9, no. 2, p. 82–91.

Krumbein, W. C. and Graybill, F. A., 1965, An introduction to statistical models in geology: New York, McGraw-Hill, 475 p.

Li, J. C. R., 1964, Statistical inference: Ann Arbor, Mich., Edwards Bros, v. 1, 658 p.

Link, R. F., and Koch, G. S., Jr., 1962, Quantitative areal modal analysis of granitic complexes: discussion of article by E. H. T. Whitten: Geol. Soc. Am. Bull., v. 73, p. 411–414.

Rand Corporation, The, 1955, A million random digits with 100,000 normal deviates: Glencoe, Illinois, Free Press, 200 p.

Rosenfeld, M. A., 1954, Petrographic variation in the Oriskany Sandstone (abs.): Geol. Soc. Am. Bull., v. 65, p. 1298–1299.

Wilks, S. S., 1962, Mathematical statistics: New York, John Wiley & Sons, 644 p.

Chapter 4

Inference

Chapters 2 and 3 are about descriptive statistics. Chapter 2 explains how observations can be summarized by an empirical frequency distribution and introduces theoretical frequency distributions and probability. Chapter 3 explains the sampling distributions of two statistics, the mean and the variance, for observations that follow the normal distribution. In this chapter, the material in Chapters 2 and 3 is extended for making *inferences* about the population mean and the population variance from the information contained in a single statistical sample. By *inference* is meant the drawing of a conclusion or conclusions about a population from a sample by inductive reasoning.

Two methods of inference are explained. The first is interval estimation to determine the range of parameter values that are consistent with the information in a sample. The second is the hypothesis test to learn whether the information in a sample is consistent with a postulated value or range of values for a parameter. In making both kinds of inferences, mistakes are inevitably made because the source of information is one or more statistics whose values fluctuate from sample to sample. Evaluation of these mistakes and their control within acceptable bounds constitutes the basic contribution of statistical theory to the analysis of data.

4.1 ESTIMATION OF POPULATION PARAMETERS

In Chapters 2 and 3, we point out that many kinds of numerical data may be represented by a theoretical frequency distribution. However, to represent data completely, we must not only choose a mathematical function but also specify parameter values. Thus, for the normal distribution, several different distributions (figs. 2.5 and 2.6) are plotted for different values of the parameters μ and σ. Now, in this chapter one of the central problems of statistics is introduced: How, assuming a theoretical frequency distribution, can the population parameters best be estimated from a sample?

A population parameter is estimated by calculating the value of a statistic from the observations in a statistical sample. Because more than one statistic can always be calculated to estimate a given parameter, it is necessary to choose among the statistics. Statistical criteria may aid in making a choice, but application of the statistical criteria depends on correspondence of data to underlying assumptions about the theoretical frequency distribution. Thus, to estimate a given parameter, no one statistic is always best. This important fact is often forgotten.

The two most important statistical criteria for choosing a statistic to estimate a given population parameter are now explained. One criterion is that a statistic should be an *unbiased estimate* of a population parameter. An unbiased estimate is one with a sampling distribution whose mean is equal to the estimated parameter. For example, the sample mean is an unbiased estimate of the population mean, but the sample standard deviation is a biased estimate of the population standard deviation (secs. 3.6 and 3.7). The *bias* δ is equal to the difference between the average value of the statistic and the value of the parameter being estimated. Thus, for the examples,

$$\delta = \mu_{\bar{w}} - \mu = 0,$$

but

$$\delta = \mu_s - \sigma \neq 0.$$

Although unbiased estimates are desirable, estimates with a small bias, for example, the sample standard deviation, may be perfectly acceptable and useful for some purposes.

The second criterion is that an estimate should be *efficient*. An *efficient estimate* is a statistic that, for a given sample size, has the smallest mean-square error of all the statistics that might be calculated. The mean-square error (M.S.E.) may be written as

$$\text{M.S.E.} = \sigma_g^2 + \delta^2,$$

where σ_g^2 is the variance of the statistic g calculated from the statistical sampling distribution of g, and δ is the bias of the statistic g. If the statistic is an unbiased estimate, δ is 0, and the preceding equation becomes

$$\text{M.S.E.} = \sigma_g^2.$$

TABLE 4.1. RELATIVE EFFICIENCIES OF THREE STATISTICS FOR ESTIMATING THE POPULATION MEAN

Name of statistic	Value of statistic	Variance of statistic	Efficiency of estimate (%)
Mean, $\bar{w}$	5	$\sigma^2/n = 0.2\sigma^2$	100
Midrange	4.5	$0.261\sigma^2$	77
Median	6	$0.287\sigma^2$	70

If the sampling distribution of the statistic is symmetrical and more or less normal and if the statistic is an unbiased estimate, the efficient estimate is likely to be the best that can be made under any circumstances.

The estimation of a parameter by several statistics is illustrated by considering three statistics that can be used to estimate the population mean μ. In section 3.3 the population mean is estimated by the sample mean $\bar{w}$, but the estimation can be made by other statistics, for example, the midrange and the median. How does one decide which estimate to use? In table 4.1 these three statistics are calculated for the fictitious data of table 3.1. In column 1 the value of the statistics is given. From table 3.1 the mean is 5. The midrange, the average of the highest and lowest observations, is $(2 + 7)/2 = 4.5$. The median, the value with half the observations larger and half smaller, is 6. The variance for the mean is equal to the population variance divided by the sample size of 5. For the midrange and the median, the variances of their sampling distributions can be obtained from the population variance by multiplying by constants tabulated by Dixon and Massey (1969, table A-8b(4), p. 488). Because from other considerations the mean is known to be the best estimate, the last column shows that the other estimates are only 70 to 77 percent as efficient.

Departures of data from the assumptions made about the theoretical frequency distribution can affect both the lack of bias and the efficiency of estimates of parameters. Sometimes a choice of estimate may be made. For example, one may choose to lose some efficiency to guarantee not to introduce bias. On the other hand, in another situation, one might choose to keep efficiency up, even though a moderate bias would be introduced. If one can find estimation procedures that are relatively insensitive to departures of the data from the underlying assumptions, he has a high degree of confidence in the conclusions drawn. But if the procedures are very sensitive to departures of data from the assumptions, then one must be wary of the results. We attempt to choose and present procedures that do not depend sensitively on departures of data from the underlying assumptions; such procedures are named *robust*. Because no procedure is completely robust, it is difficult to state quantitatively how insensitive a particular procedure is, but procedures in this book that are clearly very sensitive are pointed out.

So far only statistical criteria have been given for choosing among statistics. Actually there are other criteria based on such matters as the calculation time required, the costs of obtaining different kinds of observations, and geological judgments about how well data correspond to assumptions. Thus, when there are several statistics available to estimate a parameter, it is necessary to consider carefully which is most appropriate for a particular purpose. We sometimes present several methods, as, for example, in section 6.2, in which four methods are given for calculating a coefficient of variation for data that follow the lognormal frequency distribution. Each of these four methods has good and bad points, and to choose among them one must think not only about what the data are like internally and how they are related to the geological environment but also about how much time is to be spent in calculating and what use is to be made of the final statistic.

The choice of statistic may also be complicated by lack of the appropriate statistical theory to give criteria for choosing among different statistics. Theory is particularly lacking for the effects of departures from normality, although this field is one of active statistical research today in which better understanding can be anticipated in the next decade.

4.2   POINT AND INTERVAL ESTIMATES

In section 3.3 the methods of calculating the mean $\overline{w}$ and the variance s^2 of a statistical sample are explained. For five observations with the values 2, 4, 6, 6, and 7 the mean is 5 and the variance is 4. Both $\overline{w}$ and s^2 are point estimates, each with a single value. In sections 3.6 and 3.7 it is also shown that these statistics are unbiased estimates of the corresponding population parameters because, if all possible samples are taken, the mean of the corresponding sampling distribution is equal to the value of the parameter being estimated.

There are two disadvantages to a point estimate. The first is that a point estimate is almost sure to be wrong. Of the 9 samples of size 2 drawn from the population 2, 4, 6 (table 3.9), only three give the population mean of 4, even though the mean of all sample means is an unbiased estimate of the population mean. In the real world an estimate of driving time between two points will not be accurate to the minute; an estimate of grade of ore in a mine will not exactly check with the grade as finally mined; an estimate that a granite contains 65.1 percent quartz will be questioned by the first new analysis. The second disadvantage to a point estimate arises from the first:

the point estimate conveys no information about how wrong it is liable to be.

Because of the disadvantages of point estimates, the *confidence interval* estimate was devised to incorporate three basic attributes of a sample: the mean, the variability as measured by the standard deviation, and the sample size. In interval estimates the parameter being estimated is specified to lie between two values, called *confidence limits*, for a specified percentage of the intervals so calculated. Instead of a point estimate that a granite contains 65.1 percent quartz, a confidence interval estimate is made of the form: with a 90-percent confidence, the granite contains not less than 60.1 percent nor more than 70.1 percent quartz. Confidence intervals are of two kinds: two-sided and one-sided. A two-sided interval, like the illustration, is bounded on both sides by a calculated value. In a one-sided interval, one side is bounded by a noncalculated value, such as 0 or 100 percent, about which there is no risk of being wrong. A statement for a one-sided confidence interval is of the form: with a 90-percent confidence, the granite contains not less than 62.8 percent quartz.

The two procedures most used today for calculating interval estimates were introduced by Fisher in 1935 [1950, (reprinted) paper 25, p. 391] and by Neyman (1937). Fisher coined the name *fiducial intervals*, and Neyman used the name *confidence intervals*. At present confidence intervals are more used, although fiducial intervals are also used, particularly by British workers. In practice, although the definitions and calculations differ for some multivariate cases, the two methods usually give the same results, and in this book only confidence intervals are calculated. The differences between the two kinds of intervals are discussed by Wilks (1962, p. 370).

In order to calculate confidence limits about a mean, the standard deviation and the sample size are used. It seems to be intuitively obvious that, if the variability among observations as measured by the standard deviation is small, an estimate of the mean within narrow limits can be obtained; conversely, if the variability is large, an estimate of the mean within only broad limits is available. Moreover, it seems clear that a better estimate of the mean can be obtained with many observations than with only a few.

These concepts can now be put into formal terms. In section 3.6 on the distribution of the sample mean, it is shown that, if $\overline{w}$ is the mean of a random sample from a normal distribution with mean μ and variance σ^2, the probability is 90 percent (9 chances out of 10) that $\overline{w}$ is not more than 1.645 standard deviations larger or smaller than μ. In other words, in nine random samples out of 10, the sample mean is in the calculated interval. This relation may be written as the algebraic inequality

$$\mu + 1.645 \frac{\sigma}{\sqrt{n}} > \overline{w} > \mu - 1.645 \frac{\sigma}{\sqrt{n}}.$$

Subtracting the quantity $(\mu + \overline{w})$ from this inequality yields

$$-\overline{w} + 1.645\,\frac{\sigma}{\sqrt{n}} > -\mu > -\overline{w} - 1.645\,\frac{\sigma}{\sqrt{n}},$$

and multiplying through by -1 gives the desired result, that

$$\overline{w} - 1.645\,\frac{\sigma}{\sqrt{n}} < \mu < \overline{w} + 1.645\,\frac{\sigma}{\sqrt{n}}.$$

This interval is presented graphically for two different values of σ in figure 4.1. If σ is known, the interval may be evaluated to give a 90-percent confidence interval for μ. Such an interval is said to have a 90-percent *confidence coefficient*. If such intervals are calculated repeatedly for random samples

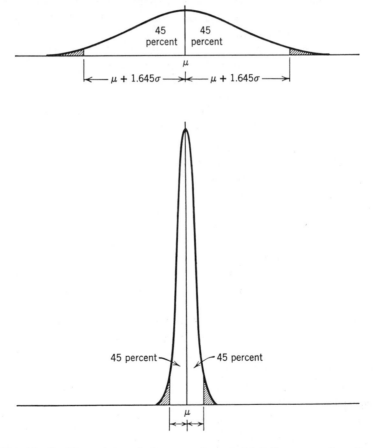

Fig. 4.1. Confidence intervals for two values of σ^2 but the same value of μ.

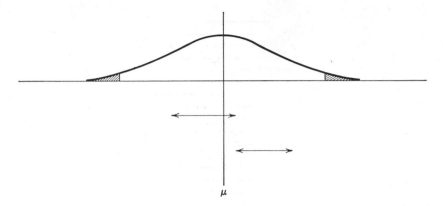

Fig. 4.2. Two confidence intervals, one containing μ and one not containing μ.

from a normal distribution, 9 out of 10 will be correct (that is, they will contain μ); whereas 1 out of 10 will be incorrect (that is, they will not contain μ). In figure 4.2, two intervals are graphed, one containing μ and one not containing μ. In figure 4.3 fifty 90-percent confidence intervals are plotted for samples of size 10 drawn from a table of random numbers from a population with a mean of $\mu = 50$, and a standard deviation of $\sigma = 100$. For these 50 intervals, 5 (or 10 percent) are incorrect as expected; they do not contain μ. Of course, in practice, whenever a confidence interval is calculated, it is unknown whether it is one of the correct ones, but the interval does give some feel for the parameter values, and the interval width gives an idea of the closeness of the estimate.

The method of calculating confidence limits outlined so far is correct, but generally it is useless because σ is seldom known. Therefore σ must usually be replaced by its estimate s. First, however, it is necessary to introduce another sampling distribution, Student's t-distribution.

4.3   STUDENT'S t-DISTRIBUTION

In section 4.2 it is pointed out that interval estimates can be calculated from the standardized normal distribution,

$$\text{s.n.d.} = \frac{\overline{w} - \mu}{\sigma_{\overline{w}}} = \frac{\overline{w} - \mu}{\sigma/\sqrt{n}},$$

with μ equal to 0 and σ equal to 1, provided that σ is known. Because σ is

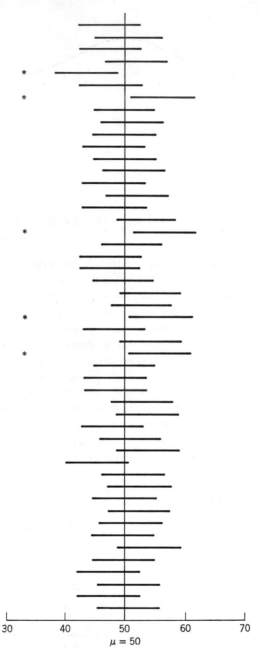

Fig. 4.3. Fifty 90-percent confidence intervals for samples of size 10 from a population with $\mu = 50$, $\sigma = 100$; σ assumed to be known.

seldom if ever known, however, it must be replaced by its estimate s, the sample standard deviation, to form the new statistic,

$$t = \frac{\bar{w} - \mu}{s/\sqrt{n}},$$

named *Student's t* after its originator, W. S. Gosset, who used the pseudonym "Student."

The frequency distribution of the statistic t is a new sampling distribution. In sections 3.6 and 3.7 the sampling distributions for the sample mean and the sample variance are illustrated for all possible samples of size 2 from a population of size 3. A similar illustration for the t-statistic is given by Li (1964, I, p. 100). The distribution of the t-statistic for normally distributed observations has been calculated mathematically.

Unlike the standardized normal distribution, which is one curve, the t-distribution is a family of curves, one for each number of degrees of freedom, as illustrated in figure 4.4 for selected numbers of degrees of freedom. Selected percentage points for different numbers of degrees of freedom are given in table A-4. To find t values for degrees of freedom not tabulated it is often accurate enough to use the value for the next smallest tabulated number of degrees of freedom. A more exact value may be obtained by linear interpolation by using the reciprocal of the number of degrees of freedom; for example, interpolation for 240 degrees of freedom can be done as follows:

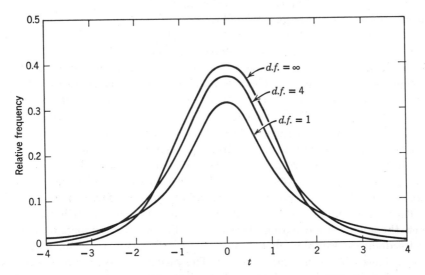

Fig. 4.4. Student's t-distributions with various degrees of freedom.

between the tabulated 5 percentage points for 120 and infinite degrees of freedom

$$1/120 = 0.0083333 \qquad t = 1.980$$
$$1/240 = 0.00416667 \qquad t = 1.970 \quad \text{(by interpolation)}$$
$$1/\infty \; = 0.00000000 \qquad t = 1.960.$$

It is convenient to explain the t-table, starting with the bottom row of numbers for an infinite number of degrees of freedom, for which s^2 is equal to σ^2, because of the following simple algebraic argument. As stated in section 3.7, the mean of the statistic

$$\frac{(n-1)s^2}{\sigma^2}$$

is $n-1$, and the variance is $2(n-1)$.

If each of these statistics is divided by the constant $(n-1)$, it is known by the rule for dividing a statistic by a constant (sec. 3.3), that the new mean is equal to the original mean divided by $(n-1)$, and the new variance is equal to the original variance divided by $(n-1)^2$. The results are that the mean of the statistic

$$\frac{s^2}{\sigma^2}$$

is 1 and the variance is $2/(n-1)$. As the number of degrees of freedom approaches infinity, the variance of s^2/σ^2 approaches 0; therefore for infinity degrees of freedom s^2 becomes equal to σ^2.

The numbers in the bottom row of the t-table are the same for corresponding percentage points of the standardized normal distribution in table A-2; for instance, for the 5-percent point t equals 1.645, meaning that 5 percent of calculated t-values with infinity degrees of freedom will be larger than 1.645. The only difference between the bottom row of table A-4 and the lower half of column 2 of table A-2 is that fewer percentage points are listed in table A-4. Only the upper *tail* of the t-distribution is tabulated. The percentage points of the lower tail are found from the symmetry of the t-distribution about 0. Thus the 95-percent point of the t-distribution has the same numeric value as the 5-percent point of the t-distribution with a negative sign. For example, the 95-percent point for t with infinity degrees of freedom is -1.645.

Thus for an infinite number of degrees of freedom the t-distribution is, in fact, the standardized normal distribution. For smaller numbers of degrees of freedom, the sampling distribution of t is still symmetrical about 0; but the variance of t becomes larger than 1. The variance becomes larger because, for an infinite number of degrees of freedom, only the numerator $\overline{w}$ varies from sample to sample; but for smaller degrees of freedom, both the numerator $\overline{w}$ and the denominator $s/\sqrt{n}$ vary from sample to sample. Whenever s as calculated from a particular sample is smaller than σ, the term $s/\sqrt{n}$ is smaller than $\sigma/\sqrt{n}$, a condition that leads to a spread-out distribution.

In the t-table the numbers become smaller in the lower part of each column because, as the number of degrees of freedom increases, the stability of the estimate s of σ in the denominator increases. The numbers in each row become larger from right to left because, to get in a larger and larger percentage of the t-distribution, we must go farther outward from the mean.

4.4 CONFIDENCE INTERVALS FOR THE POPULATION MEAN

In this section the t-statistic is used to calculate confidence intervals for the population mean μ and for the difference between two population means. The expression for a two-sided confidence interval for the population mean with σ unknown,

$$\overline{w} - t_{5\%}\frac{s}{\sqrt{n}} < \mu < \overline{w} + t_{5\%}\frac{s}{\sqrt{n}},$$

is like that with σ known, presented in section 4.2, except that σ is replaced by its estimate s, and the 5-percent point of the standardized normal distribution, 1.645, is replaced by the corresponding t-value with the appropriate number of degrees of freedom from table A-4. In table 4.2 calculations of

TABLE 4.2. EXAMPLE CALCULATIONS OF TWO-SIDED CONFIDENCE INTERVALS

	Source of data		
Calculations	Fictitious data from table 3.1	Phosphate data from table 3.4	Fresnillo mine, 2200 vein, 305-m level (% Pb)
$\overline{w}$	5	25.23	9.42
s	2	3.02	8.47
n	5	224	218
$\sqrt{n}$	2.236	14.967	14.731
$s/\sqrt{n}$	0.8945	0.2018	0.5750
d.f.	4	223	217
$t_{5\%}$	2.132	1.651	1.651
$t(s/\sqrt{n})$	1.9	0.33	0.95
μ_L	3.1	24.90	8.47
μ_U	6.9	25.56	10.37

Formulas:

$$\mu_L = \bar{w} - t_{5\%} \frac{s}{\sqrt{n}}$$

$$\mu_U = \bar{w} + t_{5\%} \frac{s}{\sqrt{n}}$$

confidence intervals are presented for three sets of data. The first two sets are the fictitious data used to introduce the calculation of mean and variance in table 3.1 and the phosphate data from table 3.4. The third set is 218 lead values from assays of mine samples taken at 2-meter intervals in the Fresnillo mine. The lead-assay data are typical of base-metal assay data in that the coefficient of variation is 0.90 in contrast to the phosphate data with a coefficient of variation of only 0.12. Although the number of observations is the same in both sets of data, the confidence intervals for the lead assays are almost three times as wide, as we note when we compare the values of $t(s/\sqrt{n})$ for the two cases.

In figure 4.3 fifty 90-percent confidence intervals for random samples of size 10 from a population with a mean of 50 and a known population variance of 100 are plotted. In figure 4.5, a comparison plot is presented, the single change being the different assumption that the population variance is unknown so that its estimate s is calculated for each sample. In figure 4.5, unlike figure 4.3, the confidence intervals are of different lengths because the value of s changes from sample to sample. Although the mean of s is nearly equal to σ, on the average the intervals are wider because the $t_{5\%}$-value for 9 degrees of freedom, 1.833, replaces the corresponding standard-normal-deviate value of 1.645. However, on the average 90 out of 100 intervals still contain the population mean μ. In the case of figure 4.5, 46 (92 percent) contain the population mean, showing good correspondence to theory. The first three misses are the same as those for the case of figure 4.3 with σ assumed known, but the fourth is different. Because s is calculated separately for each sample, exact correspondence with figure 4.3 is not to be expected, although there is some relationship because the means are the same in both cases.

In calculating the confidence intervals, a confidence coefficient of 90 percent was used. If another confidence coefficient is adopted, the value of t is different, and the confidence interval has a different width for the same data. In table 4.3, confidence intervals are calculated for the Fresnillo data using different confidence coefficients. For the confidence coefficient of 50 percent, the width of half the interval, $t(s/\sqrt{n})$, is only 0.39, but there is a 50-percent chance of making a mistake. However, for a confidence coefficient of 98 percent the chance of making a mistake is only 2 percent, but the width of half

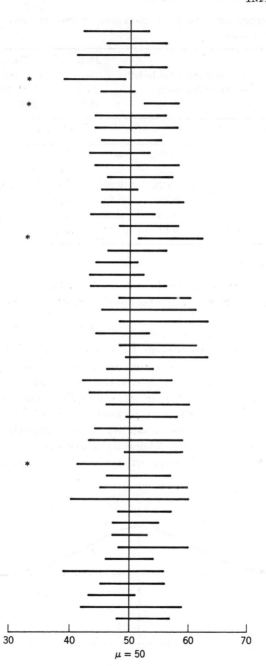

Fig. 4.5. Fifty 90-percent confidence intervals for samples of size 10 from a population with $\mu = 50$, $\sigma = 100$; σ assumed to be unknown.

TABLE 4.3. EFFECT OF CHANGING CONFIDENCE COEFFICIENT ON LENGTH OF CONFIDENCE INTERVAL

		Item	Value
Summary data from table 4.2		$\overline{w}$	9.42
for the Fresnillo mine		$s/\sqrt{n}$	0.5750
		d.f.	217

Item	Values				
Confidence coefficient	50%	80%	90%	95%	98%
Percentile of t	$t_{25\%}$	$t_{10\%}$	$t_{5\%}$	$t_{2.5\%}$	$t_{1\%}$
Value of t	0.676	1.286	1.651	1.970	2.342
$t(s/\sqrt{n})$	0.39	0.74	0.95	1.13	1.35
Lower confidence limit, μ_L	9.03	8.68	8.47	8.29	8.07
Upper confidence limit, μ_U	9.81	10.16	10.37	10.55	10.77

the interval has increased more than threefold to 1.35. Figure 4.6 illustrates that narrowing the interval provides less chance of catching the population mean.

The confidence coefficient is chosen by the investigator. We usually use a confidence coefficient of 90 percent for confidence intervals involving the mean, so that, for random samples from normally distributed populations, 90 percent of the confidence intervals contain the population mean μ. Some geologists, and most investigators in other disciplines, conventionally choose higher confidence coefficients. Our preference for the 90-percent level is made for three reasons: (a) the variability in most geological data is larger than in data from controlled laboratory experiments or manufacturing processes, (b) the likelihood always exists that the geological body or other

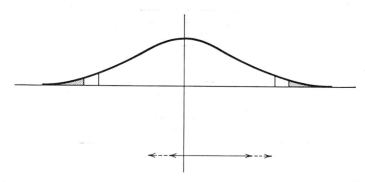

Fig. 4.6. Relation of width of confidence interval to confidence coefficient.

entity under investigation is terminated by an unexpected structural discontinuity, and (c) in applications involving valuation and other subjects of mineral economics, use of the 90-percent level is consistent with the higher risks associated with the mineral industries, risks that are illustrated by the higher interest rates in mineral industries than in manufacturing industries.

From two-sided confidence intervals we now turn to the actual calculation of one-sided confidence intervals by using the t-statistic, where the confidence statement is of the form "with a specified confidence of being correct, the population mean is not lower than a specified value," or of the form "with a specified confidence of being correct, the population mean is not higher than a specified value." The first kind of interval is named a *one-sided lower confidence interval*, and the second kind a *one-sided upper confidence interval*. In figure 4.7 a one-sided lower confidence interval is illustrated schematically. The words "one-sided" refer to the fact that only one side is calculated; actually the other side is present but is known, so there is no chance of its being incorrect. It is known to be either plus or minus infinity—or the extreme value possible, such as 0 or 100 percent or the content of the metal being estimated that is present in the pure mineral. A one-sided 90-percent lower confidence interval is calculated from the relation

$$\overline{w} - t_{10\%} \frac{s}{\sqrt{n}} < \mu < \infty,$$

and a one-sided 90-percent upper confidence interval is calculated from the relation

$$-\infty < \mu < \overline{w} + t_{10\%} \frac{s}{\sqrt{n}}.$$

In mining geology, in which it is usually more important that the ore be above cutoff grade than to set an upper limit to the grade, the one-sided

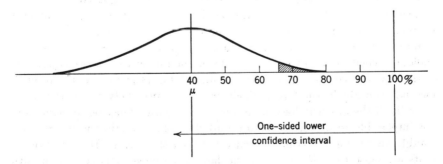

Fig. 4.7. Schematic representation of one-sided confidence interval.

TABLE 4.4. EXAMPLE CALCULATIONS OF ONE-SIDED LOWER CONFIDENCE INTERVALS

| Calculations | Source of data | | | Remarks |
	Fictitious data from table 3.1	Phosphate data from table 3.4	Fresnillo mine, 2200 vein, 305-m level (% Pb)	
$\overline{w}$	5	25.23	9.42	From table 4.2
$s/\sqrt{n}$	0.8945	0.2018	0.5750	From table 4.2
d.f.	4	223	217	
$t_{10\%}$	1.533	1.286	1.286	From table A-4
$t(s/\sqrt{n})$	1.4	0.26	0.74	
μ_L	3.6	24.97	8.68	

confidence interval is especially pertinent. In table 4.4 one-sided confidence limits are calculated for the data in table 4.2. Comparison of the one-sided confidence limits in table 4.4 to the two-sided confidence limits in table 4.2 shows that the one-sided limits are substantially higher. In estimating grade of ore, the differences may well be enough to put an estimate above rather than below a cutoff grade.

One more kind of confidence interval for the mean, the confidence interval for the difference between two population means, is now explained. The calculation is made for a two-sided interval in the fictitious illustration of table 4.5. Assume that in a study comparing average compositions of marine and fresh-water shales, a geologist has obtained the Fe analyses in table 4.5. The six marine shales (group 1) yield a mean Fe analysis of 4.883. The seven fresh-water shales (group 2) yield a mean Fe analysis of 4.857. The question is whether the population means for the two kinds of shales are evidently the same or different.

If the population mean of the marine shales is designated μ_1 and the population mean of the fresh-water shales is designated μ_2, a confidence interval can be constructed for the difference $(\mu_1 - \mu_2)$ between these two sampling processes. The interval reveals several things. First, for the chosen confidence coefficient, the interval contains the differences in the population means that are consistent with the sample data. Therefore, if the interval does not contain 0, the two population means are likely to be the same. Secondly, if the interval does contain 0, the two means may be the same, but it is impossible to be certain because all the values within the interval are suitable values for the true difference between the means. If both interval limits are close to 0, there can be no large differences between the means. How close is "close" depends on the geologist's judgment.

TABLE 4.5. EXAMPLE CALCULATION OF TWO-SIDED CONFIDENCE INTERVAL FOR DIFFERENCE OF TWO POPULATION MEANS *

Item	Statistical sample number				Combination	
	(1)		(2)			
Observa-tions:	6.9	6.5	4.8	4.7		
	3.8	3.5	3.9	6.0		
	4.9	3.7	5.9	3.7		
			5.0			
$\sum w$	29.3		34.0			
n	6		7			
$\overline{w}$	4.883		4.857		$\overline{w}_1 - \overline{w}_2$	0.026
$(\sum w)^2$	858.49		1156.00			
$(\sum w)^2/n$	143.08		165.14			
$\sum w^2$	154.25		169.84			
SS	11.17		4.70		pooled SS	15.87
d.f.	5		6		pooled d.f.	11
					s_p^2	1.4427
$1/n$	0.1667		0.1429		$1/n_1 + 1/n_2$	0.3096
					$s_p^2(1/n_1 + 1/n_2)$	0.4467
					$\sqrt{s_p^2(1/n_1 + 1/n_2)}$	0.6683
					$t_{5\%}$, with 11 d.f.	1.796
					$t_{5\%}\sqrt{s_p^2(1/n_1 + 1/n_2)}$	1.200
					μ_L	-1.174
					μ_U	1.226

Formulas:

$$s_p^2 = \frac{SS_1 + SS_2}{(n_1 - 1) + (n_2 - 1)}$$

$$\mu_L = (\overline{w}_1 - \overline{w}_2) - t_{5\%}\sqrt{s_p^2(1/n_1 + 1/n_2)}$$

$$\mu_U = (\overline{w}_1 - \overline{w}_2) + t_{5\%}\sqrt{s_p^2(1/n_1 + 1/n_2)}$$

* From Li, 1964, I, p. 151.

In geological terms, there is no evident difference in the Fe analyses of the marine and fresh-water shales. [Although fictitious data are used for illustration, the geological situation is copied from a real investigation of marine and fresh-water shales by Keith and Bystrom (1959).]

In table 4.5 a confidence interval with a 90-percent confidence coefficient is calculated for the example data. The first step is to estimate the difference $(\mu_1 - \mu_2)$ by the difference between the two sample means $(\overline{w}_1 - \overline{w}_2)$; the second is to estimate the variance of the difference $(\mu_1 - \mu_2)$. It is shown later (sec. 5.8) that if both sampling procedures are assumed to have a common variance, the variance of the difference is

$$\sigma^2\left(\frac{1}{n_1} + \frac{1}{n_2}\right).$$

Because the sample sizes are different, it is necessary to give different weights to the estimates s_1^2 and s_2^2 from the two individual samples, which are to be combined to form a *pooled* estimate of σ^2. The appropriate weights are the respective numbers of degrees of freedom, $(n_1 - 1)$ and $(n_2 - 1)$. The pooled estimate s_p^2 is calculated by adding the two SS values and dividing by the sum of the numbers of degrees of freedom. Because the number of degrees of freedom of the pooled estimate is the sum of the individual numbers of degrees of freedom, the appropriate t-value has 11 degrees of freedom. As the last step, the upper and lower confidence limits for the difference $(\mu_1 - \mu_2)$ are calculated. The conclusion may be drawn that there are no important differences between the means because the limits include 0, and moreover the upper and lower limits are not far from 0.

One- and two-sided confidence intervals are now compared and contrasted. If a two-sided confidence interval does not contain the parameter (or function of the parameter) being estimated, the parameter may be either above the upper limit of the interval or below the lower limit of the interval. However, if a one-sided confidence interval does not contain the parameter being estimated, the parameter is known to be either above or below the interval limits. With a two-sided confidence interval, it is conventional to distribute the risk so that one-half of the intervals have their high limits set too low, and the other one-half have their low limits set too high. However, in principle the risk can be distributed in any way desired, although with symmetric frequency distributions, putting one-half of the risk at either end of the interval leads to the shortest intervals. With a one-sided confidence interval, all the risk has been put into one side of the interval. Consequently, the calculated limit for a one-sided confidence interval is closer to the observed statistic than the corresponding limit for a two-sided interval. The two-sided interval is also shorter on the average than the one-sided interval, but if the investigator's interest is really in only one side, this fact is of no interest.

The distribution of risk in confidence intervals is simple for the cases given, but it can become complex. More complicated situations in apportionment of risk are considered in section 5.12 on multiple comparisons.

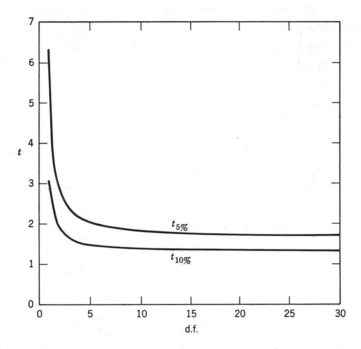

Fig. 4.8. Change in value of selected percentage points of the *t*-statistic with increasing degrees of freedom.

In conclusion it is important to note that the width of the two-sided confidence interval for the mean depends on the term

$$t \frac{s}{\sqrt{n}}.$$

Normally, it is desirable to make this term small so that the confidence interval will be narrow. The change in this term as the sample size increases is illustrated by figures 4.8 to 4.10. Figure 4.8 shows that, as the sample size increases, *t* decreases, thus making the term smaller. However, for degrees of freedom larger than about 10, the decrease is relatively small, and the value of *t* soon becomes very close to that of the standardized normal deviate. Figure 4.9 shows that, as the sample size increases, $1/\sqrt{n}$ decreases, also making the term smaller. The rate of decrease slows as the sample size gets larger, as shown by the fact that the sample size must be multiplied by 4 to cut the factor $1/\sqrt{n}$ in half. Combining these factors (fig. 4.10), one can see that the shrinkage of the average half-interval length is dominated by the

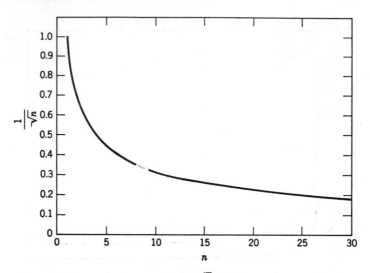

Fig. 4.9. Change in the quantity $1/\sqrt{n}$ with increasing sample size.

factor $1/\sqrt{n}$ after the sample size has reached 10. Thus, for samples of sizes larger than 10, the interval width is nearly proportional to $1/\sqrt{n}$.

As stated, one's aim usually is to narrow the confidence interval. This can be done in several ways. As shown by the graphs, n can be increased, although the advantage does not increase very fast above a sample size of about 10.

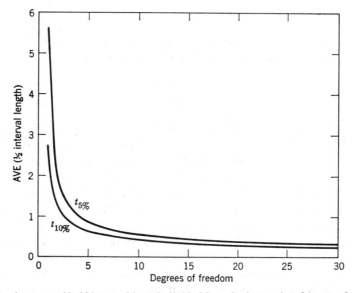

Fig. 4.10. Average of half-interval length divided by σ for increasing degrees of freedom.

More risk of being wrong can also be accepted by choosing a lower confidence coefficient. The problem can also be formulated so that a one-sided rather than a two-sided confidence interval is calculated. A final way is to decrease the size of s. Part of the variation expressed by s is natural variation inherent in the rock body or other material sampled. But part is introduced by the procedures of sampling, sample preparation, and assaying or chemical analysis. This part can be reduced by more careful methodology, as discussed in Chapter 8.

4.5   EXAMPLE ESTIMATION OF ORE GRADE WITH CONFIDENCE INTERVALS

Confidence intervals are one of the most useful statistical concepts for geology. Geologists have traditionally been inclined toward interval rather than point estimates because their data are characterized by large and changing variability. Through the confidence interval, the variability, sample size, and mean can be incorporated in a single quantitative statement. To illustrate the value of confidence limits in a specific geological situation, we turn to the problem of ore estimation from diamond-drill-hole cores from a typical vein. The procedures are readily transferable to other geological situations. As usual when one is dealing with the real world, complications arise in this analysis, complications that are treated in section 4.9.

Can useful estimates of metal content be made by boring diamond-drill holes through a vein? That is, can information from drill holes provide usable quantitative results, or are the drill holes useful only to locate vein structures? Our analysis (Koch and Link, 1964), using confidence limits calculated for typical data to answer this question, is summarized here.

The data for the example analysis are assay values from diamond-drill cores from 18 boreholes through the Don Tomás vein of the Frisco mine (see. 1.6). The Don Tomás vein (fig. 4.11) is a well-defined geologic structure that was first found on the lower levels of the mine and pinches out above level 9. When the data were obtained, development was essentially complete to level 15; drifts on each level had been driven to where the vein became too narrow or the grade became too low to warrant additional development. The entire vein is below the zones of oxidation and supergene enrichment. The Don Tomás vein is cut by the Frisco fault whose displacement is 35 meters or more. As discussed in section 5.4, because the distribution of metals is markedly different on either side, we believe this fault to be of pre-ore age. Two post-ore faults (100-S and 480-N faults) displace the vein horizontally about 14 and 11 meters, respectively, but the components of displacement in the plane of the longitudinal section are negligible.

TABLE 4.6. CONFIDENCE INTERVAL ESTIMATES FOR THE DON TOMÁS VEIN BASED ON ASSAYS FROM 18 DIAMOND-DRILL CORE SAMPLES AND 1829 DRIFT SAMPLES

| | Estimate from 18 core samples | | | Estimate from 1829 drift samples | | |
| | | Confidence limits | | | Confidence limits | |
Item	Mean	Lower	Upper	Mean	Lower	Upper
Gold content	0.27	0.09	0.45	0.28	0.26	0.30
Silver content	278.0	158.0	398.0	329.0	316.0	342.0
Lead content	6.6	3.5	9.7	8.6	8.3	8.9
Copper content	0.36	0.23	0.49	0.49	0.46	0.52
Zinc content	8.6	4.9	12.3	11.2	10.8	11.6

For this study metal content was measured in meter-percent per metric ton for base metals and in meter-grams per metric ton for precious metals. Reasons for use of these units, obtained by multiplying assay values by vein widths, are taken up in section 13.2. All drifts had been sampled by chip samples at 2-meter intervals (sec. 1.4). In all, 18 holes were bored between 1940 and 1958 at locations plotted on figure 4.11. Cores of either AX or EX size were recovered and split; half of the core in vein was assayed and the other half kept for reference. Although core recoveries were not available for all holes, they were generally about 90 percent, judging from five holes logged by one of us (GSK). Sludge was not assayed. Most holes were bored at essentially right angles to the vein, but measured widths were reduced for three holes that were not.

In table 4.6 two kinds of estimates for metal content in the Don Tomás vein are presented, one based on 1829 drift samples and the other based on 18 diamond-drill holes. The table emphasizes that, although the estimated standard deviations in each case are similar, the confidence intervals based on 1829 samples are much narrower because the sample size is so much greater. In fact, as expected, the intervals are about 10 times as wide for the data from the 18 diamond-drill holes, because

$$\sqrt{18} \approx \frac{\sqrt{1829}}{10}.$$

Thus the estimate from the large sample of drift assay values forms a standard for comparison of the less precise estimates made from assays of the drill-hole samples. Nonetheless, the less precise estimates are successful in that, for all five metals, the confidence limits include the standard means from the assays of drift samples.

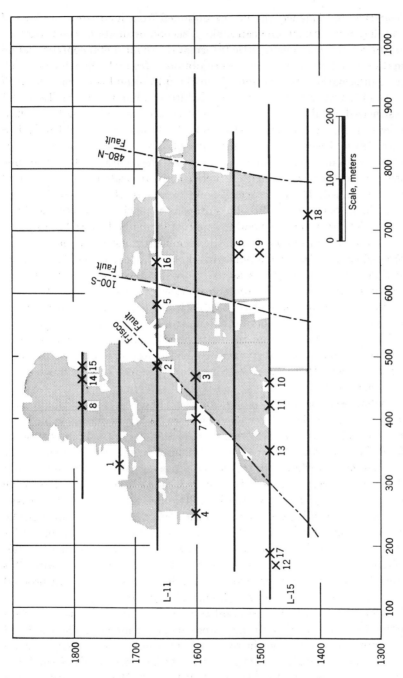

Fig. 4.11. Don Tomás vein, vertical longitudinal section. Observer looking west. Drifts shown by solid lines, other workings by gray pattern, faults by dashed line, locations of diamond drill holes by "x's" with hole numbers.

Considering, in sequence, the means obtained from increasing numbers of holes as they were drilled, one can make a second estimate of metal content for the entire vein. Obviously, little information about metal content and no information at all about variation were obtained from the first hole drilled, but more information was obtained from the first-second combinations; still more was obtained from the first-second-third combinations; etc. Table 4.7 gives estimated metal contents calculated from the first hole (line 1), the first two holes (line 2), and so forth, up to the best estimate calculated from all 18 holes (line 18), which is the same as that of table 4.6. The table illustrates that, in general, as the number of drill holes increases, the confidence limits become narrower or, in other words, that an increasingly accurate estimate of metal content is obtained. The sequence of holes from 1 to 18 corresponds to the order in which they were drilled. This order is somewhat arbitrary, and the order is not intended to correspond to one that would be followed in exploration drilling of a vein unknown from development mine workings.

It is possible from the 18 diamond-drill holes to specify confidence limits that include the metal contents actually obtained from the drift samples. Similar satisfactory limits for tonnages, based on the width data, were obtained by calculations presented in the original paper. Table 4.6 indicates that all the means of metal content calculated from diamond-drill core are lower than those obtained from the drift samples. This fact is not surprising because the holes were bored where the existence or location of the vein was most in doubt. However, when the sample variance is considered, there is no discrepancy between the two methods of sampling.

The confidence intervals based on data from the 18 diamond-drill holes are wide because the standard deviations are large, as is characteristic for base-metal veins and because the number of observations is small. Now assume that only the drill-hole data were available prior to the development of the vein through mine workings. In this hypothetical situation, a decision would have been made, based on the estimates from these data, whether to drive development workings to expose the vein. Through analysis of mining costs, based on factors including the distance of the vein from existing workings, the dip of the vein, and the width of the vein, a minimum dollar value of ore for profitable mining would have been computed. Through application of metal-price factors, the calculated grade would have been compared with this minimum value. If the grade is below the minimum value, we might proceed in one of three ways to re-estimate the metal contents.

First, a one-sided confidence interval could be computed for the same data by putting the risk entirely into the lower side of the interval. The results of such a computation are given in table 4.8. Of course this procedure could have been followed initially and might well have been preferable. Second, the confidence coefficient could have been reduced, say, to 75 percent—the results

TABLE 4.7. ESTIMATES OF METAL CONTENT FOR THE DON TOMÁS VEIN BASED ON CUMULATIVE ASSAYS OF CORES FROM 1 TO 18 DRILL HOLES BORED IN SEQUENCE

| | Point estimates of metal content | | | | | 90-percent confidence intervals | | | | | | | | | |
| | Gold content (ppm) | Silver content (ppm) | Lead content (%) | Copper content (%) | Zinc content (%) | Gold content (ppm) | | Silver content (ppm) | | Lead content (%) | | Copper content (%) | | Zinc content (%) | |
Hole no.						Lower	Upper	Lower	Upper	Lower	Upper	Lower	Upper	Lower	Upper
1	0.15	17	1.8	0.09	1.4										
2	0.20	66	2.9	0.08	7.3	0	0.48	0	377	0	10.2	0.02	0.14	0	44.4
3	0.53	172	4.2	0.09	6.8	0	1.51	0	492	0	8.4	0.05	0.13	0	16.9
4	0.40	172	4.4	0.13	6.4	0	1.04	0	355	2.0	6.9	0.04	0.22	0.5	12.2
5	0.32	316	6.2	0.24	6.1	0	0.80	0	649	2.0	10.4	0	0.50	2.0	10.2
6	0.27	277	5.4	0.22	5.3	0	0.65	8	546	1.7	9.1	0.02	0.42	1.7	8.9
7	0.41	272	6.2	0.25	5.4	0	0.84	52	491	2.8	9.5	0.07	0.42	2.4	8.3
8	0.37	242	5.7	0.24	5.3	0	0.74	49	436	2.7	8.7	0.10	0.39	2.8	7.8
9	0.33	294	6.1	0.26	5.7	0	0.66	101	487	3.4	8.8	0.13	0.39	3.4	7.9
10	0.30	265	5.5	0.24	5.1	0	0.59	87	443	2.9	8.1	0.12	0.36	2.9	7.4
11	0.27	249	5.6	0.23	5.8	0	0.54	87	411	3.2	7.9	0.13	0.34	3.5	8.1
12	0.34	249	5.4	0.26	6.1	0.06	0.61	103	396	3.3	7.5	0.15	0.37	3.9	8.3
13	0.31	250	5.8	0.26	7.6	0.06	0.56	116	384	3.7	7.9	0.16	0.36	4.3	10.9
14	0.29	253	7.7	0.33	9.6	0.05	0.52	130	377	3.9	11.6	0.18	0.48	4.9	14.3
15	0.27	239	7.3	0.33	9.3	0.05	0.49	123	356	3.7	11.0	0.19	0.47	4.9	13.7
16	0.28	241	7.0	0.32	8.9	0.08	0.49	132	349	3.6	10.5	0.18	0.45	4.7	13.1
17	0.27	236	3.7	0.35	9.0	0.07	0.46	134	338	3.4	10.0	0.21	0.49	5.1	12.9
18	0.27	278	3.6	0.36	8.6	0.09	0.46	158	398	3.5	9.7	0.23	0.49	4.9	12.3

TABLE 4.8. ONE-SIDED LOWER CONFIDENCE INTERVAL ESTIMATES FOR THE
DON TOMÁS VEIN, BASED ON ASSAYS OF CORES FROM 18 DIAMOND-DRILL HOLES

Item	Mean	One-sided lower confidence limit, 90%	One-sided lower confidence limit, 75%
Gold content	0.27	0.13	0.20
Silver content	278.0	186.0	230.0
Lead content	6.6	4.3	5.4
Copper content	0.36	0.26	0.31
Zinc content	8.6	5.8	7.1

are given in table 4.8. Reduction in the confidence coefficient could also have
been made for the two-sided interval. Third, we could improve the quality of
the data, or obtain more data. A small reduction might be made in the
estimated standard deviations by assaying all of the core instead of only
half (sec. 7.4). More holes could be drilled, although this procedure would
be expensive because a total of about 72 holes would be needed to cut the
interval widths in half through the increase in the sample size (because
$1/\sqrt{72} = 1/2\sqrt{18}$). The 72 holes would have only a small effect on the
reliability of s for the purpose of constructing one-sided confidence limits for
the mean because, with a 90-percent confidence coefficient, the t-value for 17
degrees of freedom is 1.333, and the corresponding value for 71 degrees of
freedom decreases to 1.294.

4.6 CONFIDENCE INTERVALS FOR THE POPULATION VARIANCE

The previous sections of this chapter mainly concern the calculation of
confidence intervals for a population mean. Confidence intervals can be
calculated for any other parameter for which a method has been devised. This
section briefly describes a method for calculating confidence intervals for the
population variance. One way to use confidence intervals for the variance
is in estimating the variability of a chemical element in a geological body;
another way is in estimating variability in a fossil population.

In section 3.7 it is shown that the distribution of the statistic $(n - 1)s^2/\sigma^2$
follows the chi-square distribution with $(n - 1)$ degrees of freedom. If the
value of this statistic is calculated for repeated random samples of a given
size from a population with a given variance σ^2, 95 percent of the calculated
values will be larger than the chi-square 95-percent point from table A-3, and
5 percent will be larger than the chi-square 5-percent point; that is, 90 percent

of the calculated values will fall between these two tabulated values, satisfying, for nine samples out of 10, the inequality

$$\chi^2_{95\%} < \frac{(n-1)s^2}{\sigma^2} < \chi^2_{5\%}.$$

If the reciprocal of the above inequality,

$$\frac{1}{\chi^2_{95\%}} > \frac{\sigma^2}{(n-1)s^2} > \frac{1}{\chi^2_{5\%}},$$

is multiplied by $(n-1)s^2$, the result is

$$\frac{(n-1)s^2}{\chi^2_{95\%}} > \sigma^2 > \frac{(n-1)s^2}{\chi^2_{5\%}},$$

which is a 90-percent confidence interval for the population variance σ^2.

A lower one-sided confidence interval for the population variance may be calculated, starting with the relation

$$\chi^9_{90\%} < \frac{(n-1)s^2}{\sigma^2} < \infty.$$

If the reciprocal of the above inequality,

$$\frac{1}{\chi^2_{90\%}} > \frac{\sigma^2}{(n-1)s^2} > 0,$$

is multiplied by $(n-1)s^2$, the result is

$$\frac{(n-1)s^2}{\chi^2_{90\%}} > \sigma^2 > 0.$$

As with confidence intervals for the population mean, the confidence coefficient is chosen by the investigator. We again adopt the value of 90 percent.

4.7   HYPOTHESIS TESTS

Sections 4.4 to 4.6 on confidence intervals explain ways to estimate a reasonable range of parameter values from information contained in a sample. Now, in this section on hypothesis tests a statistical method is explained to answer such questions as: Is the information contained in a sample consistent with a postulated parameter value? At the end of the section, statistical inference and hypothesis tests are contrasted with inference by confidence limits. Although confidence limits are more useful than hypothesis tests for most geological purposes, the discourse on hypothesis tests has some application and also will strengthen the reader's understanding of sampling distributions.

A hypothesis test decides which of two alternatives is correct. Such a dichotomy is often encountered in everyday life: a jury finds a defendant innocent or guilty; a doctor decides that his patient is sick or well. In each of these situations, a decision is made between only two alternatives. The decision is either right or wrong. In a trial, the decision is right if a guilty person is convicted or if an innocent person is acquitted; the decision is wrong if an innocent person is convicted or if a guilty person is acquitted. In a doctor's office the decision is right if a sick person is diagnosed to be sick or if a well person is diagnosed to be well; the decision is wrong if a well person is diagnosed to be sick or if a sick person is diagnosed to be well. Although the alternative may be made more complicated, for example, by considering the degree of guilt of a convicted person, or the degree of sickness of a person diagnosed sick, the two-choice situation is the easiest to think about and to discuss. As applied in statistics, the hypothesis test defines a more or less artificial framework for making inferences about data and arrives at the necessary dichotomy by introducing some rather artificial restraints into the problem. Just as in the practice of law or medicine, complicated social or medical situations can be forced into a simple pattern and useful results obtained, so in the practice of statistics, useful decisions can be made.

Principles and Nomenclature

In order to explain hypothesis test procedures, a highly artificial situation is introduced in which the problem is to choose between two population means, the assumption being that one is correct. An analysis is made of a set of fictitious data consisting of 25 observations drawn at random from a population with a known mean μ of 25 and a known variance σ^2 of 25. The first step is to introduce the terminology of hypothesis testing. Suppose that the problem is to decide whether a sample comes from a population with a mean μ of 25 or a mean μ of 27. The observations are assumed to follow a normal distribution with a variance σ^2 of 25, but, of course, the population mean is assumed to be unknown. Because the population mean μ is to be investigated, it is natural to base the decision on the value of the sample mean $\overline{w}$. Clearly, if the sample mean is large, the decision will tend to favor a population mean of 27, whereas, if the sample mean is small, the decision will tend to favor a population mean of 25. The problem thus resolves itself into defining a single value so that, if the sample mean is larger, the population mean of 27 is chosen; whereas if the sample mean is smaller, the population mean of 25 is chosen. Because the decision must be based on a statistic whose value fluctuates according to its sampling distribution, a wrong decision may be made. A wrong decision corresponds to one of two mistakes: the decision can be for a population mean of 25 when the true mean is 27, or the decision can be for a population mean of 27 when the true mean is 25.

The preceding general remarks are now restated formally in the technical language of hypothesis testing. The hypothesis chosen is that μ is equal to 25, which may be written

$$H_0: \mu = 25.$$

The alternative hypothesis is that μ is equal to 27, which may be written

$$H_1: \mu = 27.$$

When a choice is made between the hypotheses H_0 and H_1 on the basis of a sample, there is always a risk of making an incorrect decision. If the alternative hypothesis H_1 that μ is equal to 27 is chosen, and if the hypothesis H_0 is actually true, a true hypothesis has been rejected, and the mistake made is named a *type I error*. On the other hand, if the hypothesis H_0 that μ is equal to 25 is chosen, and if the alternative hypothesis H_1 is actually true, a false hypothesis has been accepted, and the mistake made is named a *type II error*. The names of the two types of mistakes are arbitrary and conventional.

Because the decision in a hypothesis test is based on information from a sample, there is always the probability of making a type I error. This probability, named the α *risk level* (also designated *level of significance* in some other books), is the probability of rejecting a true hypothesis H_0 and accepting the false alternative hypothesis H_1. The α risk level is chosen by the investigator; we choose an α value of 10 percent. Once the α risk level is set, the probability of making a type II error, named the β *risk level*, can be found. The β risk level is the probability of accepting a false hypothesis H_0 and rejecting the true alternative hypothesis H_1. In table 4.9, the two hypotheses and the two kinds of outcomes of a hypothesis test are listed in a table with two rows and two columns.

TABLE 4.9. HYPOTHESIS-TEST OUTCOMES

Decision of investigator	Actual status of hypothesis	
	(H_0 true; H_1 false)	(H_0 false; H_1 true)
Accept hypothesis H_0	Correct decision	Incorrect decision Type II error Percentage risk $= \beta$
Reject hypothesis H_0; accept alternative hypothesis H_1	Incorrect decision Type I error Percentage risk $= \alpha$	Correct decision

In terms of the legal analogy, in a criminal trial the jury accepts either the hypothesis H_0 that the defendant is innocent or the alternative hypothesis H_1 that the defendant is guilty. If the jury accepts either hypothesis when it is in fact true, no mistake is made. But if the jury accepts the hypothesis H_0 that the defendant is innocent, when in fact he is guilty, it has committed a type II error. On the other hand, if the jury accepts the hypothesis H_1 that the defendant is guilty, when in fact he is innocent, it has committed a type I error. Of course, in the United States and in many other countries, the intent in criminal law is to choose a low α risk level so that few innocent persons will be sent to prison—at the expense of a high β risk level, even though some criminals will thereby escape conviction.

To make a hypothesis test one must choose a statistic. For the example problem, the sample mean is a natural choice that also is statistically valid, because the sample mean is a good estimate of the population mean under a wide variety of circumstances and is the best estimate if the observations follow a normal distribution. The value of the sample mean $\bar{w}$ may be plotted as a point on a line extending from minus infinity to plus infinity. This line must be partitioned into segments, one group containing values of $\bar{w}$ for which the hypothesis H_0 is to be accepted, and the other group containing values of $\bar{w}$ for which the alternative hypothesis H_1 is to be accepted. In statistical nomenclature, these two groups of line segments are named *regions*, although in all cases discussed in this book the regions are segments of one straight line, and there is no connotation of area. In particular, the group of line segments for which the alternative hypothesis H_1 is to be accepted is named the *critical region*, and a statistic is said to lie *inside* or *outside* a critical region.

In principle the line segment could be partitioned into many pieces, some inside the critical region and some outside. What guide then can be used to define the critical region? For one thing, it is clear that the size of the β risk level should be as small as possible. Although sometimes this goal is difficult to accomplish, for the specific case under discussion it can be shown that a critical region consisting of the line segment to the right of a *cutting point*, designated Ω, minimizes the size of β, that is, the probability of making a type II error. Therefore the hypothesis H_0 is accepted if the sample mean is less than Ω, and the alternative hypothesis H_1 is accepted if the sample mean is more than Ω.

Now that the critical region has been generally defined, all that remains is to evaluate Ω. Associated with the line (fig. 4.12) on which the sample means lie is a frequency distribution representing the sampling distribution of the sample mean. Therefore associated with any line segment there is an area under the frequency curve for the *true* population mean and variance, an area corresponding to a probability. Accordingly, the evaluation of the

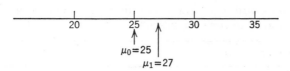

Fig. 4.12. Hypothetical means for example test of hypothesis H_0: $\mu = 25$, H_1: $\mu = 27$.

cutting point Ω is made by a calculation such that, if the hypothesis H_0 is true, the probability of making a type I error is equal to α, the area under the frequency curve that applies in this case.

Because α is chosen equal to 10 percent, in the example problem, the location of the cutting point Ω can be easily calculated since, if the hypothesis H_0 that μ is equal to 25 is true, the distribution of the sample mean has a mean of 25 and a variance equal to the population variance divided by the sample size n. Thus, if the mean is normally distributed, by using the 10-percent point from the standardized normal distribution (table A-2), the result is that

$$\Omega = 25 + 1.282\,\frac{\sigma}{\sqrt{n}}.$$

The probability that the mean of a random sample will be larger than Ω if the hypothesis H_0 is true is 10 percent. Notably, the decisions about the hypothesis to be tested, the alternative hypothesis, and the α risk level of making a type I error can all be made before any observations are obtained, or even before a sample size is decided upon.

In order to illustrate the initial test of the hypothesis H_0 that μ equals 25 against the alternative hypothesis H_1 that μ equals 27, and to illustrate subsequent hypothesis tests, an example set of fictitious data consisting of 25 observations is randomly drawn from a population with a known mean μ of 25 and a known variance σ^2, also of 25. The fictitious data may be regarded as phosphate assays, alumina determinations from igneous rocks, or other variables, as the reader prefers. As the subject is developed, it will be pretended that one or both of the two parameters are unknown and that their estimates $\overline{w}$ of 24.36 and s^2 of 16.49 are to be used in their places. In table 4.10 the original observations are listed, and the sample means are calculated.

Example Hypothesis Test. H_0: $\mu = 25$, H_1: $\mu = 27$

The first example is to perform the actual calculations for the hypothesis test introduced in the preceding subsection on principles and nomenclature. The test is of the hypothesis H_0 that μ equals 25 against the alternative hypothesis H_1 that μ equals 27. Now that the observations in table 4.10 are presumed to be at hand, the calculations for the hypothesis test can be

TABLE 4.10. OBSERVATIONS AND SAMPLE MEAN FOR
EXAMPLE HYPOTHESIS TESTS *

Observations:

29	20	21	26	32
19	29	26	27	33
22	21	21	23	25
26	20	25	27	19
30	21	21	24	22

Computations:

Item	Value
n	25
$\sum w$	609
$\overline{w}$	24.36

* From Li, 1964, I, pp. 65 and 575.

performed, the pretension being that the population variance is known but
that the population mean is unknown. In table 4.11 a rather formal procedure
is set up in a format that is followed throughout this section. The hypothesis
and alternative hypothesis are listed first. The statistic to be computed to
test the hypothesis is the sample mean $\overline{w}$. The chosen α risk level, the size of

TABLE 4.11. EXAMPLE TEST OF HYPOTHESIS H_0: $\mu = 25$, H_1: $\mu = 27$

Hypothesis:	H_0: $\mu = 25$
Alternative hypothesis:	H_1: $\mu = 27$
Statistic:	$\overline{w}$
Risk of type I error:	$\alpha = 10\%$
Critical region:	$\overline{w} > \Omega$

Computation:

Line	Item	Value	Remarks
1	σ	5	
2	n	25	
3	$\sqrt{n}$	5	
4	$\sigma/\sqrt{n}$	1	Standard error of the mean
5		1.282	10% point from table A-2
6	Ω	26.282	$25 + 1.282\sigma/\sqrt{n}$
7	$\overline{w}$	24.36	

Conclusion. Accept hypothesis H_0 that $\mu = 25$.

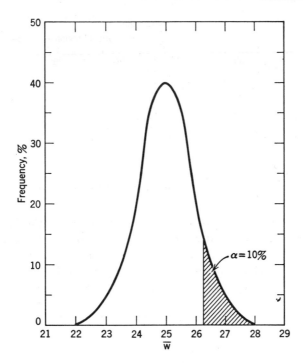

Fig. 4.13. Sampling distribution of $\bar{w}$, showing size of α risk level if H_0 is true for example test of hypothesis in table 4.11.

type I error, is 10 percent. The population standard deviation σ is assumed to be known to be equal to 5. Two other assumptions are made in this test, as in all other tests of significance in this section: (a) the major one that the observations are a random sample from the population of interest, and (b) the minor one that the theoretical frequency distribution of the observations is normal. In section 4.9 the consequences that follow when data do not fit these assumptions are considered.

If the hypothesis H_0 that μ equals 25 is true, the sampling distribution of $\bar{w}$ is that plotted in figure 4.13, because σ is known to be 5, and n is known to be 25. On computation, the value of Ω is found to be 26.282 (line 6). Because the sample mean of 24.26 is smaller than 26.282 (fig. 4.14), it is *outside* the critical region, and the conclusion is to accept the hypothesis H_0.

For the example of table 4.11 the β risk level, the size of the type II error, can be readily calculated. A type II error can be made only if the alternative hypothesis H_1 that μ equals 27 is true; whereupon the statistic $\bar{w}$ follows the sampling distribution with mean equal to 27, graphed as a solid line in

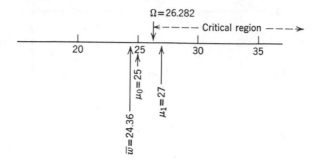

Fig. 4.14. Critical region for example test of hypothesis in table 4.11.

figure 4.15. The probability of a type II error is equal to the size of the area under the curve with mean at 27 and to the left of 26.282, the cutting point Ω. The calculations to obtain this area are given in table 4.12. In line 2, Ω, found to be 26.282 in table 4.11, is calculated to be -0.718 unit from 27. In line 3, -0.718 is divided by the standard error of the mean, 1, from line 1, to

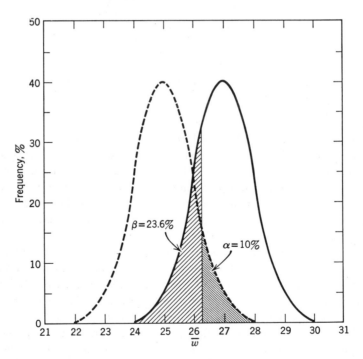

Fig. 4.15. Sampling distribution of $\bar{w}$, showing sizes of α and β risk levels for example test of hypothesis in table 4.11.

TABLE 4.12. EXAMPLE CALCULATION OF β RISK LEVEL

Alternative hypothesis: H_1: $\mu = 27$
Statistic: $\overline{w}$
Ω: 26.282

Computation:

Line	Item	Value	Remarks
1	$\sigma/\sqrt{n}$	1	From table 4.11
2	$\Omega - 27$	-0.718	Boundary of critical region minus μ, 26.282 $- 27$
3	$(\Omega - 27)/(\sigma/\sqrt{n})$	-0.718	s.n.d.
4	$100\% - \beta$	76.4%	From table A-2
5	β	23.6%	

yield -0.718. From table A-2 by interpolation, or directly from a larger table, -0.718 is found to correspond to 76.4 percent; in other words, 76.4 percent of sample means lead to the correct conclusion if the hypothesis H_1 that μ equals 27 is true. Because 76.4 percent give the correct result, $(100 - 76.4)$ or 23.6 percent give the wrong result if H_1 is true, thus giving the size of the type II error or the β risk level.

Example Hypothesis Test. H_0: $\mu = 25$, H_1: $\mu < 25$ or > 25

The test of the hypothesis H_0 that μ equals 25 against the alternative hypothesis H_1 that μ equals 27 has been discussed in detail. But usually the alternative hypothesis is less specific, and sometimes the hypothesis itself is also less specific. In the remaining examples three of the less specific cases are explained. In the first the case of a specific hypothesis with a less specific alternative hypothesis is considered. In the second, the case of a less specific hypothesis and a less specific alternative hypothesis is considered. In the last, the hypothesis test with the population variance unknown is considered.

The test of the hypothesis H_0 that μ equals 25 against the alternative hypothesis H_1 that μ is smaller or larger than 25 is now considered. Clearly, the hypothesis H_0 should be rejected if the sample mean $\overline{w}$ is too much larger or smaller than 25. Thus the critical region should consist of two line segments, one to the right of 25 and one to the left of 25, with the segment of the line in the vicinity of 25 being outside the critical region. The 10-percent α risk of making a type I error must be divided between the two line segments constituting the critical region. Although the risk need not be divided evenly, there is no reason in this example not to do so, and therefore it is divided evenly. Because the critical region is divided into two line segments that lie

TABLE 4.13. EXAMPLE TEST OF HYPOTHESIS H_0: $\mu = 25$, H_1: μ IS SMALLER OR LARGER THAN 25

Hypothesis:	H_0: $\mu = 25$
Alternative hypothesis:	H_1: $\mu < 25$ or $\mu > 25$
Statistic:	$\overline{w}$
Risk of type I error:	$\alpha = 10\%$
Critical region:	$\overline{w} < \Omega_L = 25 - 1.645\sigma/\sqrt{n}$ or
	$\overline{w} > \Omega_U = 25 + 1.645\sigma/\sqrt{n}$

Computation:

Line	Item	Value	Remarks
1	$\sigma/\sqrt{n}$	1	Standard error of the mean
2		1.645	5% point from table A-2
3	Ω_L	23.355	$25 - 1.645\sigma/\sqrt{n}$
4	Ω_U	26.645	$25 + 1.645\sigma/\sqrt{n}$
5	$\overline{w}$	24.36	

Conclusion. Accept hypothesis H_0 that $\mu = 25$.

under the tails of the assumed H_0 frequency distribution, this type of hypothesis test is named a *two-tailed* test, in contrast to the hypothesis test of the first example which is named a *one-tailed* test.

In table 4.13 the hypothesis-test calculations are performed by using the set of observations from table 4.10. The α risk level is still chosen equal to 10 percent; this area is divided equally between the two tails of the postulated distribution, as shown in figure 4.16. Each half of the critical region is defined with the use of the 5-percent (or 95-percent) point from the standardized normal distribution (table A-2). Because $\overline{w}$ lies outside the critical region, the hypothesis H_0 that μ equals 25 is accepted.

Example Hypothesis Test. H_0: $\mu \leq 25$, H_1: $\mu > 25$

The next example of a hypothesis test is a generalized version of the one-tailed test in table 4.11, in which the hypothesis H_0 that μ equals 25 is tested against the alternative hypothesis H_1 that μ equals 27. In the new example, table 4.14, the hypothesis H_0 that $\mu \leq 25$ is tested against the alternative hypothesis H_1 that $\mu > 25$. Again, the critical region is chosen to maximize the chance of rejecting the hypothesis H_0 if it is false. All the calculations are the same as those for table 4.11. However, although the α risk level is still 10 percent, the size of the α area now is 10 percent only if μ is exactly 25; otherwise the size of the α area is between 0 and 10 percent if μ is smaller than 25. The value of Ω is still 26.282 and is calculated for the case in which μ is

TABLE 4.14. EXAMPLE TEST OF HYPOTHESIS H_0: μ
IS EQUAL TO OR LESS THAN 25, H_1: μ IS LARGER
THAN 25

Hypothesis:	H_0: $\mu \leq 25$
Alternative hypothesis:	H_1: $\mu > 25$
Statistic:	$\overline{w}$
Risk of type I error:	$\alpha = 10\%$
Critical region:	$\overline{w} > \Omega$
Boundary of critical region:	26.282
$\overline{w}$:	24.36

Conclusion. Accept hypothesis H_0 that $\mu \leq 25$.

exactly 25, which corresponds to the cutting point between acceptance of the hypothesis H_0 and the alternative hypothesis H_1.

Example Hypothesis Test. H_0: $\mu \leq 22$, H_1: $\mu > 22$, σ unknown

In the preceding examples σ is assumed to be known. Actually, in practice σ must usually be replaced by its estimate s derived from the sample. An

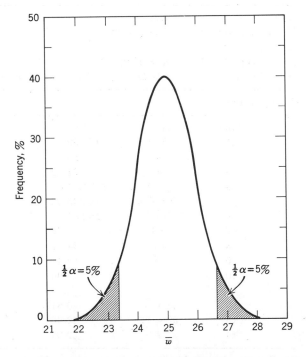

Fig. 4.16. Sampling distribution of $\overline{w}$ if H_0 is true for example test of hypothesis H_0: $\mu = 25$, H_1: μ is smaller or larger than 25.

example of the procedure is given in the test of table 4.15, where the data
from table 4.10 are used, but the assumption is made that both σ and μ are
unknown. To make a realistic problem, assume that the observations in table
4.10 are phosphate assays from a mine with a cutoff grade of 22 percent.
Because several assays in table 4.10 are below 22 percent, there is a question
whether the true grade is above 22 percent. The appropriate test is to set up
the hypothesis H_0 that $\mu \leq 22$ percent and compare it with the alternative
hypothesis H_1 that $\mu > 22$ percent. The reason for stating the hypothesis in
this form is that the risk of mining nonpaying phosphate rock is controlled by
specifying the α risk level. In this test, the t-statistic rather than the sample
mean $\overline{w}$ is chosen for the test statistic because it is necessary to introduce the
t-sampling distribution for 24 degrees of freedom in place of the standardized
normal distribution. The value of Ω is the tabulated t-value of 1.318; calculated t-values larger than 1.318 are inside the critical region. The only basis

TABLE 4.15. EXAMPLE TEST OF HYPOTHESIS $H_0: \mu$ IS
EQUAL TO OR LESS THAN 22, $H_1: \mu$ IS LARGER THAN 22,
FOR THE FIRST SET OF DATA

Hypothesis:	$H_0: \mu \leq 22$
Alternative hypothesis:	$H_1: \mu > 22$
Statistic:	t
Risk of type I error:	$\alpha = 10\%$
Critical region:	$t > \Omega \ (t_{10\%} = 1.318,$ with 24 d.f.$)$

Computations:

Line	Item	Value
1	$\sum w$	609
2	n	25
3	$\overline{w}$	24.36
4	$(\sum w)^2$	370,881
5	$(\sum w)^2/n$	14,835.24
6	$\sum w^2$	15,231
7	SS	395.76
8	d.f.	24
9	s^2	16.49
10	s	4.06
11	$\sqrt{n}$	5
12	$s/\sqrt{n}$	0.812
13	$\overline{w} - 22$	2.36
14	t	2.91

Conclusion. Reject hypothesis H_0 that $\mu \leq 22$. Accept
alternative hypothesis H_1 that $\mu > 22$.

for rejecting the hypothesis H_0 is by calculated values that are too large, never by ones that are too small. Because the computed value of t, 2.91, is larger than Ω, 1.318, and is therefore inside the critical region, the hypothesis H_0 is rejected, and the alternative hypothesis that $\mu > 22$ is accepted. In this case, a correct decision has been made because the sample is known to have been drawn from a population with μ equal to 25.

In the fictitious example of phosphate cutoff grade of table 4.15 the hypothesis test reaches a correct decision. Because the sample is actually drawn from a population with μ equal to 25, a type I error can never be made, but because the α risk level is 10 percent a type II error is necessarily made in 10 percent of the cases in which H_0 is true. An example of how type II error can be made is provided by the random sample of table 4.16, which is

TABLE 4.16. EXAMPLE TEST OF HYPOTHESIS H_0: μ is EQUAL TO OR LESS THAN 22, H_1: μ IS LARGER THAN 22, FOR THE SECOND SET OF DATA

Observations:

22	21	14	19	23
20	15	19	21	21
22	21	35	28	23
16	23	14	27	18
20	21	24	23	22

Computations:

Line	Item	Value
1	$\sum w$	532
2	n	25
3	$\overline{w}$	21.28
4	$(\sum w)^2$	283,024
5	$(\sum w)^2/n$	11,320.96
6	$\sum w^2$	11,806
7	SS	485.04
8	d.f.	24
9	s^2	20.21
10	s	4.49
11	$\sqrt{n}$	5
12	$s/\sqrt{n}$	0.898
13	$\overline{w} - 22$	-0.72
14	t	-0.802

Conclusion. Accept hypothesis H_0 that $\mu \leq 22$. (Type II error is made.)

also drawn from the population with μ equal to 25 and σ^2 equal to 25. The same hypothesis test is made as that in table 4.15, and when the calculations are performed, a t-value of -0.802 is obtained. Because this t-value is outside the critical region, the hypothesis H_0 that $\mu \leq 22$ is accepted. A false hypothesis H_0 has been accepted, and a type II error has been made. Translated into the geological situation, the mistake made has been to infer that the sampled grade is below the cutoff grade.

The Power of a Test

The *power* of a test is the probability that a hypothesis H_0 will be rejected. Because the concept of power explains how the β risk level varies and also is important for decision making (Chap. 14), an outline is given here; a particularly clear and full account is presented by Dixon and Massey (1969, pp. 263–273). The *power* of the test is of interest only when the hypothesis H_0 is in fact false (hence its name). Because in practice it is not known whether H_0 is false, the power is defined for all parameter values—both consistent with H_0, i.e., H_0 true, and inconsistent with H_0, i.e., H_0 false. If the hypothesis H_0 is in fact false, the power of a test is equal to 100 percent minus the β risk level. If the hypothesis H_0 is in fact true, the power of a test depends on whether the parameter hypothesized under H_0 is a single value or a range of values. If it is a single value, the power is equal to the α risk level; if it is a range of values, the power is equal to the α risk level only for the extreme value at the end of the range; otherwise the power is between 0 percent and the α risk level. The use of power of a test is that, if there is a choice among two or more otherwise appropriate tests, the one with the highest power is selected in order to minimize the probability of making a type II error. These rather abstract and difficult definitions are illustrated and clarified here by reference to some of the previous examples of hypothesis testing.

When the simple hypothesis H_0 that μ equals 25 is tested against the simple alternative hypothesis H_1 that μ equals 27 (table 4.11), the power is computed to be (table 4.12, line 4) 76.4 percent when H_0 is false. For the more complicated two-tailed test of the hypothesis H_0 that μ equals 25, against the alternative hypothesis H_1 that μ is less than or greater than 25, the power depends on the actual value of μ. When the power is calculated for all values of μ, it may be plotted against the values of μ to obtain a *power curve*. In figure 4.17 the solid line on the graph is the power curve for this two-tailed test for a sample size of 25. The graph shows that, if the hypothesis H_0 that μ equals 25 is true, the probability or power of rejecting a mean of 25 is 10 percent, which corresponds to the α risk level. For true means increasingly different from 25, both arms of the power curve rise because the probability β of rejecting a false hypothesis H_0, which is increasingly different

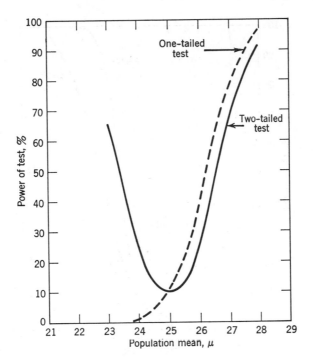

Fig. 4.17. Example power curves for one- and two-tailed test, $\mu = 25$, $\alpha = 5$ percent, $n = 25$.

from the true hypothesis, decreases. In figure 4.17 the dashed line on the graph is the power curve (also for sample size of 25) for the corresponding one-tailed test of the hypothesis H_0 that μ is equal to or less than 25 against the alternative hypothesis H_1 that μ is larger than 25. Notably, if H_0 is false, the one-tailed test is more powerful than the two-tailed test, because none of the α risk level is "spent" in the wrong tail. Therefore, if a one-tailed test is appropriate, it is always preferable to a two-tailed test.

The power of a test depends upon the sample size as well as on the type of test. Both examples in figure 4.17 are for a sample size of 25. In figure 4.18 are plots of the power curves for the second test in figure 4.17 of the hypothesis H_0 that μ is less than or equal to 25 against the alternative hypothesis H_1 that μ is greater than 25, for three different sample sizes, 10, 25, and 100. As the sample size increases, so does the power, which depends on the standard error of the sample mean, $\sigma/\sqrt{n}$. The increase from a sample size of 10 to one of 100 is striking. The importance of this relation in deciding how many mine samples to take of a given deposit is considered in Chapter 14.

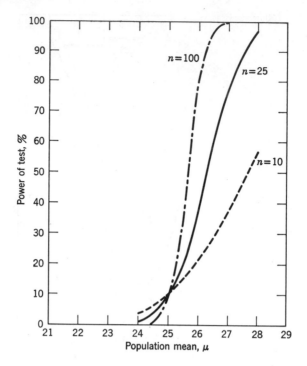

Fig. 4.18. Example power for one-tailed test, $\mu = 25$, $\alpha = 5$ percent, $u = 10$, 25, and 100.

Relation between Hypothesis Tests and Confidence Intervals

The point of view of the investigator making a hypothesis test is different from that of the investigator constructing a confidence interval, although most of the arithmetic is similar. Consider, as an example, hypothesis tests regarding the population mean contrasted with confidence intervals for the population mean. In a hypothesis test, the hypothetical mean and the estimated standard error of the mean are used to construct a critical region, and then a decision is made depending on whether the sample mean is inside or outside the critical region. On the other hand, in constructing a confidence interval, the sample mean and the estimated standard error of the mean are used to construct an interval that sets upper and lower bounds for the population mean.

In a sense, constructing a confidence interval corresponds to making an infinite number of hypothesis tests. Thus, if a confidence interval for the mean, with a certain confidence coefficient, is constructed from a set of data, it is known that, if a hypothesis test with an α risk level equal to 100 percent

TABLE 4.17. EXAMPLE TEST OF HYPOTHESIS OF EQUALITY
OF TWO POPULATION MEANS *

Hypothesis:	$H_0: \mu_1 = \mu_2$
Alternative hypothesis:	$H_1: \mu_1 < \mu_2$ or $\mu_1 > \mu_2$
Statistic:	t
Risk of type I error:	$\alpha = 10\%$
Critical region:	$|t| > t_{5\%}$ ($t_{5\%} = 1.796$, with 11 d.f.)

Computations:

Line	Item	Value	Remarks
1	$\overline{w}_1 - \overline{w}_2$	0.026	From table 4.5
2	$\sqrt{s_p^2(1/n_1 + 1/n_2)}$	0.6683	From table 4.5
3	t	0.039	

Conclusion. Accept hypothesis, $\mu_1 = \mu_2$.

* From Li, 1964, I, p. 151.

minus the confidence coefficient is made for the same set of data, all hypotheses about means falling inside the confidence interval are accepted, and all hypotheses about means falling outside the confidence interval are rejected. For example, in table 4.5 a 90-percent confidence interval is constructed for the difference between two population means of fictitious marine and freshwater shale data. The hypothesis is accepted when, in table 4.17, a test of the hypothesis H_0 that μ_1 equals μ_2 at an α risk level of 10 percent is made for the same data, because the conclusion is that μ_1 minus μ_2 equals 0, with 0 falling inside the confidence interval.

Uses for both confidence intervals and hypothesis tests in analysis of geological data are given throughout the book. In general the confidence interval is most useful when the aim is to summarize the information in a set of data to report what has been found out and how well it is known. The length of the confidence interval gives a direct indication of the precision of estimate. On the other hand, a hypothesis test is generally more useful if a specific decision is required between two alternatives. In the hypothesis test the precision is not directly indicated, although it can be obtained from the power curve.

4.8 BAYESIAN STATISTICS

The preceding sections of this book present the orthodox, classical kind of statistical inference that has dominated the thinking of most statisticians in this century. During the last few years, a different approach has been developed named Bayesian statistics, after Thomas Bayes who published a famous statistical paper in 1763. Statisticians of the Bayesian school argue that little experimental work, data gathering, and statistical analysis are performed in a vacuum and that prior information is usually available before an investigation is begun. For instance, if an investigator assays gold ore, he expects the gold content to be a few parts per million rather than a few parts per hundred; if he searches rocks for fossil birds, he expects to find only a few, not thousands. The Bayesian school argues that advantage should be taken of whatever relevant prior information is available, and it has developed formal methods to do this based on Bayes' theorem (sec. 2.5).

The principal difference between classical and Bayesian statistics is in the interpretation of the parameter under investigation. Classical statistics assumes that this parameter has a single value to be estimated, either by forming a confidence interval or by making a hypothesis test concerning a postulated single value. In contrast, Bayesian statistics assumes that the parameter has a probability distribution rather than a single value. The assumed probability distribution of the parameter is combined with the sampling distribution of the corresponding statistic to form a new modified distribution for the parameter. Thus, in Bayesian statistics, the information in the sample as summarized by a statistic modifies the investigator's postulated distribution of the parameter; whereas in the classical approach, the statistic influences the investigator's opinion about a single postulated value of a parameter or about the range of reasonable values for a parameter.

In classical statistics every distribution has a frequency interpretation, but in Bayesian statistics the distribution of the parameter may or may not be subject to a frequency interpretation. Two kinds of distributions illustrate this point. Records may show that a gold mine which has been in production for several years has had during this time an average weekly production of 20,000 ounces, with a standard deviation of 3000 ounces and a roughly normal distribution; for this distribution a clear-cut frequency interpretation is evident. On the other hand, suppose that 10 geologists hunt for fossils at an outcrop. If one of them offers an opinion that the time needed to locate the first fossil has a distribution that is normal with a mean of 3 hours and a standard deviation of 1 hour, the distribution, arrived at informally and subjectively, is not subject to a frequency interpretation. Either of these two kinds of distributions is acceptable to a Bayesian statistician.

The Use of Bayes' Theorem

As previously explained, Bayesian statistics assumes that a parameter has a probability distribution. The distribution of the parameter assumed before data have been taken is named the *prior probability distribution of the parameter*, or simply the *prior distribution*; and the distribution of the parameter calculated after the data have been collected is named the *posterior prob-ability distribution of the parameter*, or simply the *posterior distribution*. The posterior distribution is obtained from the prior distribution and the distribution of the sample statistic by applying Bayes' theorem (sec. 2.5) in the version for continuous rather than discrete distributions.

For continuous distributions Bayes' theorem states that, *if $D_0(\theta)$ is the distribution of the parameter and if $f(w, \theta)$ is the distribution of the statistic w for a given value of θ, the posterior distribution $D_1(\theta|w)$ is a conditional distribution dependent upon the observed value of the statistic that is calculated by the formula*

$$D_1(\theta|w) = \frac{f(w, \theta) D_0(\theta)}{\int f(w, \theta) D_0(\theta) d\theta}.$$

Suppose that the prior distribution of a parameter θ is normal with a mean equal to ϕ_0 and a variance equal to σ_0^2. If a random sample is then taken from another normal distribution with a mean of θ and a variance of σ^2, the distribution of the sample mean $\bar{w}$ is normal with a mean of θ and a variance of σ^2/n, where n is the sample size. Bayes' theorem may be used to calculate the posterior distribution of the parameter θ, given $\bar{w}$. The calculation indicates that the posterior distribution of θ is normal with the parameters

$$\text{Posterior mean} = \phi_1 = \frac{n\bar{w}/\sigma^2 + \phi_0/\sigma_0^2}{n/\sigma^2 + 1/\sigma_0^2},$$

and

$$\text{Posterior variance} = \sigma_1^2 = \frac{1}{n/\sigma^2 + 1/\sigma_0^2}.$$

The weighting factors to combine the prior mean ϕ_0 and the sample mean $\bar{w}$ to obtain the posterior mean ϕ_1 are the reciprocals of the variances of these two quantities, and the variance σ_1^2 of ϕ_1 is the reciprocal of the sum of the weighting factors. Such weighting by reciprocals of variances is used again and again in statistics in many contexts (e.g., sec. 5.3).

The use of Bayes' theorem can be shown numerically by considering three fictitious cases of sampling from normal populations with known variances (table 4.18). In case 1 assume that, based on long experience, the weekly production of a gold mine is known to have a mean of ϕ_0 of 20,000 ounces, with a standard deviation σ_0 of 3000 ounces. Each week the mill super-intendent forecasts current production before the final returns are in with a standard deviation σ of 1000. (This standard deviation measures how closely his forecast production agrees with the production calculated from the final returns; based on many years' forecasts, it may be considered to be a parameter.) In the week in question his estimate of w is 22,000 ounces, and

TABLE 4.18. THREE FICTITIOUS CASES OF SAMPLING TO ILLUSTRATE BAYES' THEOREM

Line	Item	Case 1	Case 2	Case 3
1	ϕ_0	20,000	20,000	20,000
2	σ_0	3,000	6,000	100,000
3	w	22,000	22,000	22,000
4	σ	1,000	1,000	1,000
5	$1/\sigma_0^2$	0.1111×10^{-6}	0.0278×10^{-6}	0.0001×10^{-6}
6	$1/\sigma^2$	1×10^{-6}	1×10^{-6}	1×10^{-6}
7	ϕ_0/σ_0^2	0.2222×10^{-2}	0.0555×10^{-2}	0.0002×10^{-2}
8	w/σ^2	2.2×10^{-2}	2.2×10^{-2}	2.2×10^{-2}
9	$\phi_0/\sigma_0^2 + w/\sigma^2$	2.4222×10^{-2}	2.2555×10^{-2}	2.2002×10^{-2}
10	$1/\sigma_1^2 = 1/\sigma_0^2 + 1/\sigma^2$	1.1111×10^{-6}	1.0278×10^{-6}	1.0001×10^{-6}
11	ϕ_1	21,800	21,945	21,999.8
12	σ_1	949	986	999.95

the problem is to estimate the value of the week's production that will be established when the final returns are in. In lines 1 to 4 of table 4.18, the parameters already introduced are listed, and in lines 5 to 10 are calculated the quantities required in the preceding equations to estimate the mean ϕ_1 and standard deviation σ_1 of this week's production. By taking into account prior information, the superintendent changes his estimate of ϕ_1 by about 10 percent.

In case 2 the initial four parameters are the same except that the standard deviation σ_0 of the prior distribution is assumed to be larger, with the value of 6000 instead of 3000. When the calculations are repeated, because the standard deviation of the previous production is large, that is, the prior production fluctuates more widely from week to week, there is little information available in the previous production to change the superintendent's estimate; thus, his estimate of 22,000 ounces is only altered to 21,945. Finally, in case 3 assume that the mine has entered a new orebody so that the variance of the prior distribution, instead of being calculated, is arbitrarily set at the very high value of 100,000, implying that it is practically unknown. Thus the superintendent's estimate is altered insignificantly to 21,999.8.

For most real problems the variance of the prior distribution is infinite or very large, and almost all of the information about the posterior distribution is contributed by the sample, as the equations and numerical results demonstrate. Thus in practice, almost no one, except the most ardent Bayesians, considers prior information to be relevant when testing hypotheses or constructing confidence intervals, although interpretation of the results of these procedures is somewhat modified.

Hypothesis testing has little meaning in a Bayesian context, as can be demonstrated by considering a Bayesian test about means. Assume that the postulated value of the population mean is the mean of the prior distribution and that the variance of the prior distribution is infinite. The posterior distribution is then centered about the observed sample mean, with a variance equal to the variance of the sample mean. The distance in standard deviation units between the prior and posterior means and the probability of an observation at a greater distance can be calculated. If this probability is small, the conclusion is that the prior mean is inconsistent with the posterior distribution; if the probability is large, the conclusion is that the prior mean is consistent. There are at least two ways to assess the meaning of a probability of an observation at a greater distance: (a) set a fixed significance level, say 5 percent, or (b) calculate the exact probability from the prior mean to infinity. Because a hypothesis test depends on a logical framework in which type I and type II errors are considered for single values for parameters rather than for a frequency distribution of parameters, this Bayesian procedure is clearly very different from a hypothesis test.

When Bayesian estimates are appropriate, some statisticians, for instance Mosteller and Wallace (1964), advocate calculating odds rather than using classical significance tests. Prior and posterior odds can be calculated to choose between two candidate sets of parameter values, and if the odds in favor of one set are very high, that set is accepted. Even though this approach has not found wide favor, it should not be ignored.

Seen from a Bayesian point of view, the interpretation of confidence intervals is changed like that of hypothesis tests. Instead of regarding an interval as either containing or not containing a single-valued parameter, the interval is interpreted as containing the central part of the posterior distribution. For instance, for a 90-percent confidence interval, the classical interpretation is that 90 percent of the intervals constructed contain the parameter being estimated, and 10 percent do not. The Bayesian interpretation is that, if the prior distribution is assumed to have an infinite variance, a 90-percent interval contains the central 90 percent of the posterior distribution.

If the variance of the prior distribution is not infinite, the percentage of the posterior distribution is not 90. Consider once again the estimated weekly production of the fictitious gold mine. If one ignores any prior information, the estimate is 22,000 ounces, with a standard deviation of 1000. A two-sided 90-percent confidence interval for θ has the bounds [22,000 $-$ 1.645 (1000), 22,000 $+$ 1.645 (1000)], or (20,355, 23,645). The constant 1.645 is the 5-percent point of the standardized normal distribution. However, if one takes for case 1 the prior information that ϕ_0 is equal to 20,000 and σ_0 is equal to 3000, the posterior distribution has as parameters ϕ_1 equal to 21,800 and σ_1 equal to 949. Between the confidence limits, the amount of area in this distribution is 91 percent. Thus 91 percent of the posterior distribution lies within the confidence interval. If this calculation is repeated for the prior distribution of case 3 with σ equal to 100,000, the amount of the posterior distribution in the confidence interval is essentially 90 percent because the variance of the prior distribution is so large that it has a negligible effect.

Unless appropriate assumptions are made about the shape of the prior distribution, the calculation of the posterior distribution can be very difficult. Therefore most Bayesians use prior distributions that yield easy calculations for the posterior distribution, as explained, for instance, by Schlaifer (1961).

Concluding Remarks

If data are fairly extensive and if the prior distribution has a large variance, classical and Bayesian statistics differ but little. Because results are only slightly modified if Bayesian theory is applied, such prior distributions are named *gentle priors*, and although an investigator's attitude toward the results is somewhat different in theory, his course of action is likely to be the same. However, if data are sparse and the prior distribution has a relatively small variance, considerably more accurate results can be obtained from a Bayesian approach—if the prior information is relevant.

Deciding whether a prior distribution is indeed relevant is liable to be the stickiest point. Different people have different prior distributions in mind because of differences in their background and position. Thus it is unlikely that the president, vice president, comptroller, and production manager of a company would have the same set of priors, and it is even improbable that their priors would agree enough to be reconciled.

4.9   INDICATIONS

Two important assumptions are made whenever a confidence interval is formed or a hypothesis test is made. The first is that the sample is random, and the second is that the observations are drawn from a normal population. These assumptions, introduced in sections 2.4 and 3.4, are repeatedly examined throughout the book. In this section their relation to *indications*, a name given to statistical results whose precise inferential validity is in doubt, is considered.

Failure of real data, even when obtained in a designed experiment, to correspond exactly to the assumptions of randomness and normality is the rule rather than the exception. Moreover, almost always the relation is uncertain between the target population and the sampled population for which the randomization is performed. Because the investigator draws inferences about the target population, these inferences must rest on geological as well as on statistical reasoning. These problems arise even in carefully designed experiments; for data obtained in undesigned experiments or from the literature, there is seldom if ever a reliable way to test their correspondence to the assumptions.

When the results of a statistical analysis, though uncertain, appear to be suggesting something about a problem, the statistical inferences may be named *indications* rather than decisions or conclusions. The term indications was first used by Tukey (1968), who provides a thoughtful discussion.

If the first assumption of randomly distributed observations is not met, great trouble can result, and even at best the reliability of any statistical analysis is always uncertain. Lack of randomness causes many problems; some interesting ones are outlined by Wallis and Roberts (1956, pp. 337–340). There is no unequivocal way to prove the randomness of statistical samples obtained in undesigned experiments. Although various tests can be applied, at best they can indicate that samples appear to behave like random samples; such tests can never prove that samples are in fact random samples.

Many geological observations, including most of the example data in this book, are demonstrably not random; yet we must use them, for they are the best we have. Even if the observations are randomly drawn from a sampled population, they are seldom if ever random with respect to the target population. Yet, because of the importance of randomness in making any statistical inferences, it is not an exaggeration to state that the two most important statistical decisions that an investigator makes are the relating of target and sampled populations and the devising of an appropriate scheme for random sampling. These matters are explored further in section 7.2.

If the second assumption, that observations are normally distributed, is not met, the exact risk levels for confidence limits and hypothesis tests are unknown. For instance, if observations actually come from the skewed distribution of figure 4.19, instead of 10 percent of the observations being more than 1.65 standard deviations greater than the mean and another 10 percent being more than 1.65 standard deviations smaller than the mean, 8 percent are actually in the left tail and 12 percent in the right. Thus a "10 percent" confidence interval based on the right tail is actually a 12-percent interval. For a hypothesis test with the critical region in one tail, the α risk level is 8 or 12 percent rather than the specified 10 percent. Of course, the investigator does not know that his probability levels are changed. Although troublesome and untidy, this distortion of the normal curve does not cause the geologist too much trouble, because with a problem in the real world the risk level is chosen arbitrarily anyway.

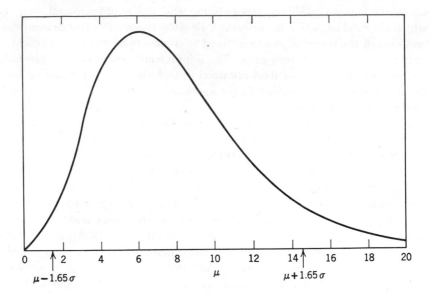

Fig. 4.19. Skewed quasi-normal distribution.

A potentially more serious problem, if the assumption of normality is not met, is that statistical procedures designed for the normal distribution may be inefficient for nonnormal distributions. Then *transformations* (Chap. 6) may be advantageously applied to observations. The attitude that one should take toward possible loss of efficiency is difficult to assess; sometimes one should be very worried, other times one can be unconcerned. The need for efficient methods is closely related to the number of observations. If observations are few and inferences are uncertain, efficiency may be very important; if observations are numerous, conclusions may be unmistakable, and efficiency may be unimportant.

As the preceding remarks suggest, failure to satisfy the assumption of randomness is much more serious than failure to satisfy that of normality. Fortunately, for a new investigation it is always possible, though not always easy, to sample randomly the sampled population. And, although few if any real data come from a population that is normal or even quasi-normal, the only consequences of failure to meet the normality assumption are distortion of the theoretical risk levels and reduction in the efficiency of estimation. As data of a certain kind accumulate, an investigator can always try to make more exact statements as he learns how skewed a distribution actually is. Even for a highly skewed distribution, such as the lognormal one (sec. 6.2), the problem is not too great.

Where to draw the line and call an inference an indication rather than a conclusion or a decision is a troublesome problem, for which judgment is the only guide.

REFERENCES

Dixon, W. J., and Massey, F. J., Jr., 1969, Introduction to statistical analysis: New York, McGraw-Hill, 638 p.

Fisher, R. A., 1950, The fiducial argument in statistical inference, *in* Contributions to Mathematical Statistics: New York, John Wiley & Sons, paper 25.

Keith, M. L., and Bystrom, A. M., 1959, Comparative analyses of marine and fresh-water shales: Penn. State Univ., Mineral Industries Expt. Sta. Bull.

Koch, G. S., Jr., and Link, R. F., 1964, Accuracy in estimating metal content and tonnage of an ore body from diamond-drill-hole data: U.S. Bur. Mines Rept. Inv. 6380, 24 p.

Li, J. C. R., 1964, Statistical inference: Ann Arbor, Mich., Edwards Bros., v. 1, 658 p.

Mosteller, Frederick, and Wallace, D. L., 1964, Inference and disputed authorship: The Federalist: Reading, Mass., Addison-Wesley, 285 p.

Mosteller, Frederick, and Tukey, J. W., 1968, Data analysis, including statistics: *in* Handbook of Social Psychology, 2nd. ed., Reading, Mass., Addison-Wesley, p. 80–203.

Neyman, J., 1937, Outline of a theory of statistical estimation based on the classical theory of probability: Royal Soc. London Philos. Trans., ser. A, v. 236, p. 333–380.

Schlaifer, Robert, 1961, Introduction to statistics for business decisions: New York, McGraw-Hill, 374 p.

Wallis, W. A., and Roberts, H. V., 1956, Statistics: a new approach: Glencoe, Illinois, Free Press, 619 p.

Wilks, S. S., 1962, Mathematical statistics: New York, John Wiley & Sons, 644 p.

Chapter 5

Analysis of Variance

In Chapters 2 to 4 selected basic statistical methods and calculation procedures are explained in detail. Comprehension of the subjects in these chapters is essential for the geologist who wishes to use statistics effectively and is also necessary for understanding the rest of this book. However, from Chapter 5 onward, only a few statistical methods are explained in detail; for other methods the reader is referred to statistics books and research papers.

Chapter 5, an introduction to the analysis of variance, is about the sources of variability in data and about the purposeful partitioning of the sum of squares of the observations into pieces, each of which can be given an interpretation in terms of geology or other discipline to which statistics are applied. Space permits only an introduction to a few of the many topics in the analysis of variance.

The following topics are taken up in this chapter. In sections 5.1 to 5.6 the one-way analysis of variance is explained, together with one modification— the nested analysis of variance. In section 5.7 the analysis of variance in randomized blocks is discussed. Section 5.8 explains linear combinations, a mathematical device necessary for the single-degree-of-freedom analysis, which is treated in sections 5.9 and 5.10. In section 5.11 the preceding sections are tied together through a discussion of linear models. Section 5.12 is about multiple comparisons of data, and section 5.13 summarizes the statistical methods introduced up to this point in the book.

Many traditional subjects in the analysis of variance are not even mentioned in this book, partly because they would expand it too much, but mainly because they pertain to the analysis of *designed experiments* (sec. 8.6),

131

in which data are collected according to a formal plan that is carefully conceived and executed. Such designed experiments are common in agriculture, in some of the physical sciences, and in some industries. For instance, a herd of 16 cows may be randomly partitioned into four different groups; each group may be fed different rations and the weights of the 16 cows may be periodically compared. Or several different brands of light bulbs may be used in a number of factories and the lives of the bulbs then contrasted. Although some geological studies are equally formal (e.g., four different rock formations might be compared by measuring paleomagnetism at six outcrops in each), most geologists do not today seem to work in this way. There are many methods for the statistical analysis of designed experiments; Cochran and Cox (1957) and Davies (1957) give excellent accounts; and computer programs are readily available for the more popular ones.

5.1   COMPARISON OF MORE THAN TWO MEANS

In sections 4.4 and 4.7 the problem of comparing agreement among means is introduced. The means are compared by constructing confidence intervals and by making a hypothesis test. When such a comparison is made, the question at once arises: How would the check for agreement in sampling be made if three or more groups of specimens had been taken? In this chapter this problem is investigated for means. Although agreement in data may also be compared by studying other parameters, the comparison of means is usually of greatest interest for geological problems.

The ideas and techniques that have been developed for comparing means have many uses and provide the basis for much of the applied statistics in any discipline. In this chapter these ideas and techniques are developed through the explanation of some highly simplified geological problems. Later, particularly in Chapters 7, 8, and 11, more interesting geological applications are taken up.

Suppose that in area A a geologist has mapped five limestone outcrops which he is uncertain whether to assign to a single formation. Suppose, further, that he believes the key indicator is the sand content of the rock, so that if he concludes that this content is the same in the five outcrops he will decide that they belong to the same formation. To investigate the sand content the geologist might take four hand specimens from each outcrop, or 20 altogether, and measure the sand content in each to obtain the data in table 5.1. In the table some statistics are calculated for each of the five samples.

TABLE 5.1. FICTITIOUS OBSERVATIONS OF SAND CONTENT CONSTITUTING STATISTICAL SAMPLES FROM FIVE OUTCROPS IN AREA A

		Statistical sample number				
Line	Item	(1)	(2)	(3)	(4)	(5)
1	Observations	9.8	9.5	9.3	15.5	7.6
2	(sand content,	7.5	7.2	10.6	8.9	11.0
3	%)	10.1	10.4	9.6	12.4	9.8
4		10.9	11.3	13.2	10.0	10.8
5	$\sum w$	38.3	38.4	42.7	46.8	39.2
6	n	4	4	4	4	4
7	$\overline{w}$	9.575	9.600	10.675	11.700	9.800
8	$(\sum w)^2$	1466.89	1474.56	1823.29	2190.24	1536.64
9	$(\sum w)^2/n$	366.72	368.64	455.82	547.56	384.16
10	$\sum w^2$	373.11	377.94	465.25	573.22	391.44
11	SS	6.39	9.30	9.43	25.66	7.28
12	d.f.	3	3	3	3	3
13	s^2	2.13	3.10	3.14	8.55	2.43

To compare k means corresponding in the fictitious illustration to the five sample means of sand percentage, one might be tempted to compare each mean with every other by two-sample confidence intervals or the t-hypothesis test. There are two difficulties. One is that as the number of comparisons becomes large the probability of making a type I error, by rejecting the true hypothesis that two population means are equal, also becomes large; for instance, if k is 10, 45 pairwise comparisons are possible. In a test with a 10-percent significance level four or five of the comparisons would on the average lead to rejecting the hypothesis, through chance alone, even though all were true. The second difficulty is that there may be too few observations in each statistical sample to yield good estimates of the population variance σ^2. These difficulties can be overcome by the method of *multiple comparisons*, an explanation of which is deferred until section 5.12, in order to introduce first the basic procedures involved in the analysis of variance.

Mathematical Model

Before solutions to the specific problem are discussed, a mathematical model (sec. 1.5) is introduced. For each outcrop, the four determinations of sand percentage are defined as a sample of four observations from a population corresponding to that outcrop. Thus there is a single sample of four observations from each of five populations. Each population has a mean μ_i

and a common variance σ^2. If the geologist concludes that the means of the five populations are equal,

$$\mu_1 = \mu_2 = \mu_3 = \mu_4 = \mu_5,$$

he would conclude that the five outcrops belong to the same formation. However, if he concludes that the means are unequal, he might be unwilling to conclude that the five outcrops belong to different formations without a further investigation of the data with a more complicated model (sec. 9.2). Thus his interpretation of the data depends on the mathematical model that he uses.

The problem, then, is to estimate each of the μ_i values and the common variance σ^2 and to devise a test of the hypothesis H_0 that

$$\mu_1 = \mu_2 = \mu_3 = \mu_4 = \mu_5.$$

The jth observation from the ith population may be represented as

$$w_{ij} = \mu_i + e_{ij},$$

where μ_i represents the mean of the population from which the observation is taken and e_{ij} represents the random fluctuation of the observation (e.g., the fourth observation from the third population is $w_{34} = 13.2$). The assumption is made for all the random fluctuations e_{ij} that their mean is 0 and that their variance is equal to the common population variance σ^2, or,

$$\mu_e = 0,$$
$$\sigma_e^2 = \sigma^2.$$

The expression for an observation may be rewritten

$$w_{ij} = \mu + (\mu_i - \mu) + e_{ij},$$

where μ is the average of the five μ_i values. If the μ_i values are all equal, the terms in parentheses $(\mu_i - \mu)$ are all 0. If they are not all equal, one measure of their difference is the variance of population means

$$\sigma_\mu^2 = \frac{(\mu_1 - \mu)^2 + (\mu_2 - \mu)^2 + (\mu_3 - \mu)^2 + (\mu_4 - \mu)^2 + (\mu_5 - \mu)^2}{4},$$

where the number 4 in the denominator is equal to the number of means minus 1.

The hypothesis H_0 that

$$\mu_1 = \mu_2 = \mu_3 = \mu_4 = \mu_5$$

may be rewritten

$$\sigma_\mu^2 = 0,$$

for, if the variance of population means is equal to zero, the means must also be equal to each other.

All of the quantities needed to test this hypothesis have now been introduced. They are the population means μ_i, the mean of population means μ, the common population variance of the random fluctuations σ^2, and the variance of population means σ_μ^2. Statistics to estimate the first three of these parameters are intuitively clear. The sample mean 9.58 from population 1 is an estimate of μ_1 (table 5.1), and similarly for the other populations. The mean of sample means 10.27 is an estimate of μ (table 5.2). The third parameter σ^2 may be estimated by the combination of individual sample variances, which is the pooled sample variance s_p^2 (sec. 4.4), equal to 3.87 (table 5.2).

If the fourth parameter, the variance of population means σ_μ^2, is equal to 0, the average value of the variance of the five sample means is $\sigma^2/5$. However,

TABLE 5.2. ONE-WAY ANALYSIS OF VARIANCE TO TEST THE HYPOTHESIS THAT THE POPULATION MEANS OF FICTITIOUS STATISTICAL SAND CONTENT FROM FIVE OUTCROPS ARE EQUAL

Hypothesis:	$H_0: \mu_1 = \mu_2 = \mu_3 = \mu_4 = \mu_5$, or $\sigma_\mu^2 = 0$
Alternative hypothesis:	H_1: means unequal, or $\sigma_\mu^2 > 0$
Statistic:	F
Risk of type I error:	$\alpha = 10\%$
Critical region:	$F > F_{10\%}$, with 4 and 15 d.f. ($= 2.36$)

Computations:

Line	Item	Value	Remarks
1	$\overline{w}$	9.575, 9.600, 10.675, 11.700, 9.800	From table 5.1
2	n	4, 4, 4, 4, 4	,,
3	s^2	2.13, 3.10, 3.14, 8.55, 2.43	,,
4	$\sum \overline{w}$	51.350	
5	k	5	
6	$\overline{\overline{w}}$	10.27	
7	$(\sum \overline{w})^2$	2636.8225	
8	$(\sum \overline{w})^2/k$	527.36450	
9	$\sum \overline{w}^2$	530.72625	
10	$SS_{\overline{w}}$	3.36175	
11	d.f.	4	
12	$s_{\overline{w}}^2$	0.84044	
13	n	4	
14	$n s_{\overline{w}}^2$	3.3616	
15	$\sum SS$	58.05	
16	$\sum (d.f.)$	15	
17	s_p^2	3.87	
18	F	0.87	

Conclusion. Accept hypothesis $H_0: \mu_1 = \mu_2 = \mu_3 = \mu_4 = \mu_5$, or $\sigma_\mu^2 = 0$.

if this quantity is larger than 0, because the five population means are unequal, the average value of the variance of the five sample means must be larger than $\sigma^2/5$. In fact, this average value is equal to $\sigma^2/5 + \sigma_\mu^2$. For the fictitious data, the variance of sample means $s_{\bar{w}}^2$ is equal to 0.8404 (table 5.2), which is nearly equal to the pooled sample variance $s_p^2/5$ of 0.7740 (table 5.2).

Thus the criterion of comparison for the population means is the ratio of two variances calculated from the k samples, both of which estimate the population variance, *provided* that the k means are equal. The first one is calculated as follows: Consider k samples, each with the same number of observations n. For each sample a sample mean $\bar{w}_i$ and a sample variance s_i^2 are calculated. Then the common population variance σ^2 can be estimated by the pooled sample variance s_p^2 obtained by the formula

$$s_p^2 = \frac{\sum (n - 1)s_i^2}{k(n - 1)}$$

which, to simplify notation, may be rewritten

$$s_p^2 = \frac{\sum \mathrm{SS}_i}{k(n - 1)},$$

where the summation extends over the k samples. The statistic s_p^2 has $k(n - 1)$ degrees of freedom.

The second of these variances may be introduced as follows: The variance of the sample mean $s_{\bar{w}}^2$ is

$$s_{\bar{w}}^2 = \frac{\sum (\bar{w}_i - \bar{\bar{w}})^2}{k - 1}$$

with $(k - 1)$ degrees of freedom, where $\bar{\bar{w}}$ is the mean of sample means. This variance is related to the underlying population variance of observations σ^2 because the variance of a sample mean $\bar{w}$ is equal to the population variance divided by the sample size, that is, to σ^2/n (sec. 3.6). Thus $s_{\bar{w}}^2$ is an unbiased estimate of σ^2/n, and $ns_{\bar{w}}^2$ is an unbiased estimate of σ^2, *provided that the k population means are equal.*

Therefore, if the population means are the same, the ratio of the two sample variances,

$$\frac{ns_{\bar{w}}^2}{s_p^2},$$

should be nearly 1. On the other hand, if the population means are different, both the value of $s_{\bar{w}}^2$ and the ratio will be larger, on the average—how much larger depends on the sampling distribution of the ratio

$$F = \frac{ns_{\bar{w}}^2}{s_p^2}.$$

The F-sampling distribution, one of the most useful in statistics, is explained in section 5.2 before the actual calculations for the analysis of variance are introduced in section 5.3.

5.2   THE F-DISTRIBUTION

The F-distribution is a family of theoretical frequency distributions for the ratio of two variances, calculated from independent random samples drawn from two populations whose observations are normally distributed and whose variances are equal. Conceptually, the distribution is obtained as follows. From the first population, with variance σ^2, all possible samples of a given size n_1 are drawn, and, for each sample, the sample variance s_1^2 with $(n_1 - 1)$ degrees of freedom is calculated. Likewise, from the second population, also with variance σ^2, all possible samples of a given size n_2 are drawn; and, for each sample, the sample variance s_2^2 with $(n_2 - 1)$ degrees of freedom is calculated. Then, every sample variance s_1^2 is divided by every sample variance s_2^2 to yield a ratio

$$F = \frac{s_1^2}{s_2^2},$$

the number of ratios being equal to the number of possible samples in the first population multiplied by the number in the second population. The frequency distribution of all possible F-ratios is the F-distribution.

In figure 5.1 frequency curves of several F-distributions are graphed. The figure illustrates several points about the F-distribution. First there is a different F-distribution for every combination of number of degrees of freedom in the numerator and denominator of the F-ratio. Second, because the F-ratio is obtained by dividing two positive numbers, its minimum value is 0, and its maximum value is infinity. Third, as the number of degrees of freedom in both numerator and denominator increase, the frequency curve becomes more nearly symmetrical because the distributions of both s_1^2 and s_2^2 become nearly symmetrical and tightly clustered around 1 because of the central limit theorem. Thus, since there are fewer large values of s_1^2 and fewer small values of s_2^2, the distribution of the ratio itself becomes nearly symmetrical. Fourth, the mean of every F-distribution with fairly large n_2 is nearly 1 and is exactly equal to $(n_2 - 1)/(n_2 - 3)$.

The F-distribution may be related to the example of section 5.1. If the ratio

$$F = \frac{n s_{\bar{w}}^2}{s_p^2},$$

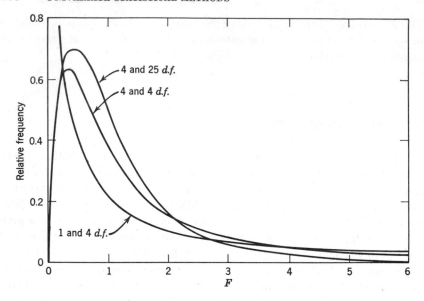

Fig. 5.1. F-curves with various degrees of freedom.

with $(k - 1)$ and $k(n - 1)$ degrees of freedom is too large, the hypothesis that the k sample means are equal is rejected. How large is too large depends upon the risk level chosen by the investigator. In figure 5.2 the F-distribution with 3 and 24 degrees of freedom is graphed. The concept is like that used for the t-test (sec. 4.7), except that the critical region is entirely in the right-hand tail. There is no reason to reject the hypothesis if the calculated F-value is in the left tail because this fact indicates that the within-sample variance is large compared with the among-sample variance. However, if F is very small (less than 0.1 for typical problems), the data are suspect because with proper randomization an F-value this small is unlikely.

The critical region of the F-test for selected α risk levels and degrees of freedom is given in table A-5. There is a separate subtable for each α risk level; the subtable for the 10-percent risk level is described, and then the differences in the other tables are mentioned. Each column corresponds to a different number of degrees of freedom in the numerator; each row corresponds to a different number of degrees of freedom in the denominator. The table entries decrease downward in each column, because the average variance of the denominator s_2^2 decreases. Likewise, except for very small degrees of freedom in the denominator, the table entries decrease toward the right, with increasing degrees of freedom in the numerator. For an infinitely large number of degrees of freedom in both numerator and denominator the value

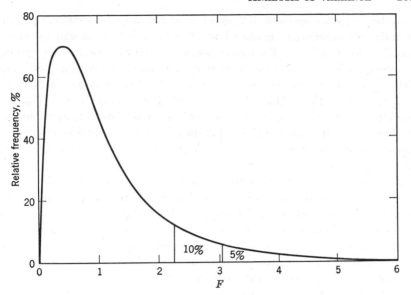

Fig. 5.2. F-curve with 3 and 24 degrees of freedom to illustrate critical regions for 5 and 10 percent risk levels.

of the F-ratio is 1, recorded in the lower right-hand corner entry, because both s_1^2 and s_2^2 become equal to the common variance σ^2 for infinitely large samples. Except for this entry, which is the same for all subtables, the subtables for the 5- and 1-percent risk levels are like those for the 10-percent subtable, except that each entry gets larger as illustrated for one case by figure 5.2.

5.3 ONE-WAY ANALYSIS OF VARIANCE

Now that the F-distribution, which is essential for interpreting the sample-variance ratio calculated in the analysis of variance, has been introduced, the one-way analysis of variance for the fictitious data of table 5.1 can be presented. In table 5.2 the hypothesis test and calculations for the data of table 5.1 are outlined as those for the t-test in table 4.15. The hypothesis H_0 that the four population means are equal is tested against the alternative hypothesis H_1 that they are unequal. If the hypothesis H_0 is true, the variance of population means σ_μ^2 is equal to 0; if the alternative hypothesis H_1 is true, σ_μ^2 is greater than 0.

In the computations (table 5.2) $s_{\bar{w}}^2$ is first obtained by the short-cut variance formula; the calculations in lines 1 to 12 of table 5.2 are similar to those in lines 5 to 13 of table 5.1. The numerator $ns_{\bar{w}}^2$ of the variance ratio is obtained by multiplying by n; the denominator s_p^2 is obtained by dividing $\sum$ SS in line 15 by $\sum$ (d.f.) in line 16. The F-ratio, obtained in line 18 as the final result, is compared with the tabulated F-value of 2.36. Being smaller than the tabulated value, the computed value 0.87 is *outside* the critical region, and the hypothesis H_0 that the five population means are equal is accepted. In geological language the five outcrops in area A are judged to belong to the same formation.

The calculations in table 5.2 are performed to demonstrate the algebra. However, when a desk calculator is used, it is generally more convenient to organize the calculations differently, in the format of table 5.3. In the preliminary calculations T is the sum total of observations in each sample. The advantage of this format is that $\sum w$ and $\sum w^2$ can be obtained in one operation by desk calculator, as can $\sum T$ and $\sum T^2$ as a check. As a demonstration of calculations the fictitious illustration of table 5.1 is reworked in table 5.4. If the sample sizes are unequal, the format of table 5.4 is generalized to that of table 5.5. In table 5.5 n_0 is equal to $k/\sum (1/n_i)$.

TABLE 5.3. SIMPLIFIED COMPUTING METHOD FOR ONE-WAY ANALYSIS OF VARIANCE

Type of sum	Preliminary calculations		
	(1) Total of squares	(2) Number of observations in sum	(3) = (1)/(2) Total of squares per observation
Grand	$(\sum T)^2$	kn	I
Sample	$\sum T^2$	n	II
Observation	$\sum w^2$	1	III

Source of variation	Analysis of variance			
	Sum of squares	Degrees of freedom	Mean square	F
Among-sample means	II − I	$k - 1$	$ns_{\bar{w}}^2$	$ns_{\bar{w}}^2/s_p^2$
Within samples	III − II	$k(n - 1)$	s_p^2	
Total	III − I	$kn - 1$		

TABLE 5.4. SIMPLIFIED COMPUTING METHOD FOR ONE-WAY ANALYSIS OF VARIANCE
APPLIED TO DATA OF TABLE 5.1

	Preliminary calculations		
	(1)	(2) Number of	(3) = (1)/(2) Total of
Type of sum	Total of squares	observations in sum	squares per observation
Grand	42,189.16	20	2109.458
Sample	8,491.62	4	2122.91
Observation	2,180.96	1	2180.96

	Analysis of variance				
Source of variation	Sum of squares	Degrees of freedom	Mean square	F	$F_{10\%}$
Among-sample means	13.45	4	3.36	0.87	2.36
Within samples	58.05	15	3.87		
Total	71.50	19			

TABLE 5.5. SIMPLIFIED COMPUTING METHOD FOR ONE-WAY ANALYSIS OF VARIANCE,
GENERALIZED FOR UNEQUAL SAMPLE SIZES

	Preliminary calculations		
	(1)	(2) Number of	(3) = (1)/(2) Total of
Type of sum	Total of squares	observations in sum	squares per observation
Grand	$(\sum T)^2$	kn	I
Sample			II $= \sum (T^2/n)$
Observation	$\sum w^2$	1	III

	Analysis of variance			
Source of variation	Sum of squares	Degrees of freedom	Mean square	F
Among-sample means	II − I	$k - 1$	$n_0 s_{\bar{w}}^2$	$n_0 s_{\bar{w}}^2 / s_p^2$
Within samples	III − II	$\sum n - k$	s_p^2	
Total	III − I	$\sum n - 1$		

TABLE 5.6. FICTITIOUS OBSERVATIONS OF SAND CONTENT CONSTITUTING STATISTICAL SAMPLES FROM FIVE OUTCROPS IN AREA B

Line	Item	Statistical sample number				
		(1)	(2)	(3)	(4)	(5)
1	Observations	14.0	4.8	9.4	3.1	15.8
2	(sand content,	16.2	7.2	13.2	8.8	16.3
3	%)	16.9	1.8	9.2	4.2	13.6
4		15.4	1.6	9.0	5.5	14.5
5	$\sum w$	62.5	15.4	40.8	21.6	60.2
6	n	4	4	4	4	4
7	$\bar{w}$	15.625	3.850	10.200	5.400	15.050
8	$(\sum w)^2$	3906.25	237.16	1664.64	466.56	3624.04
9	$(\sum w)^2/n$	976.56	59.29	416.16	116.64	906.010
10	$\sum w^2$	981.21	80.68	428.24	134.94	910.540
11	SS	4.65	21.39	12.08	18.30	4.53
12	d.f.	3	3	3	3	3
13	s^2	1.55	7.13	4.03	6.10	1.51

A Second Fictitious Illustration

For the fictitious data from area A (table 5.1) the conclusion is drawn that the five population means are equal. Consider now the fictitious data from five outcrops in area B presented in table 5.6, for which the same hypothesis H_0 that the five population means are equal is to be tested. For area B the within-sample variances are of about the same sizes as those for area A, but the variance of the five sample means $s_{\bar{w}}^2$ is 116.2, which indicates a larger variability among sample means in area B. For area B an analysis of variance (table 5.7) yields a computed F-value of 28.59, which, being larger than the tabulated F-value of 2.36, leads to accepting the alternative hypothesis H_1. In geological language the five outcrops in area B are judged to belong to different formations.

TABLE 5.7. ONE-WAY ANALYSIS OF DATA FROM FICTITIOUS AREA B

Source of variation	Sum of squares	Degrees of freedom	Mean square	F	$F_{10\%}$
Among-sample means	464.65	4	116.16	28.59	2.36
Within samples	60.95	15	4.06		
Total	525.60	19			

Remarks

In hypothesis tests with the analysis of variance, as in any hypothesis tests, whenever a hypothesis H_0 is accepted, the investigator runs the risk of making a type II error through accepting a hypothesis H_0 when it is in fact false. Sometimes this risk is verbalized by a weaseling double-negative statement, such as, "The data do not contradict the hypothesis H_0." The chance of making a type II error depends on the ratio of the population variance of the means to the population variance of the random fluctuations (σ_μ^2/σ^2) and on the number of degrees of freedom in the numerator and denominator. This ratio can be estimated, and the chance of making a type II error calculated by a method beyond the scope of this book (see Dixon and Massey, 1969, Chap. 14). Thus, when an investigator accepts a hypothesis H_0, he should be wary, but if he rejects it he is protected because the chance of type I error is set exactly.

The name *analysis of variance* describes the partitioning of the total variance into meaningful pieces. For the fictitious illustrations of areas A and B there are two pieces: one for the variation among sample means and another for the variation of observations within the individual samples. If k samples, each containing n observations, are treated as one large sample, the total sum of squares with $(kn - 1)$ degrees of freedom can be calculated for the large sample. However, the sum of degrees used to estimate the pooled sample variance s_p^2 has $k(n - 1)$ degrees of freedom, and the sum of squares used to estimate the among-sample-means variance $ns_{\bar{w}}^2$ has $(k - 1)$ degrees of freedom. Thus, as illustrated by table 5.4, the total sum of squares is equal to the sum of the among-samples sum of squares and the within-samples sum of squares, and the total number of degrees of freedom is equal to the sum of the numbers of degrees of freedom in the partitioned sum of squares.

According to Cochran's theorem (Wilks, 1962, p. 212), whenever a sum of squares can be partitioned into pieces so that the pieces add up to the total sum of squares and the degrees of freedom of the pieces add up to the total degrees of freedom, the sampling distributions of the individual pieces are independent of one another. Thus in the division into among sample sum of squares and within-sample sum of squares and the corresponding variances the two statistics $ns_{\bar{w}}^2$ and s_p^2 are independently distributed, and their ratio follows the F-distribution.

Finally, the estimation of the variance of population means σ_μ^2 is of interest, particularly if a hypothesis H_1 is accepted, as in the fictitious illustration of data from area B. The mean value of the variance of sample means $\mu_{s_{\bar{w}}^2}$ may be expressed as the sum of two components (Dixon and Massey, 1969, p. 162). The first, the square of the standard error of the mean σ^2/n, is the common

TABLE 5.8. AVERAGE MEAN-SQUARE VALUES IN ONE-WAY ANALYSIS OF VARIANCE

Source of variation	Degrees of freedom	Mean square	Average mean square
Among-sample means	$k - 1$	$ns_{\bar{w}}^2$	$\sigma^2 + n\sigma_\mu^2$
Within samples	$k(n - 1)$	s_p^2	σ^2

population variance of random fluctuations divided by the sample size. The second is σ_μ^2. Thus

$$\mu_{s_{\bar{w}}^2} = \frac{\sigma^2}{n} + \sigma_\mu^2.$$

In table 5.8 the preceding formula is related to the analysis-of-variance format of table 5.4. Table 5.8 shows the mean-square statistics that estimate the mean-square parameters. As in table 5.4, the variance of sample means $s_{\bar{w}}^2$ is multiplied by the number of sample means n, so that, in table 5.8, the average variance of sample means,

$$\frac{\sigma^2}{n} + \sigma_\mu^2,$$

is multiplied by n to yield

$$\sigma^2 + n\sigma_\mu^2.$$

Hence by subtracting the within-sample mean square from the among-sample mean square and dividing by n the desired estimate of the variance of population means σ_μ^2 is found to be

$$\frac{ns_{\bar{w}}^2 - s_p^2}{n}.$$

If $ns_{\bar{w}}^2$ is less than s_p^2, the result is negative, and it is customary to set this estimate equal to 0. Whenever an average variance is composed of several pieces, the pieces are named *variance components*, and the statistics estimating them are named *variance-component estimates*.

5.4   EXAMPLES OF ONE-WAY ANALYSIS OF VARIANCE

For studying many sets of geological data, the one-way analysis of variance is a powerful tool, even though more complicated analyses of variance, presented subsequently in this chapter, may be required to understand fully all variance components. To emphasize the value of the one-way analysis of variance two example problems are reviewed in this section in some detail.

Comparison of Mean Silver Content of a Vein on Two Sides of a Fault

The first example is a comparison of mean silver content in a vein on two sides of a fault. In the Frisco mine (sec. 1.4) the Don Tomás vein (sec. 4.5, fig. 4.11) is cut by the Frisco fault whose displacement is 35 meters or more. The original data were the assays for five metals of chip samples from 1810 stations in drifts on the vein. In order to reduce and measure local fluctuation, metal contents from successive groups of 10 samples along each drift were averaged, and the resulting means were the example data. Thus there were 181 summary points (1810/10), 63 on the south side and 118 on the north side of the fault.

In the original paper (Koch and Link, 1963) several methods for comparing the mineralization are given. However, in this book only a one-way analysis of variance for the 181 summary-point observations for silver is given as an example. The problem is formulated in the following hypothesis test:

1. *Hypothesis:* The two population means for silver content are equal on the north and south sides of the fault; that is, $H_0: \mu_n = \mu_s$.
2. *Alternative hypothesis:* $H_1: \mu_n \neq \mu_s$, or $\sigma_\mu^2 > 0$.
3. *Assumptions:* Random samples drawn from normal populations with the same variance.
4. *Risk of type I error:* $\alpha = 10$ percent.
5. *Critical region:* $k = 2$, $n_n + n_s = 181$; accordingly, degrees of freedom are 1 and 179. From table A-5, $F_{10\%}$ is 2.74.

Table 5.9 presents the one-way analysis of variance to test this hypothesis. The total variance is partitioned into two pieces: one corresponding to the variance between the north and south sides of the fault, with 1 degree of freedom; and the other corresponding to the variance within vein segments,

TABLE 5.9. ONE-WAY ANALYSIS OF VARIANCE TO TEST THE HYPOTHESIS THAT SILVER CONTENT IS THE SAME IN THE DON TOMÁS VEIN ON BOTH SIDES OF THE FRISCO FAULT

Source of variation	Sum of squares	Degrees of freedom	Mean square	F	$F_{10\%}$
Between north and south sides of Frisco fault	1,726,043	1	1,726,043	34.7	2.74
Within vein segments	8,895,472	179	49,695		

with 179 degrees of freedom. Because it is larger than the tabled $F_{10\%}$-value, the calculated F-value of 34.7 is inside the critical region, and the alternative hypothesis H_1 is accepted, that $\mu_n \neq \mu_s$, or σ_μ^2 is greater than 0. In geological terms the evidence supports a difference in silver mineralization.

Had these silver values been obtained through a designed experiment, this analysis would be one step toward the objective of comparing mineralization on the two sides of the Frisco fault. Some or all of the following variables would have been investigated: the five metals of commercial value, other chemical elements or substances, and physical properties such as rock competency (perhaps reflecting recemented and therefore stronger rock on one side of the fault but not on the other). Target and sampled populations would have been defined, and random statistical samples would have been taken on either side of the fault. If criteria for comparing mineralization were established in enough detail, appropriate sample sizes might have been determined from the estimates of underlying variance derived from preliminary sampling. Such a designed experiment is an ideal to strive for.

Because most of the Don Tomás vein had been mined out, sampling in a designed experiment would have been impossible; therefore, the best available data were the assays obtained in the regular course of mining. In such a situation, which today seems to be more the rule than the exception in geology, the suitability of a particular statistical analysis is always debatable. Even though the question is not further discussed at this point, the issue is raised to warn the reader. With so large a calculated F-value, a difference in mean silver content is clearly established. Had the calculated value been smaller, a more refined statistical analysis might have been worthwhile. The decision would have been made in view of the objectives of the entire study and the total pattern of data.

It should be noted that instead of the one-way analysis of variance, the t-test (sec. 4.7) could have been applied to these data. This test would yield a computed value of $t = \sqrt{F} = \sqrt{34.7}$, and, of course, the same conclusion that the means are unequal would have been drawn.

Comparison of Stope and Development Data from a South African Gold Mine

The second one-way analysis of variance is of gold-assay data from the City Deep mine, Central Witwatersrand, South Africa. The overall problem (Koch and Link, 1966), parts of which are discussed in Chapter 16 of this book, was to determine gold distribution and variability in a typical South African gold mine. The data were assays from a 1000-square-foot block of ground that had been fully developed and stoped out in workings extending throughout most of this section of the City Deep mine. The assays, supplied by City Deep, Ltd., consisted of 503 gold-assay values from all development samples taken in drifts and a raise, and 1536 assays from all stope samples.

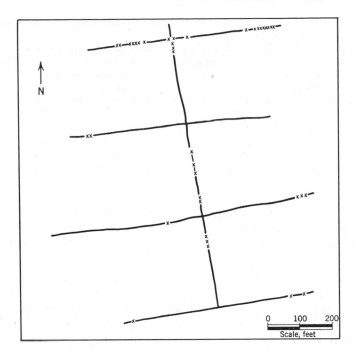

Fig. 5.3. Locations of the 39 of 503 development points accounting for 50 percent of the total in.-dwt gold content, City Deep mine, South Africa.

Figure 5.3 shows that the four parallel drifts and the raise gave good coverage for the development sampling and that the pattern of gold assays is erratic, with 50 percent of the gold content in the development samples concentrated at 39 out of 503 sample points.

We examined the development and stope data separately to learn how well they agreed. Because the purpose of taking the development samples was to predict grade of gold ore to be mined in the stopes, good agreement between the two sets of assay values was desirable for a successful prediction. Table 5.10 lists summary assay data for the stope and development samples. For these data the inch-pennyweight (in.-dwt) variable is the appropriate measure of gold content (sec. 13.2). Table 5.10 shows that the development means far overestimate the grade of gold actually mined in the stopes and also that both development and stope in.-dwt values are highly variable, with coefficients of variation of about 2. In order to learn whether the two sets of data are consistent within limits of expected statistical fluctuation, the one-way analysis of variance of table 5.11 was performed. The hypothesis H_0 that the two population means for the statistical samples of development and stope

TABLE 5.10. SUMMARY ASSAY DATA OF SAMPLES FROM 1000-FOOT-SQUARE BLOCK, CITY DEEP MINE, SOUTH AFRICA

Unit	Stope data, 1536 samples			Development data, 503 samples		
	Mean	s	C	Mean	s	C
Dwt	12.00	31.31	2.61	22.58	71.33	3.16
Inches	20.84	9.47	0.45	18.20	10.29	0.57
In.-dwt	192.40	374.70	1.95	273.46	607.80	2.22

Note: Dwt, pennyweights; In.-dwt, inch-pennyweights; s, standard deviation; C, coefficient of variation.

assays are equal was tested against the alternative hypothesis H_1 that the population means are different. This analysis has 1 degree of freedom for variation between stope and development assays and 2037 degrees of freedom (503 + 1536 − 2) for within-group variation. Because the calculated F-value of 12.7 is larger than the tabled $F_{10\%}$-value of 2.71, we accepted the alternative hypothesis H_1 and concluded that the two sets of data are inconsistent. As before, the t-test could have been used, as there are only two sets of means. In section 16.1, in which the discussion of these interesting data is continued, we suggest ways to reconcile the stope and development values.

TABLE 5.11. ONE-WAY ANALYSIS OF VARIANCE TO COMPARE DEVELOPMENT AND STOPE MINERALIZATION, CITY DEEP MINE

Source of variation	Sum of squares	Degrees of freedom	Mean square	F	$F_{10\%}$
Between stope and development	2,489,747	1	2,489,747	12.7	2.71
Within groups	400,963,342	2037	196,647		

5.5   ANALYSIS OF VARIANCE OF NESTED DATA

In the preceding sections of this chapter we have explained the analysis of variance for the simplest one-way case and have given two examples of the many uses in geology. Now, in this section the analysis of variance of nested data, also named hierarchical analysis of variance, is introduced. The method is explained and in section 5.6 application in geology is discussed. The computing details are omitted; the reader requiring them is referred to Li (1964, I, chap. 18).

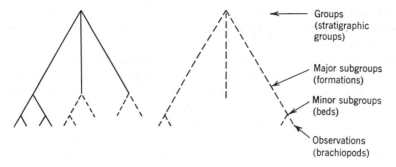

Fig. 5.4. Relation to one another of 60 fictitious observations.

The names *nested* and *hierarchical*, being descriptive, indicate that in this form of the analysis of variance the observations are partitioned into subgroups within larger groups. The method is explained by a fictitious illustration (fig. 5.4) analyzed in the format of table 5.12 for a set of 60 observations classified into five groups, with three major subgroups within each group, two minor subgroups within each major subgroup, and two observations within each minor subgroup. Any desired geological interpretation may be placed on the classification; for instance, the five groups might be stratigraphic groups, the three major subgroups might be formations, the two minor subgroups might be beds, and the observations might be sizes of brachiopods.

For these fictitious data a typical observation w_{ijk} may be written

$$w_{ijk} = \mu + (\mu_i - \mu_j) + (\mu_j - \mu_k) + (\mu_k - \mu) + e_{ijk},$$

where μ is the general mean for all of the data, μ_i is the mean of minor subgroups within major subgroups, μ_j is the mean of major subgroups within groups, μ_k is the mean of groups, and e_{ijk} is the random fluctuation. There would be 30 minor subgroup means μ_i; their averages taken two at a time would determine the 15 major subgroup means μ_k. In turn, the averages of

TABLE 5.12. FICTITIOUS ILLUSTRATION OF NESTED ANALYSIS OF VARIANCE FORMAT

Source of variation	Sum of squares	Degrees of freedom	Mean square
Among groups		4	
Among major subgroups		10	
Between minor subgroups		15	
Within minor subgroups		30	

the 15 major subgroup means taken three at a time would determine the five group means μ_k. Finally, the average of the five group means would determine the value of μ.

The differences $(\mu_i - \mu_j)$ measure the variation of minor subgroup means within a major subgroup; the differences $(\mu_j - \mu_k)$ measure the variation of major subgroup means within a group; the differences $(\mu_k - \mu)$ measure the variation among groups. The random fluctuation e_{ijk} is defined as having a mean of 0 and the same variance for all observations.

Through the model, the degrees of freedom are partitioned. In all there are $5 \times 3 \times 2 \times 2 = 60$ observations (fig. 5.4). Because there are two observations in each minor subgroup, there is one degree of freedom associated with the variation within each minor subgroup. Thus, with 30 minor subgroups, there are altogether 30 degrees of freedom for within-minor-subgroups variation. Because there are two minor subgroups within each of the 15 major subgroups, there are 15 degrees of freedom for the variation of minor subgroups within major subgroups. Because there are three subgroups within each of the five groups, there are $2 \times 5 = 10$ degrees of freedom for the variation of the major subgroups within groups. Finally, because there are five groups, there are 4 degrees of freedom for variation among groups. The sum of degrees of freedom from all levels is 59, the total number of degrees of freedom in 60 observations $(30 + 15 + 10 + 4 = 59)$.

In the fictitious illustration of table 5.12 the appropriate F-value to test whether the major subgroups differ among themselves has 10 and 15 degrees of freedom. Moreover, because both the degrees of freedom and the sum of squares are additive, a hypothesized difference in the groups themselves can be tested by an F-test with 4 and 10 degrees of freedom. A nesting scheme, like that illustrated by the fictitious illustration, can be carried to any number of meaningful levels. The number of degrees of freedom for any level is always the number of degrees of freedom that can be calculated for observations at that level.

5.6   AN EXAMPLE OF NESTED ANALYSIS OF VARIANCE

The statistically designed experiment for sampling ore in the Homestake mine, South Dakota, by diamond-drill-hole core, introduced in section 3.4, provides an excellent example of the analysis of variance of nested data. Our purpose was to investigate several levels of variability in gold distribution: among drill holes, among 1-foot intervals, within 1-foot intervals, and among furnace runs. As shown by figure 3.1, five holes were bored, one pair in one crosscut, and the other three in an adjacent crosscut. The core was sampled

Table 5.13. Summary gold assay data from five diamond-drill holes, Homestake mine, South Dakota

Drill hole	Number of observations	Mean gold assay (oz/ton)	Standard deviation	Coefficient of variation
A	58	0.52	1.45	2.79
B	58	2.19	7.60	3.47
C	24	2.85	8.27	2.90
D	39	0.50	1.16	2.32
E	40	0.69	1.57	2.28
All holes	219	1.25	4.93	3.94

in a random design, and selected geological samples were assayed. With the assay values in hand, the first obvious step was to compute a mean and standard deviation for the individual holes and for all holes (table 5.13). Although all of these statistics are distinctly different, those for holes A, D, and E are rather similar, as are those for holes B and C. The extremely high coefficients of variation reflect the erratic gold mineralization, particularly as observed in these small samples of 1-inch-long cylinders of core.

First of all, taking into account the variability, we wanted to learn if the differences among these five sample means are large enough to indicate a difference in the underlying population means, or if the observed differences merely stem from sample fluctuation. The appropriate analysis of variance, a comparison of among-hole and within-hole variation, is presented in table 5.14A. The measure of within-hole variation is necessarily the among-foot variation rather than the among-observation variation or the among-sample-pair variation that might be expected, because the smallest scale of unrestricted random sampling is at 1-foot intervals. In statistical terminology the sampling is *blocked* within holes at 1-foot intervals, because the *constraint*, or requirement, was made that two pairs of samples be taken for every 1-foot interval. An average assay value was available for each 1-foot interval, the average of the four observations (or fewer, in the few intervals where a potential observation was lost in drilling or assaying). Thus there were 4 degrees of freedom to estimate the variability among the five holes and 51 degrees of freedom to estimate the variability within holes. The hypothesis H_0 that the five population means were equal was tested against the alternative hypothesis H_1 that they were unequal. Because the computed F-value of 0.89 was smaller than the tabled $F_{10\%}$-value of 2.07, it is outside the critical region, and the conclusion is drawn that the population means are the same.

TABLE 5.14. THREE ONE-WAY ANALYSES OF VARIANCE TO INVESTIGATE GOLD MINERALIZATION IN DIAMOND-DRILL CORE FROM THE HOMESTAKE MINE

A. Analysis of variance to compare among-hole and within-hole variation (as measured by among-foot variation)

Source of variation	Sum of squares	Degrees of freedom	Mean square	F	$F_{10\%}$
Among-hole	177.883	4	44.471	0.89	2.07
Among-foot	2536.107	51	49.728		

B. Analysis of variance to compare among-foot and within-foot intervals (as measured by between-sample pair variation)

Source of variation	Sum of squares	Degrees of freedom	Mean square	F	$F_{10\%}$
Among-foot intervals	2536.107	51	49.728	1.71	1.45
Between-sample pairs	1511.398	52	29.065		

C. Analysis of variance to compare within-sample-pair and between-sample-pair variation

Source of variation	Sum of squares	Degrees of freedom	Mean square	F	$F_{10\%}$
Between-sample pairs	1511.398	52	29.065	2.90	1.37
Within-sample pairs	1083.141	108	10.029		

We next wanted to investigate the variability on a smaller scale, comparing the variation from foot to foot within holes with the variation within each foot. Again the sampling within each foot is blocked at the scale of sample pairs, so that the appropriate measure is the between-sample-pair variation. The hypothesis H_0 that the variation among 1-foot intervals is no larger than that within 1-foot intervals is tested against the alternative hypothesis H_1 that it is larger. The analysis is performed in table 5.14*B*. As in table 5.14*A*, there are 51 degrees of freedom to measure among-foot variability, and there are 52 degrees of freedom to measure between-sample-pair variability, because there were 52 1-foot intervals with 2 pairs of assays (in the others one or more observations were lost). Because the computed F-value of 1.71 is larger than the tabled $F_{10\%}$-value of 1.45, the decision is to accept the alternative hypothesis H_1 that the variation among 1-foot intervals is larger than that within 1-foot intervals.

TABLE 5.15. NESTED ANALYSIS OF VARIANCE OF GOLD ASSAY DATA FROM THE HOMESTAKE MINE

Line	Source of variation	Sum of squares	Degrees of freedom	Mean square	F	$F_{10\%}$
1	Among hole	177.883	4	44.471	0.89	2.07
2	Among foot intervals	2536.107	51	49.728	1.71	1.45
3	Between sample pairs	1511.398	52	29.065	2.90	1.37
4	Within sample pairs	1083.141	108	10.029		
5	Total	5308.529	215			

Finally, we examined the variation on the smallest scale sampled, within 1-foot intervals, comparing the variation within the paired 1-inch cylinders with that among the paired cylinders. In table 5.14C, the hypothesis H_0 that the variation among sample pairs is no larger than that within sample pairs is tested against the alternative hypothesis H_1 that the variation among sample pairs is larger. As before, the degrees of freedom to estimate the between-sample-pair variation are 52; the degrees of freedom to estimate within-sample-pair variation are 108, because there are 108 pairs of samples (again subtracting lost observations). Because the computed F-value of 2.90 is larger than the tabled $F_{10\%}$-value of 1.37, it lies inside the critical region, and the decision is to accept the alternative hypothesis H_1 that the within-sample-pair variation is smaller than the among-sample-pair variation.

Clearly, the three one-way analyses of variance in table 5.14 are related to one another. The relationship is exhibited by combining the three analyses of variance into a nested analysis of variance in table 5.15. In table 5.15, both the total sum of squares and total degrees of freedom in line 5 have been partitioned into the 4 pieces for each on lines 1 to 4. Certain problems in this analysis are discussed in section 5.11. The Homestake example analysis has been introduced purposely to show that analysis of real data, even that obtained in a designed experiment, is not likely to be clean cut.

5.7 RANDOMIZED BLOCKS

In this section, a statistical procedure named *randomized block* is explained, first by the simplest case of the *t-test of paired observations*, and then by the general procedure for observations that are in groups larger than pairs. Through the randomized-block procedure, the experimental material (perhaps rock units or hand specimens for chemical analysis) is arranged in homogene-ous groups named *blocks*, and the experimental treatments (perhaps measure-ments of grain size by different techniques or chemical analyses by different methods) are applied randomly to the experimental material in each block. Because the randomization is confined within homogeneous groups and because all treatments are applied in each group, most of the among-group variation is eliminated. If the experimental material is not arranged in blocks, any variation in it produces additional random fluctuation.

A randomized block is a simple *experimental design*, a subject developed in sections 8.6 and 12.5. Through an experimental design, as the name im-plies, a scientific investigation is performed as a statistically designed experiment in order to obtain maximum information from the effort expended.

Application of the *t*-Statistic to Paired Observations

The *t*-test on paired observations is explained by analysis of the fictitious data of table 5.16. Assume that an investigator has nine rock specimens that he wishes to analyze for CaO by two chemical methods. He may prepare two pulverized fractions suitable for chemical analysis from each rock specimen. He may then randomly choose one fraction from each specimen to be analyzed by the first method and the other fraction to be analyzed by the second method. In this way the 18 observations in columns 1 and 2 of table 5.16 might be obtained.

In order to learn whether the two chemical methods furnish consistent results, a confidence interval may be formed for the difference between the population means for the two methods by assuming that the observations are two samples, each of size 9 from populations of infinite numbers of potential analyses. If the confidence interval includes zero, the conclusion is that the two methods are consistent; if it does not, the conclusion is that the two methods are inconsistent. Instead, the hypothesis H_0 that the two population means are equal could be tested against the alternative hypothesis H_1 that they are unequal.

Because, as calculated in table 5.16, the confidence interval for these data extends from -0.833 to 1.455, including zero, the investigator concludes that the two chemical methods are consistent. However, he has made a major blunder in data analysis by ignoring the fact that he had nine pairs of observations from nine rock samples and has treated them as though they were merely 18 unpaired observations. Therefore experimental error (random fluctuation, e) has been increased because the variation in CaO of the rocks from sample to sample has been included in it. As demonstrated next, this uncontrolled variability can readily be eliminated by the proper statistical procedure, with no increase in experimental cost.

TABLE 5.16. FICTITIOUS ILLUSTRATION, COMPARING TWO METHODS OF CHEMICAL ANALYSIS FOR CaO BY THE t-STATISTIC

Data:

	CaO (%)		
	(1)	(2)	
	10.4	10.6	
	9.9	9.7	
	9.1	8.8	
	9.6	8.9	
	8.5	8.4	
	7.4	6.8	
	8.1	7.9	
	6.6	6.3	
	7.2	6.6	

Calculations:

$\sum w$	76.8	74.0	
n	9	9	
$\overline{w}$	8.533	8.222	$0.311 = \overline{w}_1 - \overline{w}_2$
$(\sum w)^2$	5898.24	5476.00	
$(\sum w)^2/n$	655.36	608.44	
$\sum w^2$	669.16	625.56	
SS	13.80	17.12	
$SS_1 + SS_2$		30.92	
$n_1 + n_2 - 2$		16	
s_p^2		1.9325	
$1/n_1 + 1/n_2$		0.2222	
$\sqrt{s_p^2(1/n_1 + 1/n_2)}$		0.6553	
$t_{5\%}$ (with 16 d.f.)		1.746	
$t\sqrt{s_p^2(1/n_1 + 1/n_2)}$		1.144	
μ_L		-0.833	
μ_U		1.455	

An incisive way to perform the statistical analysis is outlined in table 5.17. The essential difference from table 5.16 is that now the nine analyses by the two chemical methods are recognized as being *paired*, so that the two analyses of duplicate samples are compared directly with one another. In statistical terminology the two chemical methods are named *treatments*, and the nine

TABLE 5.17. FICTITIOUS ILLUSTRATION, COMPARING TWO METHODS OF CHEMICAL ANALYSIS FOR CaO BY THE t-STATISTIC APPLIED TO PAIRED OBSERVATIONS

Data:

	CaO (%)		Difference, (1) − (2)
	(1)	(2)	(3)
	10.4	10.6	− 0.2
	9.9	9.7	0.2
	9.1	8.8	0.3
	9.6	8.9	0.7
	8.5	8.4	0.1
	7.4	6.8	0.6
	8.1	7.9	0.2
	6.6	6.3	0.3
	7.2	6.6	0.6

Calculations:

	(1)	(2)	(3)
$\sum w$	76.8	74.0	2.8
n	9	9	9
$\bar{w}$	8.533	8.222	0.31111
$(\sum w)^2$			7.84
$(\sum w)^2/n$			0.8711
$\sum w^2$	669.16	625.56	1.52
SS			0.64889
s^2			0.081111
s			0.2848
$\sqrt{n}$			3
$s/\sqrt{n}$			0.09493
$t_{5\%}$ (with 8 d.f.)			1.860
$t(s/\sqrt{n})$			0.176
μ_L			0.135
μ_U			0.487

pairs are named *blocks*. As explained in the next subsection, any number of treatments and blocks can be accommodated; the t-statistic applied to paired observations is only the special case for two treatments.

Through application of the t-statistic to the paired observations of table 5.17, a confidence interval is formed for the population mean of the differences between the paired observations. The interpretation of the confidence interval is that, because of experimental error, no one would expect the two chemical methods always to yield exactly the same result. But if the two methods are consistent, one of them should not, on the average, yield higher results than the other. Because the confidence interval for the paired differences extends from 0.135 to 0.487 and does not include zero, the conclusion is that the two methods do not agree. By recognizing that the observations are paired, and by statistically analyzing the experimental data accordingly, one obtains more precise information. The next step would be to find out which chemical method is better or whether one contains a constant bias that can readily be removed by calculation.

Randomized-Block Design with More than Two Treatments

The statistical procedure for reducing unexplained variability by blocking may be readily generalized for more than two treatments. As already explained for paired observations, in any collection of data it is always desirable to organize the observations in several blocks for the purpose of making the amount of unexplained variability as small as possible. The observations are arranged to be as homogeneous as possible within blocks, although the different blocks may be very heterogeneous. Then each treatment is randomly applied within each block. Although the within-block variability is then small, the among-blocks variability may be large, and the randomized-block design allows the effect of the among-block variability to be removed from the statistical analysis.

The general principles of the randomized-block design may best be explained with the aid of an illustration. Suppose that 20 chemical laboratories are to analyze three different rocks for lime. For a comparison of the performance of the laboratories, each laboratory is considered a treatment, and each rock is considered a block. From each rock 20 pulverized fractions are prepared for analysis and are assigned at random to the 20 laboratories. This procedure yields 60 (20 × 3) chemical analyses or observations.

Even if the 3 rocks have different mean lime contents, the performances of the 20 laboratories may be compared by removing the effect of the different means. The model for an observation w_{LR} is

$$w_{LR} = \mu + (\mu_L - \mu) + (\mu_R - \mu) + e_{LR},$$

where μ is the mean of all observations, μ_L is the mean of the observations for a single laboratory, μ_R is the mean of the observations for a single rock, and e_{LR} is the random fluctuation. The average values of μ_L and μ_R are identical and are equal to μ. The random fluctuation e_{LR} is the deviation of an observation from the average value for laboratory L and rock R. As in other models, the assumption is made that the mean of e_{LR} is 0 and that the variance of e_{LR} is the same for all values of L and R.

In this linear model the term $(\mu_L - \mu)$ is named the *laboratory effect*, and the term $(\mu_R - \mu)$ is named the *rock effect*. In words, the model may be written as

w = general mean + laboratory effect + rock effect + random error.

If the rock effect were not taken into account, the linear model would be

w = general mean + laboratory effect + random error,

and the random error would be larger, if, as generally happens, the lime contents of the rocks were different on the average. Other consequences of this model are that all laboratories are assumed to have about the same random fluctuation and to measure the same deviation from the average of the various rocks.

The calculations for a randomized-block design may be explained by the use of an example from Li (1964, I, p. 244). The data in table 5.18 are observed pH readings from the top, middle, and bottom of six core samples of soil; the problem is to determine if the average reading varies with soil depth. The two bottom rows and the right-hand column of the table list summary quantities needed for the analysis of variance calculations. The next-to-bottom row lists the three totals of the top, middle, and bottom pH readings;

TABLE 5.18. READINGS OF pH FROM THE TOP, MIDDLE, AND BOTTOM OF SIX CORE SAMPLES OF SOIL. (FROM LI, 1964, I, p. 244.)

Core sample no.	Top	Middle	Bottom	Core sample totals, T_r
1	7.5	7.6	7.2	22.3
2	7.2	7.1	6.7	21.0
3	7.3	7.2	7.0	21.5
4	7.5	7.4	7.0	21.9
5	7.7	7.7	7.0	22.4
6	7.6	7.7	6.9	22.2
Location totals, T_t	44.8	44.7	41.8	
Grand total, G				131.3

each total is the sum of the six pH readings from the individual core samples. The right-hand column lists the six totals from the six core samples; each total is the sum of the three pH readings from a single core sample. The entry in the bottom row, the total of all 18 observations, is also the sum of the three totals in the next-to-bottom row, or the sum of the six totals in the right-hand column.

The analysis-of-variance calculations for the randomized-block design in table 5.19 are similar to those for the completely randomized experiment explained in table 5.3; the only difference is the addition to table 5.19 of the lines associated with the replications. The top half of the table contains the preliminary calculations. Column 1 lists the totals of squares for the grand total of all observations squared, G^2; the sum of squares of the replication totals, $\sum T_r^2$; the sum of the squares of the treatment totals, $\sum T_t^2$; and the sum of the squares of all the observations, $\sum w^2$. In column 2 the number of observations per sum is recorded; the entries in column 3 are obtained by dividing the entries in column 1 by those in column 2. The bottom half of table 5.19, the analysis-of-variance format, shows how the sums of squares

TABLE 5.19. COMPUTING METHOD FOR RANDOMIZED BLOCK

	Preliminary calculations		
Type of sum	(1) Total of squares	(2) Number of observations in sum	(3) = (1)/(2) Total of squares per observation
Grand	G^2	kn	G^2/kn (I)
Replication	$\sum T_r^2$	k	$\sum T_r^2/k$ (II)
Treatment	$\sum T_t^2$	n	$\sum T_t^2/n$ (III)
Observation	$\sum w^2$	1	$\sum w^2$ (IV)

	Analysis of variance			
Source of variation	Sum of squares	Degrees of freedom	Mean square	F
Replication	II − I	$n - 1$		
Treatment	III − I	$k - 1$		
Error	IV − III − II + I	$(k - 1)(n - 1)$		
Total	IV − I	$kn - 1$		

are obtained from the elements labeled I, II, III, and IV in column 3 of the top half of the table.

Table 5.20 gives, in the format of table 5.19, the details of the calculations for the soil sample data in table 5.18. For these data, the hypothesis H_0 is that the pH in the top, middle, and bottom soil samples is the same; or, in other words, that there is no treatment effect. Because the calculated F-value to test this hypothesis is 31.77, which is larger than the tabled value of 2.92 for a 10-percent α risk level, the hypothesis is rejected, and the alternative hypothesis H_1 that there is a treatment effect is accepted. To explore the nature of the treatment effect further would require use of the multiple-comparison techniques discussed in section 5.12.

The analysis of variance in table 5.20 can also be used to investigate whether the six places sampled have, on the average, different soil pH values. The hypothesis H_0 that the soil is the same at the six places, that is, no replication effect exists, is compared with the alternative hypothesis H_1 that the soil is different. Because the calculated F-value to test this hypothesis is 6.43, which is larger than the tabled F-value of 2.52 at the 10-percent α risk level, the alternative hypothesis H_1 that the soil is different is accepted.

TABLE 5.20. COMPUTING METHOD FOR RANDOMIZED BLOCK APPLIED TO DATA OF TABLE 5.18

Type of sum	Preliminary calculations		
	(1) Total of squares	(2) Number of observations in sum	(3) = (1)/(2) Total of squares per observation
Grand	17,239.69	18	957.7606
Replication	2,874.75	3	958.2500
Treatment	5,752.37	6	958.7283
Observation	959.37	1	959.3700

Source of variation	Analysis of variance				
	Sum of squares	Degrees of freedom	Mean square	F	$F_{10\%}$
Replication	0.4894	5	0.09788	6.43	2.52
Treatment	0.9677	2	0.48385	31.77	2.92
Error	0.1523	10	0.01523		
Total	1.6094	17			

5.8   LINEAR COMBINATIONS

A linear combination is a convenient mathematical device for summarizing various observations or statistics. The explanation of linear combinations here is entirely mathematical and is introduced at this point because the material is needed in section 5.9.

A *linear combination* q of n observations w_i is defined as a sum of terms,

$$q = c_1 w_1 + c_2 w_2 + \cdots + c_n w_n,$$

where the quantities c_i are known constants. Alternatively, the above formula may be written in summation notation as

$$q = \sum c_i w_i,$$

where the summation extends from $i = 1$ to $i = n$. A familiar example of linear combination is the sample mean, where the known constants are all equal and are the reciprocal $(1/n)$ of the sample size. Thus, the sample mean may be written

$$q = \frac{1}{n} w_1 + \frac{1}{n} w_2 + \cdots + \frac{1}{n} w_n = \overline{w},$$

or, in summation notation,

$$q = \sum \frac{1}{n} w_i = \overline{w}.$$

If one observation w_i is drawn independently and randomly from each of k populations, the mean and variance of a linear combination q formed from these observations can be calculated by a theorem illustrated by Li (1964, I, p. 246). The mean μ_q of the linear combination q is

$$\mu_q = c_1 \mu_1 + c_2 \mu_2 + \cdots + c_k \mu_k,$$

or

$$\mu_q = \sum c_i \mu_i.$$

The variance σ_q^2 of the same linear combination is

$$\sigma_q^2 = c_1^2 \sigma_1^2 + c_2^2 \sigma_2^2 + \cdots + c_k^2 \sigma_k^2,$$

or

$$\sigma_q^2 = \sum c_i^2 \sigma_i^2.$$

For the linear combination q equal to the sample mean $\overline{w}$, in which all the observations are drawn from a single population, the constant c was shown to be the reciprocal of the sample size $(1/n)$ in the formula

$$q = \sum \frac{1}{n} w_i = \overline{w}.$$

Similarly, if the reciprocal of the sample size is substituted in the formula for μ_q above, then because $k = n$, the result obtained is

$$\mu_q = \sum \frac{1}{n} \mu = \mu = \mu_{\bar{w}},$$

that is, the mean of this linear combination is the mean of sample means. Substituting the reciprocal of the sample size in the formula for σ_q^2 above, we obtain

$$\sigma_q^2 = \sum \left(\frac{1}{n}\right)^2 \sigma^2 = \frac{\sigma^2}{n} = \sigma_{\bar{w}}^2,$$

that is, the variance of this linear combination is the variance of the mean.

Another important linear combination represents the difference between two sample means, $\bar{w}_A - \bar{w}_B$. Again the constant multiplier is the reciprocal of the sample size; for the observations w_A,

$$c_i = \frac{1}{n_A},$$

where

$$i = 1, 2, \ldots, n_A,$$

and, for the observations w_B,

$$c_i = -\frac{1}{n_B},$$

where

$$i = n_A + 1, n_A + 2, \ldots, n_A + n_B.$$

Then the linear combination q is equal to

$$q = \left(\frac{1}{n_A} w_1 + \frac{1}{n_A} w_2 + \cdots + \frac{1}{n_A} w_{n_A}\right)$$
$$- \left(\frac{1}{n_B} w_{n_A+1} + \frac{1}{n_B} w_{n_A+2} + \cdots + \frac{1}{n_B} w_{n_A+n_B}\right),$$

or, in summation notation,

$$q = \sum_{i=1}^{n_A} \frac{1}{n_A} w_i - \sum_{i=n_A+1}^{n_A+n_B} \frac{1}{n_B} w_i = \bar{w}_A - \bar{w}_B.$$

The mean of the linear combination is

$$\mu_q = \left(\frac{1}{n_A} \mu + \frac{1}{n_A} \mu + \cdots + \frac{1}{n_A} \mu\right) - \left(\frac{1}{n_B} \mu + \frac{1}{n_B} \mu + \cdots + \frac{1}{n_B} \mu\right),$$

or, in summation notation,

$$\mu_q = \sum \frac{1}{n_A} \mu - \sum \frac{1}{n_B} \mu = 0,$$

showing that if the two samples are drawn from the same population the mean difference of sample means is 0. The variance of the linear combination is

$$\sigma_q^2 = \left(\frac{1}{n_A^2} \sigma^2 + \frac{1}{n_A^2} \sigma^2 + \cdots + \frac{1}{n_A^2} \sigma^2 \right)$$
$$+ \left(\frac{1}{n_B^2} \sigma^2 + \frac{1}{n_B^2} \sigma^2 + \cdots + \frac{1}{n_B^2} \sigma^2 \right),$$

or, in summation notation,

$$\sigma_q^2 = \sum \frac{1}{n_A^2} \sigma^2 + \sum \frac{1}{n_B^2} \sigma^2 = \sigma^2 \left(\frac{1}{n_A} + \frac{1}{n_B} \right)$$

showing that, if the two samples are drawn from the same population, the variance of the difference of sample means is the sum of the variances of the individual sample means.

A *contrast* is a linear combination for which the sum of constant multipliers $\sum c_i$ is equal to 0. Contrasts, the most important kind of linear combinations, are illustrated by the last example and discussed in detail in section 5.9.

5.9 SINGLE DEGREE OF FREEDOM

In sections 5.1 and 5.3 the one-way analysis of variance is illustrated with fictitious observations of sand content from five outcrops. Now, in this section, contrasts are formed by using the method of the *single degree of freedom* to analyze similar data more incisively. Three different contrasts are used before the method of the single degree of freedom is related to the one-way analysis of variance of section 5.3. The meaning of *single degree of freedom* is explained near the end of this section.

Assume that a geologist is making a geochemical survey for selenium, which is suspected of killing cattle in Wyoming rangeland. Assume further that the geologist wishes to compare two methods of sampling for selenium— method A, taking single hand specimens of outcrops, and method B, taking chip samples across the entire outcrop surface. Finally assume that two formations, numbered 1 and 2, are to be sampled, by taking four groups of observations, each seven in number. Accordingly, the groups of observations

TABLE 5.21. DESIGNATIONS FOR FOUR POPULATIONS
OF FICTITIOUS SELENIUM DETERMINATIONS

Formation	Method of sampling	
	A	B
1	μ_1 $(1, A)$	μ_2 $(1, B)$
2	μ_3 $(2, A)$	μ_4 $(2, B)$

come from four populations, with means μ_1 to μ_4, as defined in table 5.21. With this specification of the four statistical samples, three hypotheses about the population means may be tested with contrasts.

The first hypothesis H_0 to be tested is that the two formations have the same mean selenium content. For sampling method A the two formations can be compared by forming the hypothesis that $\mu_1 = \mu_3$, which can be rearranged to the form $\mu_1 - \mu_3 = 0$. Likewise, for sampling method B, the two formations can be compared by the hypothesis that $\mu_2 = \mu_4$, or $\mu_2 - \mu_4 = 0$. These two hypotheses can be added to form the desired hypothesis H_0 that $\mu_1 - \mu_3 + \mu_2 - \mu_4 = 0$, which can be rearranged to the form $\mu_1 + \mu_2 - \mu_3 - \mu_4 = 0$. The alternative hypothesis H_1 is that the sum of the population means is not equal to 0; that is, $\mu_1 + \mu_2 - \mu_3 - \mu_4 \neq 0$. In geological terms the alternative hypothesis H_1 is that the selenium contents of the two formations are different.

Because each of the four population means μ_1 to μ_4 is estimated by a corresponding statistic $\overline{w}_1$ to $\overline{w}_4$, the quantity $\overline{w}_1 + \overline{w}_2 - \overline{w}_3 - \overline{w}_4$ is a statistic that estimates the quantity $\mu_1 + \mu_2 - \mu_3 - \mu_4$. Both quantities are linear combinations whose constant multipliers are 1, 1, -1, and -1. Because for both quantities the sum of constant multipliers is 0, the linear combinations are contrasts (sec. 5.8). To show clearly the role of the constant multipliers, one may write the contrast involving the parameters as follows:

$$\tau = (1)\mu_1 + (1)\mu_2 + (-1)\mu_3 + (-1)\mu_4,$$

where τ is a new symbol defined by the equation, and the contrast involving the statistics may be written

$$q = (1)\overline{w}_1 + (1)\overline{w}_2 + (-1)\overline{w}_3 + (-1)\overline{w}_4.$$

The contrast q formed in the preceding paragraph is essential for testing the hypothesis H_0, which now may be written $\tau = 0$, and also for calculating a confidence interval for τ. Both methods are presented, but the hypothesis test is presented first because it is directly related to the one-way analysis of variance discussed in section 5.3.

Hypothesis Tests with Single Degree of Freedom

Because the appropriate hypothesis test is an F-test for which suitable variances must be found, some preliminary material is presented. From section 5.8, the variance of the contrast q is

$$\sigma_q^2 = \sum c_i^2 \sigma_i^2.$$

The quantities σ_i^2 are the variances of the four sample means, and each is equal to the square of the standard error of the mean; that is,

$$\sigma_i^2 = \sigma_{\bar{w}}^2 = \frac{\sigma^2}{7},$$

where 7 is the number of observations taken in each group. Therefore,

$$\sigma_q^2 = (1)^2 \frac{\sigma^2}{7} + (1)^2 \frac{\sigma^2}{7} + (-1)^2 \frac{\sigma^2}{7} + (-1)^2 \frac{\sigma^2}{7},$$

or, factoring out $\sigma^2/7$,

$$\sigma_q^2 = \left(\sum c_i^2\right) \frac{\sigma^2}{7} = 4\frac{\sigma^2}{7}.$$

Each of the $\bar{w}$-values follows a normal distribution; the contrast q also follows a normal distribution with mean equal to τ and variance equal to $4\sigma^2/7$. Because a normally distributed quantity minus its mean divided by its standard deviation is normally distributed with a mean of 0 and a variance of 1 (sec. 3.6), the statistic

$$\frac{q - \tau}{\sqrt{4\sigma^2/7}}$$

follows a normal distribution with a mean of 0 and a variance of 1. If the hypothesis H_0 is true, τ is equal to 0, and the statistic may be rewritten

$$\frac{q - 0}{\sqrt{4\sigma^2/7}}.$$

Since the parameter σ^2 is generally unknown, it is replaced, as usual, with its statistic, the pooled sample variance s_p^2. Then the new quantity,

$$t = \frac{q - \tau}{\sqrt{4s_p^2/7}},$$

TABLE 5.22. FICTITIOUS SELENIUM CONTENTS FROM FOUR FORMATIONS

Line	Item	Statistical sample number			
		(1)	(2)	(3)	(4)
1		4.7	5.0	6.7	7.5
2		5.5	7.3	3.3	5.0
3		3.8	5.4	5.7	4.9
4	Observations	5.2	5.3	5.4	5.7
5	selenium, ppm	5.6	5.8	4.5	5.6
6		4.9	6.8	4.2	6.4
7		6.7	8.2	6.4	5.5
8	$\sum w$	36.4	43.8	36.2	40.6
9	n	7	7	7	7
10	$\overline{w}$	5.200	6.257	5.171	5.800
11	$(\sum w)^2$	1324.96	1918.44	1310.49	1648.36
12	$(\sum w)^2/n$	189.28	274.06	187.21	235.48
13	$\sum w^2$	194.08	282.66	196.28	240.32
14	SS	4.80	8.60	9.07	4.84
15	d.f.	6	6	6	6
16	s^2	0.80	1.43	1.51	0.81

$$s_p^2 = (s_1^2 + s_2^2 + s_3^2 + s_4^2)/4 = 1.14$$

follows the t-distribution rather than the standardized normal distribution, as explained in section 4.3. The t-distribution has the 24 degrees of freedom used to calculate s_p^2. Because $t^2 = F$ (sec. 5.13), the squared statistic,

$$F = \frac{(q - \tau)^2}{4s_p^2/7} = \frac{7(q - \tau)^2}{4s_p^2},$$

follows the F-distribution with 1 and 24 degrees of freedom. If the hypothesis H_0 is true, τ is equal to 0, and the statistic becomes

$$F = \frac{7q^2}{4s_p^2}.$$

Now that the method has been explained, the numerical calculations can be performed by using the fictitious data from table 5.22. In table 5.23 the test of the hypothesis that there is no difference in selenium contents of formations 1 and 2 is set up in the same form as that in section 4.7. In lines 1 to 4 are entered the four sample means from table 5.22, the constant multipliers, and the products of the means and the multipliers. In line 5 the value of q is calculated to be 0.486; in line 6 is entered the value 1.14 for the

TABLE 5.23. SINGLE-DEGREE-OF-FREEDOM TEST OF THE HYPOTHESIS
THAT THERE IS NO DIFFERENCE IN THE SELENIUM CONTENTS OF THE
TWO FORMATIONS

Hypothesis:	$H_0: \mu_1 + \mu_2 - \mu_3 - \mu_4 = 0$
Alternative hypothesis:	$H_1: \mu_1 + \mu_2 - \mu_3 - \mu_4 \neq 0$
Statistic:	F
Risk of type 1 error:	$\alpha = 10\%$
Critical region:	$F > F_{10\%}$ ($F_{10\%} = 2.93$, with 1 and 24 d.f.)

Computation:

Line	Sample number	$\overline{w}$	c	$c\overline{w}$	Remarks
1	1	5.200	1	5.200	Values of $\overline{w}$ from
2	2	6.257	1	6.257	table 5.22
3	3	5.171	-1	-5.171	
4	4	5.800	-1	-5.800	
5	$q = \sum c\overline{w} = 0.486$				
6	$s_p^2 = 1.14$				From table 5.22
7	$F = \dfrac{7q^2}{4s_p^2} = \dfrac{7(0.486)^2}{4(1.14)} = 0.363$				

Conclusion.   Accept hypothesis H_0.

pooled variance, obtained from table 5.22; in line 7, values are substituted in
the formula to yield an F-value of 0.363. Because this F-value is outside the
critical region, the hypothesis H_0 is accepted.

Two other single-degree-of-freedom hypothesis tests for the same data are
now presented. The first is that the selenium content estimated by the two
methods of sampling is the same. Through reasoning similar to that employed
to form the previous hypothesis, the two hypotheses that $\mu_1 - \mu_2 = 0$ and
that $\mu_3 - \mu_4 = 0$ are added to form the desired hypothesis H_0 that
$\mu_1 - \mu_2 + \mu_3 - \mu_4 = 0$. In table 5.24 calculations parallel to those in table
5.23 are performed. In the initial data in lines 1 to 4 the only difference is that
the constant multipliers in lines 2 and 3 have opposite signs from those in the
preceding table. Because the calculated F-value of 4.364 is inside the critical
region, the hypothesis H_0 is rejected, and the alternative hypothesis H_1, that
the two methods of sampling yield different estimates of selenium content, is
accepted.

The last hypothesis to be tested is that the sampling was done consistently
in the two formations, rather than sampling method A being applied differently in formation 1 than in formation 2 or sampling method B being applied

Table 5.24. Single-degree-of-freedom test of the hypothesis
that there is no difference in sampling methods

Hypothesis:	$H_0: \mu_1 - \mu_2 + \mu_3 - \mu_4 = 0$
Alternative hypothesis:	$H_1: \mu_1 - \mu_2 + \mu_3 - \mu_4 \neq 0$
Statistic:	F
Risk of type I error:	$\alpha = 10\%$
Critical region:	$F > F_{10\%}$ ($F_{10\%} = 2.93$, with 1 and 24 d.f.)

Computation:

Line	Sample number	$\overline{w}$	c	$c\overline{w}$
1	1	5.200	1	5.200
2	2	6.257	-1	-6.257
3	3	5.171	1	5.171
4	4	5.800	-1	-5.800

5	$q = \sum c\overline{w} = -1.686$
6	$s_p^2 = 1.14$
7	$F = \dfrac{7q^2}{4s_p^2} = \dfrac{7(-1.686)^2}{4(1.14)} = 4.364$

Conclusion. Reject hypothesis H_0. Accept alternative hypothesis H_1.

differently in formation 1 than in formation 2. Because the relation, if any, between a sampling method and a formation is investigated, the effect sought is named an *interaction*. If there is no interaction, the result of sampling formation 1 by method A plus the result of sampling formation 2 by method B should be equal, on the average, to the result of sampling formation 2 by method A plus the result of sampling formation 1 by method B (table 5.21). Thus, the hypothesis H_0 is that $\mu_1 + \mu_4 = \mu_2 + \mu_3$ or, rearranging, H_0 is that $\mu_1 - \mu_2 - \mu_3 + \mu_4 = 0$. In table 5.25 the hypothesis H_0 is tested against the alternative hypothesis H_1 that this contrast is not equal to 0. When, in table 5.25, calculations parallel to those in tables 5.23 and 5.24 are performed, the calculated F-value of 0.281 is outside the critical region, and, therefore, the hypothesis H_0 of no interaction is accepted.

In each of the tables 5.23 to 5.25 an F-statistic was calculated from the formula

$$F = \frac{7q^2}{4s_p^2}.$$

TABLE 5.25. SINGLE-DEGREE-OF-FREEDOM TEST OF THE HYPOTHESIS THAT TWO SAMPLING METHODS ARE BEING APPLIED CONSISTENTLY IN TWO FORMATIONS

Hypothesis:	$H_0: \mu_1 - \mu_2 - \mu_3 + \mu_4 = 0$
Alternative hypothesis:	$H_1: \mu_1 - \mu_2 - \mu_3 + \mu_4 \neq 0$
Statistic:	F
Risk of type I error:	$\alpha = 10\%$
Critical region:	$F > F_{10\%}$ ($F_{10\%} = 2.93$, with 1 and 24 d.f.)

Computation:

Line	Sample number	$\overline{w}$	c	$c\overline{w}$
1	1	5.200	1	5.200
2	2	6.257	-1	-6.257
3	3	5.171	-1	-5.171
4	4	5.800	1	5.800
5		$q = \sum c\overline{w} = -0.428$		
6		$s_p^2 = 1.14$		
7		$F = \dfrac{7q^2}{4s_p^2} = \dfrac{7(-0.428)^2}{4(1.14)} = 0.281$		

Conclusion.   Accept hypothesis H_0.

If the term $7q^2/4$ is replaced by the new quantity Q^2, the formula becomes

$$F = \frac{Q^2}{s_p^2}.$$

The quantity Q^2 is named the *single-degree-of-freedom statistic*; the name indicates that it has 1 degree of freedom.

With the single-degree-of-freedom statistic Q^2, the relation of the three contrasts can be given to the one-way analysis of variance for these data given in table 5.26. The one-way analysis of variance tests the hypothesis H_0 that $\mu_1 = \mu_2 = \mu_3 = \mu_4$. If this hypothesis is true, the mean selenium content is the same, regardless of sampling method or rock formation. Thus, this hypothesis is much less incisive than those formulated with the single degree of freedom.

In table 5.27 (an analysis of variance incorporating the three contrasts) lines 1, 5, and 6 are copied from table 5.26. The original F-value in line 1, calculated by dividing the mean square in line 1 by the mean square in line 5, is still valid, as is the conclusion that there is no difference among the sample

Table 5.26. Simplified computing method for one-way analysis of variance applied to data of table 5.22

| | Preliminary calculations | | |
Type of sum	(1) Total of squares	(2) Number of observations in sum	(3) = (1)/(2) Total of squares per observation
Grand	24,649.00	28	880.32
Sample	6,202.20	7	886.03
Observation	913.34	1	913.34

| | Analysis of variance | | | | |
Source of variation	Sum of squares	Degrees of freedom	Mean square	F	$F_{10\%}$
Among-sample means	5.71	3	1.903	1.67	2.33
Within-sample	27.31	24	1.14		
Total	33.02	27			

means if the specification of different formations and methods is not made. In line 2 the numbers are derived from table 5.23. The sum of squares, with 1 degree of freedom, is equal to Q^2, and, because F has already been calculated, Q^2 may be obtained by solving the formula

$$F = \frac{Q^2}{s_p^2}$$

Table 5.27. Analysis of variance of fictitious selenium data, incorporating three contrasts

Line	Source of variation	Sum of squares	Degrees of freedom	Mean square	F	$F_{10\%}$
1	Among-sample means	5.71	3	1.903	1.67	2.33
2	Between formations	0.41	1	0.410	0.36	2.93
3	Between methods	4.98	1	4.980	4.36	2.93
4	Formation-method interaction	0.32	1	0.320	0.28	2.93
5	Within-sample	27.31	24	1.140		
6	Total	33.02	27			

for Q^2, with the result that

$$Q^2 = Fs_p^2 = 0.363 \times 1.14 = 0.41.$$

In lines 3 and 4 are given the other sums of squares, obtained in the same way from tables 5.24 and 5.25. Notably, the sums of squares for the single-degree-of-freedom comparisons of lines 2, 3, and 4 add up to the sum of squares in line 1; moreover, the degrees of freedom in lines 2 to 4 add up to the 3 degrees of freedom in line 1. By the single-degree-of-freedom method both the sums of squares and the degrees of freedom have been partitioned.

The three single-degree-of-freedom values and Q^2 values add up to the sum of squares with 3 degrees of freedom because the contrasts are *orthogonal*. For equal sample sizes orthogonal contrasts are those in which the sums of the products of the multipliers, taken two at a time in order, add up to 0; that is,

$$\sum c_i c_i' = 0,$$

where the c_i are the coefficients of one of the contrasts, and the c_i' are the coefficients of the other contrast. For instance, in table 5.28 the multipliers from tables 5.23 and 5.24 are listed in two rows. Multiplying the numbers in each column, one column at a time, yields the quantities on the bottom line, which sum to 0.

For the three hypothesis tests with a single degree of freedom the sums of squares add up to the total sum of squares for among-sample means, and the degrees of freedom add up to the degrees of freedom for the among-sample means. Therefore, because of Cochran's theorem (sec. 5.3), these three F-tests are independent insofar as their numerator is concerned. However, some dependence is introduced because of the common denominator, the within-sample mean square. If there is a 10-percent α risk of type I error in each of the three tests, there is at most a 30-percent (3 $\times$ 10 percent) risk of type I error for all three tests combined, and the conclusions of the three tests are essentially independent. Such a joint risk for several tests combined is named the *error rate*.

TABLE 5.28. ILLUSTRATION OF MULTIPLYING
CONSTANTS FOR ORTHOGONAL CONTRASTS

Source	Multiplying constants			
Table 5.23	1	1	-1	-1
Table 5.24	1	-1	1	-1
Product	1	-1	-1	1

The single-degree-of-freedom method should be applied to only a few hypotheses and only to those chosen before the data are collected. First, the technique should be applied to only a few hypotheses because the error rate grows proportionally as the number of single-degree-of-freedom tests increases. If, for example, there are 10 means, 45 single comparisons can be made by contrasts, taking two means at a time. If all of these comparisons are made at a 10-percent risk level, an average of four or five would be rejected merely by chance even if all were true. Second, the single-degree-of-freedom method should be applied only if the hypotheses are chosen before the data are collected in order to ensure honest conclusions. Otherwise, if the data are inspected and a test is then made, the inspection will usually lead to something interesting that a test will "confirm"; for example, if the smallest sample mean in a group of 10 means is compared with the largest, the hypothesis of equality of means is almost sure to be rejected, regardless of the true situation. When, because of one or both of these conditions, the single-degree-of-freedom method is unsuitable, the method of multiple comparisons, explained in section 5.12, should be used.

In the fictitious illustration of sampling selenium there are four means and seven observations per mean. In general, with k means and n observations per mean, the general hypothesis H_0 that $\sum c_i \mu_i$ is equal to 0 is tested against the alternative hypothesis H_1 that $\sum c_i \mu_i$ is not equal to 0. As usual, the i subscripts designate the constant multipliers and means, from the first term $c_1 \mu_1$ to the last, for which i is equal to k. The sum $\sum c_i$ must be equal to 0; that is, it must be a contrast. To test the hypothesis we can calculate the single degree of freedom by the formula

$$Q^2 = \frac{nq^2}{\sum c_i^2},$$

where, as in the fictitious illustration,

$$q = \sum c_i \overline{w}_i.$$

The F-statistic with 1 and $k(n-1)$ degrees of freedom is

$$F = \frac{Q^2}{s_p^2}.$$

The modified method to apply if the numbers of observations in the samples are different is given in section 5.10.

Confidence Intervals with Single Degree of Freedom

In all tests involving contrasts, the hypothesis is that the appropriate linear combination of the population means is equal to 0, that is, that τ is equal to 0. Rather than making a hypothesis test, one can calculate confidence limits for τ by giving reasonable bounds for this value. The principle is like that used to calculate confidence limits for the difference between two means (sec. 4.4); and, as before, if the confidence limits are narrow and include 0, the decision is that the contrast cannot be very different from 0. A detailed explanation for the fictitious illustrations is given after the computations are explained.

Confidence limits for τ may be constructed because the statistic

$$ t = \frac{q - \tau}{\sqrt{(\sum c_i^2)s_p^2/n}}, $$

introduced earlier in this section, follows the t-distribution. In table 5.29, confidence intervals are calculated for the three cases previously examined by hypothesis tests. In lines 1 to 3, the values of q, s_p^2, and n obtained from tables 5.23 to 5.25 are entered. In line 4 is entered the sum of the squared values of the constant multipliers, equal to 4 in each case. After the calculations are performed, the upper and lower confidence limits for τ, τ_L, and τ_U are obtained (lines 10 and 11).

TABLE 5.29. CONFIDENCE INTERVALS FOR THREE CASES OF FICTITIOUS SELENIUM DATA

		Source of data		
Line	Notation	Case I: formation 1 vs. formation 2, table 5.23	Case II: method A vs. method B, table 5.24	Case III: interaction, table 5.25
1	q	0.486	-1.686	-0.428
2	s_p^2	1.14	1.14	1.14
3	n	7	7	7
4	$\sum c_i^2$	4	4	4
5	$(\sum c_i^2)s_p^2/n$	0.651	0.651	0.651
6	$\sqrt{(\sum c_i^2)s_p^2/n}$	0.807	0.807	0.807
7	d.f.	24	24	24
8	$t_{5\%}$	1.711	1.711	1.711
9	$t\sqrt{(\sum c_i^2)s_p^2/n}$	1.381	1.381	1.381
10	τ_L	-0.90	-3.07	-1.81
11	τ_U	1.87	-0.30	0.95

The confidence intervals in table 5.29 may be interpreted as follows. Because all of the intervals have a 90-percent confidence coefficient, they are correct in 90 percent of the cases for which they are calculated. For case I the interpretation is that reasonable bounds for *twice* the difference in selenium content of the two formations are 1.87 and -0.90. The bounds are for twice the difference rather than for the simple difference because τ is defined by the equation

$$\tau = \mu_1 + \mu_2 - \mu_3 - \mu_4,$$

and therefore the mean associated with formation 1 is

$$\frac{\mu_1 + \mu_2}{2}$$

and that associated with formation 2 is

$$\frac{\mu_3 + \mu_4}{2}.$$

The corresponding interval for the simple difference in selenium content between formations is 0.94 to -0.45, obtained by dividing the values in lines 1 and 9 of the table by 2. Of course, a different contrast could have been formulated initially to yield the desired intervals directly, but this formulation would have obscured the relation to the hypothesis test.

For cases II and III of table 5.29 the same considerations apply as for case I. In case II, after the factor of 2 is removed as before, the reasonable bounds for the difference in selenium content obtained from the two methods of sampling are -0.15 and -1.54, indicating a real but small difference in selenium content. In case III, in which the factor of 2 is also removed, the reasonable bounds for the difference in interaction are 0.48 and -0.90, indicating that if there is an interaction effect it is small, that is, less than 1 percent.

The general formulation of a confidence interval for a contrast is as follows: The contrast τ has been defined by the formula

$$\tau = \sum c_i \mu_i,$$

where $\sum c_i = 0$ and q, the statistic that estimates τ, has been defined by the parallel formula

$$q = \sum c_i \bar{w}_i.$$

Thus, by analogy with section 4.4, the boundaries of the confidence interval for τ are

$$q - ts_p \sqrt{\frac{\sum c_i^2}{n}} \le \tau \le q + ts_p \sqrt{\frac{\sum c_i^2}{n}}.$$

Appraisal of the Single-Degree-of-Freedom Method

The single-degree-of-freedom method is one of the most valuable in applied statistics. In section 5.10, example analyses are presented, and many other applications will occur to the reader. To develop the method more fully would expand this book too much. A particularly full account of the method is given by Li (1964, I, p. 252, etc.), who emphasizes its use. The main advantage of the single-degree-of-freedom procedure is that, where it is appropriate, more incisive results can be achieved than with a general F-test. Its disadvantage is that a too liberal use of the method or an application to "interesting results" discovered by an inspection of the data after they are gathered can lead to an error rate that is too large. The multiple-comparison procedures discussed in section 5.12 overcome this objection but lack the incisiveness of the single-degree-of-freedom procedures.

5.10 EXAMPLES OF ANALYSIS WITH SINGLE DEGREE OF FREEDOM

In this section two example analyses with the single degree of freedom are presented with data from the Fresnillo and Homestake mines. The Fresnillo problem is a new one; the Homestake problem is a refinement of the problem of section 5.6, in which the single-degree-of-freedom method leads to a more incisive analysis of the same data.

Single Degree of Freedom Applied to Fresnillo Mine Data

In the Fresnillo mine the 2137 vein (fig. 1.4) splits northwestward into two branches, designated the 2137 vein and the 2137 Footwall vein (fig. 5.5). It was of interest to learn whether in mineralization one of the two branches was more like the 2137 vein before the junction than the other. The purpose of the analysis was twofold: to aid in interpretation of the geologic history of the vein systems by segregating different types of mineralization and to aid mine planning. As usual, we present only part of the analysis from the original paper (Koch and Link, 1967).

In order to make the desired comparison, 200 meters of vein exposed in each of three drifts extending outward from the junction on each level were selected. On each level one of the three drifts followed the 2137 Footwall vein; of the other two, one followed the 2137 vein northwest of the junction (overlapping the 2137 Footwall vein), and the other followed the 2137 vein southwest of the junction. Data were taken from drifts on the 605- to 830-meter levels, which had been sampled at 2-meter intervals.

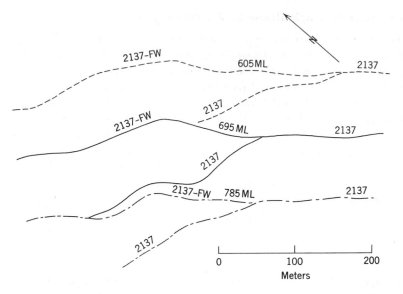

Fig. 5.5. Composite plan of drifts following 2137 and 2137 Footwall veins near their junction, Fresnillo mine.

In order to make an analysis of variance comparing the two segments of the 2137 vein northwestward and southeastward of the vein junction with one another and comparing the combined segments of the 2137 vein with the one segment of the 2137 Footwall vein, the single-degree-of-freedom method was used. In table 5.30 the calculations of the single-degree-of-freedom quantity to compare the two 2137 vein segments with one another are given for the silver-content data. The items in lines 1 to 3 of this table correspond to those in lines 1 to 4 of table 5.23 which is for the fictitious data. There are changes in table 5.30 that make the calculations a little more complex than those in table 5.23. Because the sample sizes are different, the quantity n is no longer a constant multiplier; rather the quantity $\sum (c_i^2/n_i)$ must be evaluated by summing the values in the last column of lines 1 to 3 to yield the result in line 4. For unequal sample sizes

$$Q^2 = \frac{q^2}{\sum (c_i^2/n_i)}.$$

If the n_i quantities are all equal, this formula reduces to the previous formula,

$$Q^2 = \frac{nq^2}{\sum c_i^2}.$$

TABLE 5.30. CALCULATION OF THE SINGLE-DEGREE-OF-FREEDOM QUANTITY TO COMPARE TWO SEGMENTS OF THE 2137 VEIN, FRESNILLO MINE, WITH ONE ANOTHER

Line	Vein segment	n	$\overline{w}$	c	$c\overline{w}$	c^2	c^2/n
1	2137 vein southeast of junction	469	161.24286	1	161.24286	1	0.00213220
2	2137 vein northwest of junction	458	167.55109	-1	-167.55109	1	0.00218341
3	2137 Footwall vein	411	128.91070	0	0	0	0

4 $$\sum \left(\frac{c_i^2}{n_i}\right) = 0.00431561$$

5 $\quad q = \sum c_i \overline{w}_i = 6.30823$

6 $\quad q^2 = 39.79377$

7 $\quad Q^2 = 9221$

In lines 5 and 6 q and q^2 are calculated, and finally Q^2 is calculated in line 7.

In table 5.31 parallel calculations to those in table 5.30 are performed to compare the combined segments of the 2137 vein with the one segment of the 2137 Footwall vein.

Finally, in table 5.32 the two Q^2 values from tables 5.30 and 5.31 are entered in the mean-square column. In table 5.32 the F-value comparing the two segments of the 2137 vein is only 0.2; therefore we conclude that silver mineralization in these two segments is the same. On the other hand, because the F-value of 8.6, comparing the 2137 and 2137 Footwall veins, is inside the critical region, we conclude that silver mineralization in the two veins is different.

Repeating this analysis for gold, lead, and zinc data, we (Koch and Link, 1967) found that, except for zinc, mineralization in the 2137 Footwall vein is different from that in the 2137 vein, so that the main vein northwestward from the junction can be distinguished from the branch vein through analysis of the assay data.

In the fictitious illustrations of the single-degree-of-freedom method (sec. 5.9) the contrasts are orthogonal. In the Fresnillo example the contrasts appear to be orthogonal because, in both tables 5.30 and 5.31, $\sum c_i c_i' = 0$

TABLE 5.31. CALCULATION OF THE SINGLE-DEGREE-OF-FREEDOM QUANTITY TO COMPARE THE 2137 AND 2137 FOOTWALL VEINS, FRESNILLO MINE

Line	Vein segment	n	$\bar{w}$	c	$c\bar{w}$	c^2	c^2/n
1	2137 vein southeast of junction	469	161.24286	1	161.24286	1	0.00213220
2	2137 vein northwest of junction	458	167.55109	1	167.55109	1	0.000218341
3	2137 Footwall vein	411	128.91070	−2	−257.82140	4	0.00973236

4	$\sum \left(\dfrac{c_i^2}{n_i}\right) = 0.01404797$
5	$q = \sum c_i \bar{w}_i = 70.97255$
6	$q^2 = 5037.10285$
7	$Q^2 = 358{,}564$

(table 5.33). However, the two corresponding sums of squares do not add up to the sum of squares with the 2 degrees of freedom because, for unequal sample sizes, the requirement for orthogonality is that

$$\sum \frac{c_i c_i'}{n_i} = 0.$$

Because this requirement is not met, the comparisons for the two single degrees of freedom are not orthogonal; consequently the two tests are not quite statistically independent. Therefore, instead of being exactly 20 percent

TABLE 5.32. ANALYSIS OF VARIANCE TO ASSESS METAL CONTENT NEAR THE JUNCTION OF THE 2137 AND 2137 FOOTWALL VEINS, FRESNILLO MINE

Source of variation	Sum of squares	Degrees of freedom	Mean square	F	$F_{10\%}$
Among-vein segments	367,045	2	183,523	4.4	
2137 vein southeast of junction vs. 2137 vein northwest of junction	9,221	1	9,221	0.2	2.71
2137 Footwall vein vs. 2137 vein	358,564	1	358,564	8.6	2.71
Within-vein segments	55,234,376	1335	41,374		

TABLE 5.33. MULTIPLYING CONSTANTS TO COM-
PARE MINERALIZATION IN THE 2137 AND 2137
FOOTWALL VEINS, FRESNILLO MINE

Source	Multiplying constants		
Table 5.30	1	-1	0
Table 5.31	1	1	-2
Product	1	-1	0

(twice 10 percent), the total error rate is between 10 and 20 percent. However, for practical purposes, these two particular single-degree-of-freedom tests may be considered to be independent. A general discussion of to what extent the dependence in such tests can be ignored is beyond the scope of this book; the interested reader is referred to Scheffé (1959, p. 89).

Single Degree of Freedom Applied to Homestake Mine Data

For the Homestake data studied in section 5.6 a more incisive analysis may be made to test whether the mineralization in the two crosscuts (fig. 3.1) is different. To do this the original table 5.15 is expanded by adding several new lines in the new table 5.34. The entries on these two lines partition the among-hole variability into a piece with 1 degree of freedom associated with the between-crosscut variability and a piece with 3 degrees of freedom

TABLE 5.34. NESTED ANALYSIS OF VARIANCE, INCLUDING SINGLE-DEGREE-OF-FREEDOM INTERPRETATION, OF GOLD ASSAY DATA FROM THE HOMESTAKE MINE, SOUTH DAKOTA

Line	Source of variation	Sum of squares	Degrees of freedom	Mean square	F	$F_{10\%}$
1	Total	5308.529	215			
2	Among holes	177.883	4	44.471	1.83	1.97
3	Between crosscuts	75.353	1	75.353	3.10	2.73
4	Within crosscuts	125.433	3	41.811	1.72	2.11
5	Within holes	5130.646	211	24.316		
6	Among-foot intervals	2536.107	51	49.728	3.07	1.32
7	Within-foot intervals	2594.539	160	16.216		
8	Between-sample pairs	1511.398	52	29.065	2.90	1.37
9	Within-sample pairs	1083.141	108	10.029		

associated with the variability among holes within a single crosscut. The table shows that the variability between crosscuts is larger than that within crosscuts, because the calculated F-value of 3.10 is larger than the tabled F-value of 2.73.

The total of the two sums of squares in lines 3 and 4 is larger than the sum of squares on line 2 because, as for the Fresnillo example, the contrasts are not orthogonal. They are not orthogonal because the numbers of observations in the various holes are different.

An alternative way to compare gold in the two crosscuts is to construct confidence limits for the difference between the population means of the assays from the two crosscuts. This difference may be estimated through the contrast

$$q = \tfrac{1}{3}(\overline{w}_A + \overline{w}_B + \overline{w}_C) - \tfrac{1}{2}(\overline{w}_D + \overline{w}_E),$$

where $\overline{w}_A$, $\overline{w}_B$, and $\overline{w}_C$ are mean gold assays from the three drill holes in the first crosscut, and $\overline{w}_D$ and $\overline{w}_E$ are mean gold assays from the two drill holes in the second crosscut. The limits of a 90-percent confidence interval for this difference are -0.46 and 2.98 ounces per ton. Although this estimate includes 0, the interval is quite wide; therefore a substantial difference in gold in the two drifts may exist.

5.11 LINEAR MODELS

In the discussion on soil failure (sec. 1.5) we introduce the concept of models in the example of the simple shear apparatus. In the explanation of the statistical model for shear

$$s = K_1 + (\tan K_2)p + e,$$

several concepts are introduced and discussed. In particular, an observation s depends on the values of variables and parameters explicitly defined in the model and also on implicit variables. Dependence on normal pressure is represented by the variable p; dependence on soil type is represented by the parameter values K_1 and K_2. On the other hand, dependence on temperature, soil structure, and other factors is not explicitly represented in the equation but lumped together in the random fluctuation e.

In Chapters 2 and 4 models are used, although, for simple exposition, they are not discussed explicitly. In section 5.1 models are introduced of the form

$$w_{ij} = \mu + (\mu_i - \mu) + e_{ij},$$

and in sections 5.5 and 5.7 similar but more complicated models are developed.

The model for soil failure in section 1.5 and those introduced previously in this chapter are special cases of the class of models named *linear models*. These models are highly important for several reasons. First, they are fundamental for an understanding of the analyses of variance in this chapter. Also they are fundamental to an understanding of more complicated analyses of variance: trend analysis, in Chapter 9, and multivariate analysis, in Chapter 10. In this section, after general ideas about models are discussed in some detail, the term *linear model* is defined. Then linear models are related to nonlinear models, and some other concepts are introduced.

As discussed in section 1.5, the purpose of a model, whether physical, geological, or mathematical, is to obtain a simple representation of a natural phenomenon useful for gaining insight into it. Thus, through the simple shear apparatus and the accompanying mathematical equation, shear stress is related to three variables: the cohesiveness of the soil, the pressure of the soil, and the coefficient of internal friction. By measuring maximum shear stress and pressure, the coefficients of internal friction can be calculated. Alternatively, shear stress can be calculated if pressure and the two coefficients are known.

Illustration of Mathematical Models

In this subsection functions are defined and discussed; then various kinds of mathematical models, mostly linear, are illustrated.

In the mathematical model for soil failure, the variables are related to one another by a calculating rule or a translation device named a *function*; for instance, the equation

$$s = K_1 + (\tan \theta)p$$

is a linear model in the parameters K_1 and $\tan \theta$ that may be written as

$$s = f(p, K_1, \theta),$$

where f represents the function that encompasses the rules of addition and multiplication and the calculating device for obtaining the tangent of the angle θ. In the general equation,

$$w = f(x_1, x_2, \ldots, x_k, K_1, K_2, \ldots, K_j),$$

w is the dependent variable, the x's are independent variables, the K's are parameters (constants), and f denotes the function comprising the procedures for obtaining the value of w, given the values of the x's and K's.

Functions can be classified by their mathematical properties. If each variable appears to the first degree in a single term, the function is called a *linear function*. The single variable to be calculated, w, is named the *dependent variable*, and the one or more variables, x's, used to calculate it, are named *independent variables*. The graph of a linear function with only one independent variable is a straight line; for example, the graph for the shear test, figure 1.9. Similarly, the graph of a linear function with two independent variables is a plane (sec. 9.4), and that of a linear function with more than two independent variables is a hyperplane (sec. 9.6, fig. 9.15).

In order to illustrate further the meaning of functions, several examples are instructive. For the soil-failure example the relation between shear stress and pressure for a particular soil can be written

$$w = f(x, K_1, K_2),$$

where the dependent variable w is shear stress; the independent variable K_1 is the pressure; and the independent variable K_2 is the coefficient of internal friction. The above equation can be written explicitly as

$$w = K_1 + K_2 x,$$

or, alternatively, because K_2 equals tan θ, as

$$w = K_1 + (\tan \theta)x.$$

Notably, the designation of variables as independent or dependent depends only on how the function is written, for if the above equation is "solved" for x the result obtained is

$$x = \frac{-K_1}{K_2} + \frac{w}{K_2},$$

where x has become the dependent variable and w has become the independent variable.

A second example is the calculation of the specific gravity of a sandy limestone composed entirely of quartz and calcite. Because the specific gravities and proportional weights of the two minerals are known, the specific gravity can be estimated through the functional relationship

$$w = \theta_1 x_1 + \theta_2 x_2,$$

where w is the specific gravity of the rock, θ_1 is the specific gravity of the quartz, x_1 is the proportion of quartz, θ_2 is the specific gravity of calcite, and x_2 is the proportion of calcite. The equation appears to be a linear function with two independent variables and two parameters. It is actually only a function with one independent variable and two parameters, because x_1 and x_2 sum to a constant 100 percent and therefore are not independent variables. The example illustrates the principle that, in determining the number of independent variables, it is necessary to be certain that they are independent and not directly related (chap. 11).

A third example of a functional relationship is afforded by the problem of sampling two beds of phosphate rock, each with a different but uniform phosphate content throughout. Following the approach used in section 5.1, one may write the model as

$$w_i = \mu + (\mu_i - \mu),$$

where w_i is an observation in the ith bed, μ_i is the phosphate content of the ith bed, and μ is the average of the μ_i's. Alternatively, the model may be written in terms of a functional relationship involving other variables explicitly as

$$w = \mu + (\mu_1 - \mu)x_1 + (\mu_2 - \mu)x_2,$$

where w is the observed phosphate content, μ and μ_i have the same meaning as before, and x_1 and x_2 are artificial variables set equal to $(1, 0)$ if the first bed is sampled and to $(0, 1)$ if the second bed is sampled.

The second model appears awkward and the notation needlessly complicated. However, for some purposes, for instance, regression analysis (Chap. 9), the second model may be more suitable. It is introduced here simply to show that more than one model always may be devised for a particular physical situation. The second model can also be written as

$$w = \mu + \beta(x_1 - x_2),$$

where μ is the average phosphate content of the two beds, β is half the difference between the two beds, and x_1 and x_2 have the same meaning as before. The virtue of this approach is that the difference in the average between the two beds is given by a single number, for, if $\beta = 0$, there is no difference between the two beds.

A final example is afforded by the rate of a chemical reaction which may be written

$$w = \alpha e^{-\beta/x},$$

where w is the reaction rate, α and β are parameters, and x is the temperature in degrees Kelvin. Although this equation is a nonlinear model, it may be rewritten

$$w' = \theta - \beta x',$$

where $w' = \ln w$ and $x' = 1/x$.

In terms of the new variables w' and x' and the new parameters θ and β the relationship is linear. The example illustrates that linear models can be used to express many physical relationships provided that the appropriate parameters and variables are chosen, even though the first parameters and variables that come to mind suggest a nonlinear model. However, it is important to choose variables to which a clear physical meaning can be ascribed. This subject is further developed in section 6.3 on transformations.

All the examples of functional relationships are for deterministic models. But each can be converted to a statistical model by adding a random fluctuation e to the right-hand side of the equation, just as the deterministic model for the soil-failure model in section 1.5,

$$s = K_1 + (\tan \theta)p,$$

is converted to the statistical model,

$$s = K_1 + (\tan \theta)p + e.$$

Definition of a Linear Model

A mathematical model that allows an observation w to be expressed as a function of one or more variables and parameters has a functional form that may or may not be linear. However, if the function is linear, a linear model is defined by the general equation

$$w = \alpha + \sum \beta_i x_i + e,$$

where α is a general constant, the x_i terms are independent variables with associated parameters β_i, and e is the random fluctuation. To comprehend this definition, one must recognize that, although the independent variables x_i may themselves be nonlinear functions of other variables, the model is nonetheless a linear model in terms of the parameters α and β_i; for example, the equation

$$w = 3 + 6 \cos 3z_1 + \sin 7z_2 + e$$

is a linear model, because it can be rewritten

$$w = 3 + 6x_1 + x_2 + e,$$

where $x_1 = \cos 3z_1$, and $x_2 = \sin 7z_2$. The ordinary polynomial equation is a linear model; for example, the polynomial equation

$$w = 4 + 2z + 5z^2 + 7z^3 + e$$

can be rewritten

$$w = 4 + 2x_1 + 5x_2 + 7x_3 + e,$$

where $x_1 = z$, $x_2 = z^2$, and $x_3 = z^3$.

Usually numerical values are unknown for the parameters constituting the constant term α and for the coefficients β_i of the independent variables x_i. Then, in order to estimate these parameters, corresponding statistics must be calculated and their frequency distributions determined. The procedure is as follows: Instead of describing an observation w as being drawn at random from a population with mean μ and variance σ^2, it is set equal to the population mean μ plus a random fluctuation e; that is,

$$w = \mu + e.$$

By definition, the random fluctuation e has the following properties:

1. Its mean is equal to 0; that is, $\mu_e = 0$.
2. Its variance is equal to the population variance; that is $\sigma_e^2 = \sigma^2$.
3. It is statistically independent, from observation to observation.
4. It follows a normal frequency distribution.

In order to make this definition of an observation w useful, it must be related to the preceding material by remembering that μ may be written as a linear function of as many independent variables and parameters as is appropriate; that is,

$$\mu = \alpha + \sum \beta_i x_i.$$

For instance, the discussion of the example for the two phosphate beds in the preceding subsection is an illustration of this procedure for two statistical samples.

Interpretation of Analysis of Variance with Linear Models

The fictitious illustration of the geologist sampling two formations for selenium by two different methods (sec. 5.9) serves to demonstrate the relationship between the analysis of variance and linear models. The mathematical structures imposed on the various population means in sections 5.1, 5.5, and 5.7 have all been examples of linear models, although without the detailed exposition of this section. The two sets of notations can be reconciled through the device of artificial variables, which are introduced only to accomplish this purpose and to show that the models discussed in the analysis of variance are indeed linear.

In section 5.9 two formulations of the fictitious illustration of selenium sampling are made, first in a single-degree-of-freedom analysis, and second in the simpler but less incisive one-way analysis of variance. Now, first the one-way analysis of variance formulation is related to artificial variables. In this formulation it is assumed that the geologist was sampling four populations, without connecting them through a further mathematical structure. For a one-way analysis of variance (sec. 5.1) the model for an observation is written

$$w_{ij} = \mu + (\mu_i - \mu) + e_{ij},$$

where i designated the population being sampled and j designated the jth observation from the ith population. This model may be rewritten as follows, where the principal difference in notation is the introduction of the artificial variables x_i, which play essentially the same role as the subscripts i and j in the preceding formulation:

$$w = \mu + \beta_1 x_1 + \beta_2 x_2 + \beta_3 x_3 + \beta_4 x_4 + e,$$

where

1. The general mean μ is the average of μ_1 to μ_4.
2. $\beta_1 = \mu_1 - \mu$, $\beta_2 = \mu_2 - \mu$, $\beta_3 = \mu_3 - \mu$, $\beta_4 = \mu_4 - \mu$.
3. The variables x_i are *artificial variables*, with no intrinsic meaning. They are introduced to designate the population from which an observation comes so that if an observation comes from the first population, $x_1 = 1$, and x_2 to $x_4 = 0$; if an observation comes from the second population, $x_2 = 1$, and x_1, x_3, and $x_4 = 0$; etc.

Several properties of this linear model may be pointed out. If all of the population means μ_i are equal, all of the coefficients β_i are equal to 0. In the analysis-of-variance format of table 5.8 the average within-sample mean square is equal to σ_e^2. Therefore, if the variance of the coefficients β_i is designated σ_μ^2, the average among-sample mean square of that analysis is simply $\sigma^2 + n\sigma_\mu^2$, where n is the number of observations in each sample. The coefficients β_i measure the variation among the means. Because the sum of these coefficients is 0, only three independent parameters are associated with these coefficients, and there are three degrees of freedom for these parameters. The fourth parameter is μ, the general mean of the observations.

For these same data, one can describe another statistical linear model that corresponds to the single-degree-of-freedom interpretation in section 5.9. Three new coefficients β_i and six new artificial variables x_i can be defined as follows:

1. β_1 is the difference associated with the formations.
2. β_2 is the difference associated with the methods.
3. β_3 is the difference associated with the interactions.

Then the linear model

$$w = \mu + \beta_1(x_1 - x_2) + \beta_2(x_3 - x_4) + \beta_3(x_5 - x_6) + e$$

can be written, where the artificial variables x_i have the values in table 5.35. In the previous model there were only four independent parameters. In this new model there are also four independent parameters that appear explicitly; the three β coefficients that are associated with variations among the means account for the three degrees of freedom.

Two explicit linear models using artificial variables have been devised for the two different interpretations of the fictitious selenium data. Many other

Table 5.35. Values of the artificial variables x_i for a linear model corresponding to the single-degree-of-freedom interpretation of a fictitious illustration of sampling for selenium

Formation	Method	
	A	B
1	$x_1 = 1$	$x_1 = 1$
	$x_2 = 0$	$x_2 = 0$
	$x_3 = 1$	$x_3 = 0$
	$x_4 = 0$	$x_4 = 1$
	$x_5 = 1$	$x_5 = 0$
	$x_6 = 0$	$x_6 = 1$
2	$x_1 = 0$	$x_1 = 0$
	$x_2 = 1$	$x_2 = 1$
	$x_3 = 1$	$x_3 = 0$
	$x_4 = 0$	$x_4 = 1$
	$x_5 = 0$	$x_5 = 1$
	$x_6 = 1$	$x_6 = 0$

models could be associated with this or any other set of data. Often, an appropriate model is hard to choose; usually the choice is determined by the investigator's purpose. The same set of data can be therefore interpreted in a variety of ways simply by using different models. One of the virtues in designing an experiment before gathering data lies in obtaining insight into what model might be appropriate. In analysis of data from undesigned experiments, the choice of an appropriate model and of which variables to consider can lead to formidable problems that may be difficult if not impossible to resolve.

Variance Components

Through a study of quantities named variance components, the results of an analysis of variance can be examined in more detail. Variance components identify the contribution to the observed variability of various terms or groups of terms in the linear model. The analysis of table 5.26 yields an analysis-of-variance table whose essential elements are presented in table 5.36. The last column of the table shows average mean-square values which arise from the fact that the within-sample variance is the basic fluctuation defined

TABLE 5.36. VARIANCE COMPONENTS FOR ANALYSIS OF VARIANCE OF TABLE 5.26

Source of variation	Degrees of freedom	Mean square	Average mean square
Among-sample means	3	$7s_{\bar{w}}^2$	$\sigma^2 + 7\sigma_\mu^2$
Within sample	24	s_p^2	σ^2

to be σ^2. The among-sample mean square measures $\sigma^2 + 7\sigma_\mu^2$, because the average value of $s_{\bar{w}}^2$ is $\sigma^2/7 + \sigma_\mu^2$, in general, where

$$\sigma_\mu^2 = \frac{\sum \beta^2}{3} = \frac{\sum (\mu_i - \mu)^2}{3}.$$

The β values are those in the first linear model, without the single-degree-of-freedom interpretation associated with the analysis.

Variance components are variances such as σ^2 and σ_μ^2 that identify the sizes of the sources of variability. In the example cited σ^2 represents the basic variability, whereas σ_μ^2 represents the variation introduced by the variance in the β_i values of the linear model, that is, in the variance of the four population means. For any analysis of variance, the variance structure of each line in the analysis-of-variance table can be identified. Although the components are not too difficult to recognize, the calculation of their coefficients is often complicated, too complicated to discuss in this book. The interested reader is referred to Dixon and Massey (1969, p. 153). However, once the variance structure is recognized, the sizes of the different pieces can be estimated and sometimes afford a guide to action (sec. 8.1).

Next, the essential elements of the more complicated analysis of variance (table 5.27) for the fictitious illustration of consistency of selenium sampling can be considered. This analysis is presented in table 5.37. As before, the values for σ_m^2, σ_f^2, and σ_i^2 come from the β_i values in the model. When this

TABLE 5.37. VARIANCE COMPONENTS FOR THE ANALYSIS OF VARIANCE OF TABLE 5.27

Source of variation	Degrees of freedom	Average mean square
Among-sample means	3	$\sigma^2 + 7\sigma_\mu^2$
Between methods	1	$\sigma^2 + 7\sigma_m^2$
Between formations	1	$\sigma^2 + 7\sigma_f^2$
Method-formation interaction	1	$\sigma^2 + 7\sigma_i^2$
Within sample	24	σ^2

more complicated analysis is compared with the previous simple analysis, it is evident that

$$\sigma_\mu^2 = \frac{(\sigma_m^2 + \sigma_f^2 + \sigma_i^2)}{3}.$$

The comparison makes clear the power of the single-degree-of-freedom method used in the more complicated analysis; that is, if only one of the variances on the right side of the equation is relatively large, its chance of showing up is greater in the second analysis, whereas in the first analysis its effect, being diluted by the other smaller variances, is less.

Variance components may be identified in the linear model used for the analysis of Homestake assay data (sec. 5.10). In this analysis, the following sources of variability are identified:

1. Between holes.
2. Among-foot intervals within holes.
3. Between pairs within 1-foot intervals.
4. Within pairs.

With these sources of variability recognized, each observation w may be expressed in the linear model

$$w = \mu + (\mu_i - \mu) + (\mu_j - \mu_i) + (\mu_k - \mu_j) + e,$$

where
μ = general mean,
$(\mu_i - \mu)$ = effects associated with particular holes,
$(\mu_j - \mu_i)$ = effects associated with particular 1-foot intervals,
$(\mu_k - \mu_j)$ = effects associated with particular pairs,
e = random fluctuation.

The purpose of the linear model for the Homestake data is to study the amount of variation among the various distances. The analysis of variance (sec. 5.10, table 5.15) can be used for this study because, for the model chosen, it can be shown that the average mean squares of the analysis-of-variance table (table 5.15) are those given in table 5.38. The numerical values

TABLE 5.38. AVERAGE MEAN-SQUARE VALUES OF VARIANCE COMPONENTS FOR ANALYSIS OF VARIANCE OF HOMESTAKE MINE DATA IN TABLE 5.15

Line	Source of variation	Degrees of freedom	Mean square	Average mean square
1	Among hole	4	44.471	$\sigma^2 + 2\sigma_p^2 + 4\sigma_f^2 + 43\sigma_h^2$
2	Among-foot intervals	51	49.728	$\sigma^2 + 2\sigma_p^2 + 4\sigma_f^2$
3	Between-sample pairs	52	29.065	$\sigma^2 + 2\sigma_p^2$
4	Within-sample pairs	108	10.029	σ^2

TABLE 5.39. NUMERICAL VALUES OF VARIANCE COMPONENTS FOR ANALYSIS OF VARIANCE OF HOMESTAKE MINE DATA IN TABLE 5.15

Type of variance component	Estimate of variance component
Among hole, σ_h^2	0.0†
Among foot, σ_f^2	5.1
Between-sample pairs, σ_p^2	9.5
Within-sample pairs, σ^2	10.0

† Although the estimated variance component is negative, it is set equal to 0.0 because the parameter σ_h^2 is nonnegative.

calculated for these three quantities are given in table 5.39. Of course, the F-test in the analysis of variance is used to decide whether the positive estimates represent real effects or are merely consequences of statistical fluctuations. Finally, it must be emphasized that the Homestake linear model is only one of many that might have been made for these data but is the most sensible one that we can think of.

Estimation of Parameters

There are several ways to estimate the parameters in a model. In this subsection we consider two of them, but not in any detail. The reader interested in a full discussion is referred to Wilks (1962).

One way to estimate the parameters is the least-squares method, wherein the statistics are chosen so that the quantity

$$\sum (w - \mu)^2$$

is minimized. The calculation is straightforward, if the model is linear, because the calculations required for a solution involve only simultaneous linear equations. If the model has k parameters, the sum of squares of the observations can be broken into two parts: (a) a sum of squares with k degrees of freedom, which is the variability "explained" by the k parameters, and (b) a sum of squares with $(n - k)$ degrees of freedom, which is the unexplained variability. If the second quantity is divided by $(n - k)$, an estimate of the variance σ^2 of the fluctuations is obtained. These remarks may be summarized in the equation

$$\sum w^2 = \text{SS due to parameters} + \text{residual SS}.$$

This subject is expanded and an illustration is given in section 10.3.

The second way to estimate the parameters in a linear model is the maximum-likelihood method in which the random fluctuations are assumed to follow a particular frequency distribution. Once the assumption is made, the maximum-likelihood method can be used to obtain estimates of the parameters. The theoretical frequency distributions of the n observations and the k parameters are combined to form a joint theoretical frequency distribution, whose equation is named the *likelihood function L*, where

$$L = f(w_1, \ldots, w_n, \beta_1, \ldots, \beta_k).$$

Values of the parameters β_1 to β_k are chosen to maximize this function by setting the k partial derivatives with respect to β_i equal to 0 and solving the k simultaneous equations that result. Details of the procedure are given by Hoel (1962, p. 58).

The two ways of estimating the parameters in a linear model may be compared. The least-squares method requires only the assumption that the fluctuations are independent and have equal variance. If the assumption is made that the observations follow a normal distribution with equal variance, three consequences follow:

1. Statistics calculated by the maximum-likelihood and least-squares methods are identical.

2. The sums of squares associated with the parameters are independently distributed from the sums of squares associated with the fluctuations.

3. The statistics are normally distributed.

Choice of Model

For several reasons a simple model is usually better than a complicated one. One reason is that with complicated models one is seldom certain about how to interpret the results of an analysis, because it is difficult to give a physical meaning to all of the terms of the model.

Another reason is that with a simple model there are fewer parameters to estimate. Since at least one observation is needed to estimate each parameter, with a simple model fewer observations are required, which may be an important consideration if the number of observations is limited because few were collected (as in the analysis of old data) or because few can be collected (as in the analysis of new but expensive data).

Because 1 degree of freedom is required to estimate each parameter in a linear model, the number of degrees of freedom remaining to estimate the *residual variance*, that is, the variance of the fluctuations, is $(n - k)$, where n is the number of observations and k the number of parameters. If a suitable model can be devised, it is desirable to have about 30 degrees of freedom remaining to estimate residual variability. If, for instance, there are fewer than 10 parameters to estimate, having 30 degrees of freedom yields a result only 10 percent less stable than an infinite number of degrees of freedom. With fewer than 5 degrees of freedom, an estimate of residual variability is extremely unreliable. This concept is further explained in section 9.4.

Nonlinear Models

In this section only linear models have been discussed. In principle, the mathematical model can be arbitrary; that is, the measured variable w can depend on independent variables and on parameters through a nonlinear functional relationship. But most mathematical models actually used are linear, even though many physical situations are nonlinear, so that the linear model used to describe them is only an obvious approximation. Linear models are used because the mathematical and statistical techniques available at present do not cope with nonlinear models very well.

Fortunately, any functional relationship can be approximated, at least for some limited range of the variables, by a linear equation; therefore linear models can be used in practice. If nonlinear relations are an essential part of any situation, they must be handled on a case-by-case basis. One must keep in mind that approximations must be used and that predictions, especially extrapolations, beyond the range for which data are available are liable to be very risky.

5.12   MULTIPLE COMPARISONS

In the analysis of variance a general hypothesis test is made, for example, a test of the hypothesis H_0 that the population means of several samples are equal. If hypothesis H_0 is rejected in favor of the alternative hypothesis H_1 that the population means are unequal, it is often interesting to inspect the individual sample means more closely.

The best way to make this inspection is to test, by using the method of the single degree of freedom, one or more predetermined specific hypotheses, perhaps that two sample means from samples having something geologically in common are from the same population. But, as stated in section 5.9, if more than a few comparisons are made with the single-degree-of-freedom method, the error rate, nearly as great as the combined α error, becomes too large. However, it is desirable to inspect the data for unexpected results.

The method of *multiple comparisons*, whose outstanding attribute is that the error rate is set at a constant level regardless of the number of hypotheses tested or confidence limits formed, allows one to carry out such hindsight analyses with the error rate controlled. In this section an illustrative example

TABLE 5.40. FICTITIOUS POTASH OBSERVATIONS IN 10 TRAVERSES

Traverse number	Observations					Sample mean, $\bar{w}$
1	6.4	6.8	6.8	6.0	7.6	6.72
2	2.4	2.5	2.3	3.3	2.5	2.60
3	3.0	1.4	2.4	3.7	1.6	2.42
4	4.3	4.3	4.1	5.7	4.6	4.60
5	6.0	6.6	7.2	6.7	6.2	6.54
6	2.8	3.1	3.7	4.0	2.8	3.28
7	2.2	3.3	3.4	0.4	3.5	2.56
8	5.0	5.7	4.2	3.3	5.1	4.66
9	7.4	7.1	6.4	7.8	5.2	6.78
10	2.7	4.6	6.4	5.6	3.0	4.46

is first given; then the name multiple comparisons is defined, and the calculations are explained. Finally the method is related to the method of the single degree of freedom.

As a fictitious example, suppose that a geologist wishes to investigate potash content in a gneiss exposed in a linear belt of outcrops. He naturally makes traverses at right angles to the foliation. If he makes an arbitrary number of 10 traverses and takes an also arbitrary number of five geological samples from each traverse, he may obtain the 50 fictitious potash observations listed in table 5.40, together with the sample means for the 10 traverses.

With the data in table 5.40 at hand, the first obvious step is to make a one-way analysis of variance to test the hypothesis that the 10 population means associated with the 10 traverses are the same. In table 5.41 this analysis yields an F-value of 17.8, which because it is inside the critical region, leads to acceptance of the alternative hypothesis that the 10 population means are unequal. Inspection of the 10 sample means certainly suggests

TABLE 5.41. ANALYSIS OF VARIANCE FOR FICTITIOUS POTASH DATA FROM 10 TRAVERSES

Source of variation	Analysis of variance				
	Sum of squares	Degrees of freedom	Mean square	F	$F_{10\%}$
Among traverses	137.50	9	15.2775	17.8	1.7929
Within traverses	34.30	40	0.8575		

that some population means are likely to be a good deal larger than the others. In some circumstances, with a predetermined hypothesis, a single-degree-of-freedom test might be reasonable; for example, based on additional geological information, such as changes in rock structure or texture, one could test a hypothesis that the population mean of the first traverse is not equal to the mean of the other traverses, but, if a general overall comparison of all 10 traverses is wanted, the appropriate method is that of multiple comparisons.

The purpose of the following analysis is to segregate the various traverses into groups, so that the traverses within a group will be more or less alike, whereas those in different groups will be somewhat different. With this aim in mind, the means can be ordered according to their sizes and differences can be calculated. Then the multiple-comparison procedure is used to find the size of the largest difference that can be expected between two sample means with a common population mean. If the observed difference is larger than this specified difference, the two sample means can be considered to come from different populations.

The procedure is this. For any contrast (sec. 5.8) a quantity called an *allowance* is calculated, a number that may be interpreted in two ways. The first interpretation is that an allowance is the boundary of the critical region for testing the hypothesis that the population-mean value of the contrast is equal to 0. The second interpretation is that an allowance is the quantity to be added to, and subtracted from, the numerical value of the contrast to form a confidence interval for the population mean of the contrast.

Several uses of allowances are illustrated:

1. To form groups of traverses.
2. To compare any two means to see if they are from the same population.
3. To compare any two *groups* of means to see if they are from the same population.
4. To identify, within a group, means that are extremely large or small.

Two ways to compute allowances have been formulated, and because each is better for some situations, both ways are described.

One way to compute an allowance is described as follows by Tukey (1951): For any desired contrast, the allowance is obtained by multiplying three quantities together. The first is the estimated standard error of the mean, $\sqrt{s_p^2/n}$. The second is the sum of the positive c_i quantities. The third is the critical value of the studentized range from table A-6. For the fictitious illustrative data the calculations are given in table 5.42, in which the contrast is the simple difference between any two means,

$$q = \bar{w}_i - \bar{w}_j.$$

TABLE 5.42. TUKEY MULTIPLE-COMPARISON PROCEDURE APPLIED
TO FICTITIOUS POTASH DATA

Line	Notation	Value	Calculation
1	k	10	From table 5.40
2	n	5	From table 5.40
3	s_p^2	0.8575	From table 5.41
4	s_p^2/n	0.1715	
5	$\sqrt{s_p^2/n}$	0.4140	
6	$\sum (\text{positive } c_i)$	1	
7	$q_{10\%}[k, k(n-1)]$	4.32	From table A-6
8	A_T	1.79	

In lines 2 to 5 the estimated standard error is calculated, recalling that s_p^2 is the estimate of s^2. For simple differences, the sum of the positive c_i values must be 1, as recorded in line 6. The value entered in line 7 is from table A-6, for $k = 10$, and 40 degrees of freedom [$k(n-1) = 10 \times 4$]. The numbers in lines 5 to 7 are multiplied together to obtain the allowance A_T in line 8.

The results, listed in table 5.43 and plotted in figure 5.6, may be interpreted as follows. Traverses 9, 1, and 5 form a clearly defined group designated group 1, whose traverse means are larger than any others. A second group, group 2, is composed of traverses 8, 4, 10, and 6, which have means of intermediate size. Group 2 lies between group 1 and group 3 and consists of traverses 6, 2, 7, and 3. Because traverse 6 belongs to both group 2 and group 3, it may represent an intermediate group between these two groups or may belong to one of them. More data are needed to resolve this question.

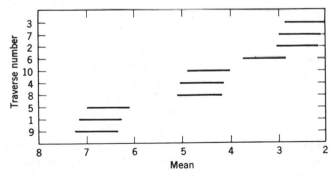

Fig. 5.6. Allowances and traverse groups defined by the Tukey multiple-comparison method.

TABLE 5.43. GROUPS OF TRAVERSES DEFINED BY THE TUKEY MULTIPLE-COMPARISON PROCEDURE

Traverse number	Mean	Difference between adjacent means	Grouping (numbers are traverse numbers)	Group number
9	6.78		915	
		0.06		
1	6.72		915	1
		0.18		
5	6.54		915	
		1.88		
8	4.66		84106	
		0.06		
4	4.60		84106	2
		0.14		
10	4.46		84106	
		1.18		
6	3.28		84106273	
		0.68		
2	2.60		6273	3
		0.04		
7	2.56		6273	
		0.14		
3	2.42		6273	

Another way to compute an allowance is described by Scheffé (1953). For any desired contrast four quantities are multiplied together. The first is the estimated standard error of the mean, $\sqrt{s_p^2/n}$, as for the Tukey method. The second is the sum of the c_i *squared* quantities, $\sum c_i^2$. The third is the critical value of $F_\alpha[k - 1, k(n - 1)]$. The fourth is $(k - 1)$. For the fictitious illustrative data, the calculations are given in table 5.44 for the same contrast as that in table 5.42. In lines 2 to 5 the standard error is estimated, as before. Line 6 is $\sum c_i^2$, which must be 2 for any simple difference. The F-value in line 7 is obtained from table A-5. Line 8 gives the value of $(k - 1)$. Line 9 gives the product of the numbers in lines 4, 6, 7, and 8. Finally, line 10 gives the square root of the values in line 9, which is A_S.

The results of the analysis by the Scheffé method are listed in table 5.45. Not one of the traverses or their combinations forms a well-defined group. Although there is some evidence of a group of traverses with high means and another with low means, three traverses (8, 4, and 10) belong to both groups, and if one tries to define a group starting with these

TABLE 5.44. SCHEFFÉ MULTIPLE-COMPARISON PROCEDURE APPLIED TO FICTITIOUS POTASH DATA

Line	Notation	Value	Calculation
1	k	10	
2	n	5	
3	s_p^2	0.8575	
4	s_p^2/n	0.1715	
5	$\sqrt{s_p^2/n}$	0.414	
6	$\sum c_i^2$	2	
7	$F_{10\%}[k - 1, k(n - 1)]$	1.79	From table A-5
8	$k - 1$	9	
9	A_S^2	5.526	Product of lines 4, 6, 7, 8
10	A_S	2.35	

three traverses all the other traverses are included in the single group thus defined.

What has been done by both methods may be formalized as follows. For all possible pairs of means the hypothesis H_0 that their population means are identical has been tested, and those means have been grouped together for which H_0 is accepted. Alternatively, the interpretation may be made that the confidence interval for the difference among all possible pairs of means has been formed, and those means have been grouped together for which the interval contains 0.

Whenever a hindsight evaluation of data is made, and the method of multiple comparisons is therefore appropriate, one must decide whether his primary interest is in simple or in complex contrasts. If primarily in simple contrasts, the Tukey method is adopted and is used for all contrasts, with the most incisive results obtained for simple contrasts. If, however, one is primarily interested in complex contrasts, the Scheffé method is chosen and again used for all contrasts, with the most incisive results obtained for complex contrasts. Whichever method an investigator adopts must be used for all contrasts formed.

Only for illustration are both the Tukey and Scheffé methods used on the fictitious data. For the particular analysis made the Tukey method is better because simple contrasts are of interest. With the Tukey method distinct groups are formed. Had the sample data been used in a complex contrast, for example, in comparing the first five traverses in a group with the second five traverses in another group, the Scheffé method would have been preferable. As stated, with real data only one method can be used for all contrasts.

The allowances calculated by either the Tukey or the Scheffé method have an interpretation similar to that placed on confidence intervals (sec. 4.4).

TABLE 5.45. GROUPS OF TRAVERSES DEFINED BY THE SCHEFFÉ MULTIPLE-COMPARISON PROCEDURE

Traverse number	Mean	Difference between adjacent means	Grouping (numbers are traverse numbers)		
9	6.78		9158410	91584106273	
		0.06			
1	6.72		9158410	91584106273	
		0.18			
5	6.54		9158410	91584106273	
		1.88			
8	4.66		9158410	91584106273	84106273
		0.06			
4	4.60		9158410	91584106273	84106273
		0.14			
10	4.46		9158410	91584106273	84106273
		1.18			
6	3.28			91584106273	84106273
		0.68			
2	2.60			91584106273	84106273
		0.04			
7	2.56			91584106273	84106273
		0.14			
3	2.42			91584106273	84106273

Recall that confidence intervals have the property of being correct in a specified percentage of cases, for example, 90 percent of the times that they are tried. Similarly, if an allowance is calculated with a 10-percent error rate, all intervals based on the allowance will be correct for 9 sets of data out of 10. Although the number of contrasts and allowances that can be calculated for each set of data is unlimited, these results still apply. Thus the concept of error rate has been extended from that of single intervals to sets of intervals for a given data set. Therefore, unlike the confidence-interval procedure, no error rate can be associated with a particular interval calculated with an allowance.

5.13   RELATIONSHIPS AMONG BASIC STATISTICAL DISTRIBUTIONS

At this point the basic ways to make statistical inferences by forming confidence intervals and by performing hypothesis tests have been introduced. Most of the remainder of the book explains how to apply these methods to

$d.f._1$ / $d.f._2$	1	2	3	....	∞
1					
2					
3	t^2				
⋮					
∞	u^2			$^2/d.f._1$	1

Fig. 5.7. Arrangement of F-table.

geological problems. Little new fundamental statistical material is introduced, although special principles and techniques are explained.

The most important sampling distributions in statistics have been introduced: the normal distribution, the t-distribution, the chi-square distribution, and the F-distribution. Although there are innumerable other distributions, some discussed in this book, these four play the central role in statistics, and it is important to realize that they are closely interrelated.

The relationship is best explained in terms of the F-distribution. Because the chi-square distribution is that of SS/σ^2 and $SS = (n - 1)s^2$, dividing the chi-square statistic by its degrees of freedom yields a ratio of s^2/σ^2, which is an F-statistic with $(n - 1)$ and infinity degrees of freedom; that is,

$$\frac{SS}{(n - 1)\sigma^2} = \frac{(n - 1)s^2}{(n - 1)\sigma^2} = \frac{s^2}{\sigma^2}.$$

Thus, if the entries in the chi-square table (table A-3) are divided by their corresponding degrees of freedom, the bottom row of the F-table (table A-5) is obtained as shown diagrammatically in figure 5.7.

If a variable t follows a t-distribution with a particular number of degrees of freedom, the variable t^2 follows an F-distribution with 1 as well as the same particular number of degrees of freedom—because, when squared, the numerator of the t-statistic forms an estimate of the variance with one degree of freedom; whereas, when squared, the denominator forms an estimate of the variance with the certain number of degrees of freedom. Finally, the fact that a t-distribution with infinity degrees of freedom is the normal distribution (sec. 4.3) shows that the entry in the F-table with 1 and infinity degrees of freedom is the square of the corresponding entry in the table for the normal distribution. When squared, the upper 2.5-percent point of the standardized normal distribution becomes the upper 5-percent point of the F-distribution, because the negative lower 2.5-percent point becomes positive when squared.

REFERENCES

Cochran, W. G., and Cox, G. M., 1957, Experimental designs: New York, John Wiley & Sons, 611 p.

Davies, O. L., 1958, Statistical methods in research and production: New York, Hafner Publishing Co., 390 p.

Dixon, W. J., and Massey, F. J., Jr., 1969, Introduction to statistical analysis: New York, McGraw-Hill, 638 p.

Hoel, P. G., 1962, Introduction to mathematical statistics: New York, John Wiley & Sons, 427 p.

Koch, G. S., Jr., and Link, R. F., 1963, Distribution of metals in the Don Tomás vein, Frisco mine, Chihuahua, Mexico: Econ. Geology, v. 58, p. 1061–1070.

——, 1966, Some comments on the distribution of gold in a part of the City Deep mine, Central Witwatersrand, South Africa, *in* Symposium on mathematical statistics and computer applications in ore valuation: South African Inst. of Mining and Metall., Proc., p. 173–189.

——, 1967, Geometry of metal distribution in five veins of the Fresnillo mine, Zacatecas, Mexico: U.S. Bur. Mines Rept. Inv. 6919, 64 p.

Li, J. C. R., 1964, Statistical inference, v. 1: Ann Arbor, Mich., Edwards Bros., 658 p.

Scheffé, Henry, 1953, A method for judging all contrasts in the analysis of variance: Biometrica, v. 40, p. 87–104.

Tukey, J. W., 1951, Quick and dirty methods in statistics, Pt. 2, Simple analyses for standard designs: Am. Soc. Quality Control, 5th Ann. Conf. Proc., p. 189–197.

Wilks, S. S., 1962, Mathematical statistics: New York, John Wiley & Sons, 644 p.

Chapter 6

Distributions and Transformations

Many observed frequency distributions are closely approximated by the theoretical normal distribution; for instance, some distributions of human heights, lives of light bulbs, and weighing errors are more or less normal. Partly because of this correspondence but mainly because the normal distribution is mathematically tractable, most of the statistical theory and methods of inference have been devised for it. But many distributions of geological observations are nonnormal; in fact one of the more common distributions is skewed with a long tail to the right, including distributions of gold and other trace elements (secs. 6.2 and 15.3, Chap. 16). Another common distribution of geological data is a mixed distribution made up of two or more distributions (sec. 6.5).

In this chapter several nonnormal distributions are introduced and methods of working with them explained. The reader is already familiar with one method of working with them: relating the nonnormal distribution to the normal distribution by means of normal approximations. Other methods involve learning a small amount of new statistics and some new tables. We also discuss the use of a *transformation*, a function of an observation that defines a new observation. There are problems involved in working with nonnormal distributions, and the validity of the results depends upon the details of the methods chosen.

Some examples of nonnormal distributions of geological data are introduced in sections 6.1 and 6.2. Then, in section 6.3, some transformations that may be used to normalize or nearly normalize data are explained. Once the data are normalized, hypothesis tests can be made and confidence

201

intervals constructed for the transformed observations. Next follows a discussion of the delicate and complicated questions: Are the results garnered from these transformed observations useful to the geologist's purpose and can they be interpreted in a manner compatible with the original variables? Special graph papers to make transformations visually meaningful and to help the geologist visualize a distribution are discussed in section 6.4. The last section (sec. 6.5) treats mixed-frequency distributions.

6.1 SOME DISCRETE DISTRIBUTIONS

So far the frequency distributions considered have been *continuous distributions* of observations derived, in principle, by measuring; thus they can have any values, not necessarily integers, in a given range. An example of a continuous observed distribution is one of metal assays lying in the range from 0 to 100 percent. An example of a continuous theoretical distribution is the normal distribution. Now introduced in this section are *discrete distributions* of observations derived, in principle, by counting; thus they are integers. Examples of discrete distributions are the number of mines in each township in the western United States, the number of assays more than three standard deviations from the mean on different mine levels, and the number of occurrences of an index fossil in different sedimentary environments.

Actually, in practice the formal distinction between discrete and continuous distributions is seldom important because one kind can be and often is transformed into the other. Thus a list of carbonate percentages can be classified into those above and those below a value defining limestone. By assigning the integer 1 to those above and the integer 0 to those below, one can transform a continuous observation, the carbonate percentage, into a discrete observation. Although information is lost in the process, the loss is not important if one's only interest is in defining limestone. On the other hand, a continuous distribution may be used to represent discrete observations, as in the normal approximation to the discrete Poisson distribution.

The Binomial Distribution

Suppose that the outcome of a trial, such as tossing a coin or classifying a rock as igneous or not igneous, is described as a dichotomy, that is, as one or the other of only two events. One event of the dichotomy is arbitrarily named a *success*, and the other is named a *failure*. Thus a simple "yes, no" classification results. The binomial distribution is the discrete distribution of the number of successes that occur in a fixed number of trials, where the outcomes are statistically independent, and the probability of success is the same for each trial.

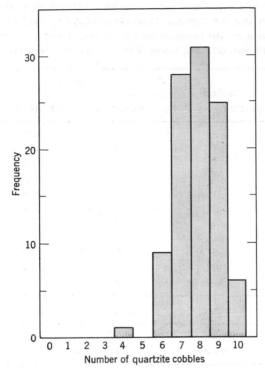

Fig. 6.1. Number of quartzite cobbles in each of 100 samples of 10 cobbles from a gravel deposit on the Gros Ventre river, Wyoming (data from R. Flemal, personal communication, 1967).

The binomial distribution may be introduced by consideration of a small set of data collected by R. Flemal (personal communication). In a study of the geologic environment of gold he counted the number of quartzite cobbles in each of 100 samples of 10 cobbles from a gravel deposit on the Gros Ventre River, Wyoming. The counts of from 0 to 10 quartzite cobbles are given in table 6.1 as a frequency distribution and in figure 6.1 as a histogram. The problem is to estimate the proportion of quartzite cobbles in the gravel. Since it is known intuitively that the number of quartzite cobbles in each sample estimates the proportion of quartzite cobbles in the entire gravel deposit, the only question that remains is how closely.

Under the assumptions that the cobbles can be classified as quartzite and nonquartzite and that the quartzite ones are randomly distributed throughout the gravel, the cobble counts follow the binomial distribution, the formula for which is

$$P(w) = \frac{n!}{w!\,(n-w)!}\ \theta^w(1-\theta)^{n-w},$$

Table 6.1. Observed and theoretical binomial
frequency distribution of quartzite cobbles in
each of 100 samples of 10 cobbles from a gravel
deposit on the Gros Ventre River, Wyoming *

Number of quartzite cobbles	Frequency	
	Observed	Theoretical
0	0	0.0
1	0	0.0
2	0	0.0
3	0	0.1
4	1	0.8
5	0	3.3
6	9	10.3
7	28	21.7
8	31	30.0
9	25	24.7
10	6	9.1

Item	Observed	Theoretical
Mean	7.87	7.87 †
Variance	1.28	1.68

* Data from R. Flemal, personal communication, 1967.
† Assuming true proportion of pebbles is 0.787.

where $P(w)$ is the probability of obtaining w successes in n items, n is the number of items in a sample (10 cobbles in the example), w is the number of successes (number of quartzite cobbles in the example), and θ is the proportion of successes in the population (proportion of cobbles that are quartzite in the area in this example). The mean of this distribution is $n\theta$, and the variance is $n\theta(1 - \theta)$.

Table 6.1 gives the empirical distribution obtained by Flemal and also the mean and variance obtained in the usual way (sec. 3.3). In column 2 the theoretical binomial distribution for the same mean of 7.87 and the theoretical variance calculated by the previous formula are given [$7.87(1 - 0.787) = 1.68$]. Comparison of the two variances shows that the empirical distribution is a little more compressed than the theoretical one because the cobbles were not chosen at random, but the agreement is fairly good. As explained at the end of this section, a chi-square test to compare the two distributions yields a chi-square value of 6.39, which is less than the tabled value of 9.24 for the

10-percent point of the chi-square distribution. Thus the good agreement indicated by the variance is corroborated.

The binomial distribution has only the single parameter θ, from which both the mean and variance are calculated. If both $n\theta$ and $n(1 - \theta)$ are large enough, say 5 or greater, the binomial distribution is satisfactorily approximated by the normal distribution. Because the statistic

$$\frac{\overline{w} - n\theta}{\sqrt{n(\theta)\,(1 - \theta)}}$$

follows the normal distribution with mean zero and variance 1, it can be used to find confidence limits for the mean and variance (Li, 1964, I, p. 459). For Flemal's data, the estimate of the mean proportion of quartzite cobbles in the gravel is 0.787, and 90-percent confidence limits for the population proportion of quartzite cobbles are 0.766 and 0.808.

In many disciplines, including engineering and the social sciences, there are natural dichotomies; for instance, in quality control of manufacturing certain parts may be grouped into batches of equal size that are tested for good or bad quality. However, geological data tend to be continuous, and there is less occasion to use the binomial distribution. Geological data come mostly from cobble counts in conglomerates, gravels, glacial tills, and the like, and from some paleontological studies.

Consequently, although the binomial distribution is the first subject considered in many statistics books and the mathematics is simple, a complete explanation is not provided in this book because it is only marginally pertinent to geology and because the statistics in this book is not developed from the principles of probability. The reader who needs to work with the binomial distribution can readily follow the explanation in a standard text, for instance, that by Hoel (1962, p. 85, etc.), and will find especially helpful the thorough explanation by Li (1964, I, pp. 443–487), relating the binomial and normal distributions. If there is more than a dichotomy, for instance, if several kinds of cobbles are to be counted rather than only quartzite versus nonquartzite cobbles, the *multinomial distribution*, also treated by Hoel and Li, may be used.

The Poisson Distribution

The number of occurrences of any one item, whether a physical object or an event, may be counted in a group of objects (which may be areas, volumes, time intervals, or something else, usually of equal size or duration). The *Poisson distribution* arises when the occurrences are randomly distributed in the objects being investigated. If, for instance, equal-sized cookies were manufactured from a very large batch of raisin dough, the number of raisins per cookie would follow the Poisson distribution, provided that the raisins were mixed at random in the dough; otherwise the Poisson distribution would not be followed. As a geological example, if oil fields are distributed randomly in the world, the frequency distribution of the number of oil fields per unit area is Poisson.

The equation for the Poisson distribution is

$$f(w) = \frac{\mu^w}{w!}\, e^{-\mu},$$

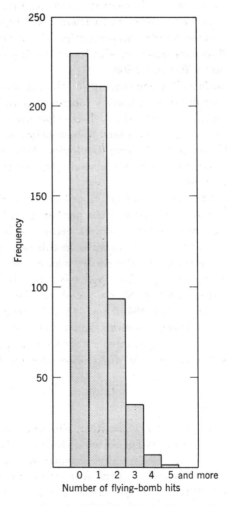

Fig. 6.2. Number of flying-bomb hits in the south of London during World War II, according to areas of 1/4 square kilometer each (after Feller, 1957, p. 150).

Table 6.2. Observed and theoretical Poisson distribution of number of flying-bomb hits in the south of
London during World War II*

Number of flying-bomb hits per $\frac{1}{4}$ sq km	Observed	Theoretical
0	229	226.7
1	211	211.4
2	93	98.5
3	35	30.6
4	7	7.2
5 and more	1	1.6
Summary items:		
Mean	0.932	0.932†
Variance	0.960	0.932
Total number of cells	576	
Total number of bombs	537	

* After Feller, 1968, p. 161.
† Assuming mean number of bombs per cell is 0.932.

where w is an observation and μ stands for both the mean and variance, which are the same for a Poisson distribution. The observation w may be any nonnegative integer, i.e., 0, 1, 2, 3, ... The sample mean $\overline{w}$ is the estimate of μ. If the sample mean is fairly large, say larger than 5, the Poisson distribution is closely approximated by the normal distribution, and 90-percent confidence limits for μ can be obtained from the relation $\overline{w} \pm 1.645\sqrt{\overline{w}}$. As an example of Poisson-distributed data, Feller (1968) cites the number of flying-bomb hits in the south of London during World War II, counted in areas of $\frac{1}{4}$ square kilometer each. (Table 6.2 and figure 6.2 give the results.) The observed and theoretical distributions are tabulated and a chi-square test (explained at the end of this section) shows that the distribution is Poisson. Two facts should be mentioned: doubtless all or nearly all the flying bombs that fell were noticed, and most people at the time believed that the hits were clustered.

As a geological example of the Poisson distribution, table 6.3 and figure 6.3 list the number of meteorites per square degree in Nebraska, counted from data tabulated by Mason (1962). The table gives both the observed distribution and the theoretical distribution for the same mean of 1.35 meteorites per square degree. We thought that, neglecting the small change in area of a square degree from north to south, the number of meteorites per square degree should be random and therefore Poisson. However, the right-hand columns in the table show that, for the entire United States and for two

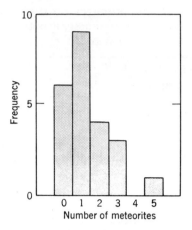

Fig. 6.3. Number of meteorites per square degree in Nebraska (date from Mason, 1962, p. 227).

selected subareas, the distribution is not Poisson because the estimated variance is too much larger than the estimated mean. For these subdivisions and for some others that we investigated, the trouble is that too many cells contain no meteorites.

The point is that these distributions are not of meteorites that fell but rather of meteorites that were found. No doubt many meteorites were not found because they are hard to spot in some areas, such as mountainous or wooded areas, or because no one was looking. The distribution is probably Poisson in Nebraska because the state is reasonably well populated, is flat, and has large plowed areas of fine soil. Not quite as bad as not looking, but also disruptive, is looking too hard; in an area of one square degree in Texas, champion meteorite finders have turned up twelve! The moral of this story is that, to find a Poisson distribution in geology, one probably has to think of a rare event that is observed every time it happens, like a White House wedding. Aside from major earthquakes, we have not been able to think of any.

The Negative Binomial Distribution

The negative binomial distribution is related to the Poisson distribution in its functional form, which is

$$f(x) = \frac{(k + x - 1)!}{x! \, (k - 1)!} \, (1 + \theta)^{-k} \left(\frac{\theta}{1 + \theta}\right)^{x},$$

where x is zero or a positive integer, and k and θ are two parameters. Having two parameters, the negative binomial distribution is somewhat more flexible

TABLE 6.3. FREQUENCY DISTRIBUTIONS OF METEORITES IN NEBRASKA, THE ENTIRE UNITED STATES, AND TWO OTHER SELECTED PARTS OF THE UNITED STATES*

Number of meteorites per square degree	Frequency				
	Nebraska				
	Observed	Theoretical	U.S.	Central U.S.	Central U.S. except N. and S. Dakota
0	6	5.98	818	69	27
1	9	8.05	179	45	34
2	4	5.43	83	30	29
3	3	2.44	23	10	10
4	0	0.82	19	12	12
5	1	0.28	12	8	8
6	0	0	5	2	2
7	0	0	1	1	1
8	0	0	2	1	1
9	0	0	1	1	1
10	0	0	0	0	0
11	0	0	0	0	0
12	0	0	1	1	1
Summary items:					
Mean	1.35	1.35†	0.55	1.51	2.05
Variance	1.60	1.35	1.36	3.63	4.11
Total number of cells	23	23	1144	180	126
Total number of meteorites	31	31	624	271	258

* Data from Mason, 1962, p. 227.
† Assuming mean number of meteorites per square degree is 1.35.

than the Poisson distribution in fitting empirical distributions. Certain mixtures of data, that are Poisson distributed with different means, can produce a negative binomial distribution. Therefore, there is some theoretical justification for using a negative binomial distribution for data that may have arisen from mixtures of Poisson distributions.

The population mean and variance for the negative binomial distributions may be estimated by the formulas

$$\theta = \frac{s^2 - \bar{x}}{\bar{x}},$$

and

$$k = \frac{\bar{x}^2}{s^2 - \bar{x}}.$$

That these estimates are inefficient is unimportant for the geologist because they are used mostly for curve fitting, which is seldom of geological interest.

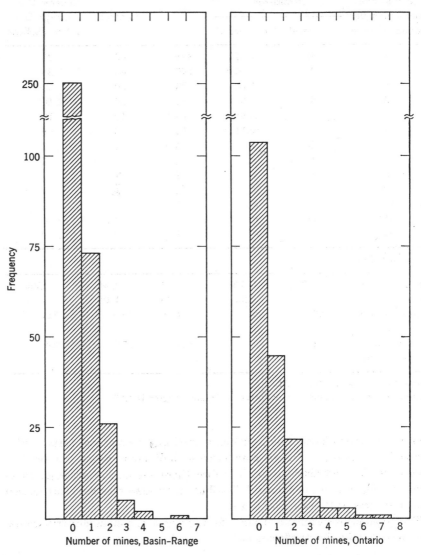

Fig. 6.4. Number of mines in cells each 1000 square kilometers in area in parts of the Basin and Range province, United States and Ontario, Canada (data from Slichter, 1960, p. 574).

Efficient estimates, if required, may be obtained by a more complicated calculation procedure explained by Haldane (1941).

Table 6.4 and figure 6.4 present two sets of data compiled by Slichter (1960). Griffiths (1966) later showed that these data follow a negative binomial distribution. The data are areal distributions of mines in the Basin and Range province in the western United States and Precambrian intrusive and sedimentary areas of Ontario, Canada. In order to obtain the counts in table 6.4, a grid with cells each 1000 square kilometers in area was superimposed on maps of the areas, and the mines in each cell were counted. The table gives the estimated parameters p and k and shows that the agreement between the observed and theoretical distributions is good. One interpretation resulting from the fact that the distributions of mines follow the negative binomial rather than the Poisson distribution is that the mines are not randomly distributed. It then becomes evident that some geological process must have been involved in the distribution. (Section 12.3 describes the situation more fully.)

TABLE 6.4. OBSERVED AND THEORETICAL NEGATIVE BINOMIAL DISTRIBUTIONS OF MINES IN PARTS OF THE BASIN AND RANGE PROVINCE, UNITED STATES, AND ONTARIO, CANADA *

Number of mines per 1000 sq km, n	Frequency, f	
	Observed	Theoretical
A. Basin and Range (154 mines, area 357 units)		
0	250	249.8
1	73	74.7
2	26	22.6
3	5	6.9
4	2	2.1
5	0	.6
6	1	.2
7	0	.1

Estimated parameters for negative binomial distribution:

p	k
0.443	0.975

B. Precambrian intrusive and sedimentary areas of Ontario (147 mines, 185 units)

Number of mines per 1000 sq km, n	Frequency	
	Observed	Theoretical
0	104	104.4
1	45	44.4
2	22	19.8
3	6	8.9
4	3	4.1
5	3	1.9
6	1	.8
7	1	.4
8	0	.2

Estimated parameters for negative binomial distribution:

p	k
0.867	0.917

* From Slichter, 1960, p. 574.

Chi-Square Test

Several theoretical frequency distributions are introduced in this chapter. The question arises of how to compare the agreement of an observed frequency distribution with a theoretical frequency distribution. The *chi-square test*, which is based on the chi-square distribution (sec. 3.7), is the usual way to make this comparison. As always, whenever the hypothesis of agreement is accepted, the risk of type II error, accepting a false hypothesis, must be recognized. Therefore, one can never be certain that a particular set of empirical observations necessarily follows a particular theoretical frequency distribution.

The chi-square test may be explained by applying it to the cobble data introduced at the beginning of this section. First, the observed frequencies O are listed (table 6.1). Then the theoretical frequencies T are obtained from the formula for the binomial distribution,

$$P(w) = \frac{n!}{w!\,(n-w)!}\,\theta^w (1-\theta)^{n-w},$$

by substituting numerical values and multiplying by the number of samples.

For instance, for $w = 9$ and an assumed value of $\theta = 0.787$, the theoretical frequency of 24.7 is obtained by the formula:

$$T = 100\left(\frac{10!}{9!\,1!}\right)(0.787)^9(0.213) = 24.7.$$

The entries for 0 to 5 cobbles in table 6.1 are combined to obtain a theoretical frequency of 4.2. (Conventionally, each cell entry must have a theoretical frequency of at least 5, but the next entry for size 6 is not included because the resulting value of 14.5 would be too large. For details see Dixon and Massey, 1969, p. 238.)

The next step is to calculate the chi-square statistic, which is

$$\chi^2 = \sum_{i=1}^{k} \frac{(O_i - T_i)^2}{T_i}.$$

The number of degrees of freedom of this statistic is equal to the number of cells minus 1 minus the number of fitted parameters. For the cobble example numerical values are substituted in this formula to obtain the value of chi-square, as follows:

$$\chi^2 = \frac{(1 - 4.2)^2}{4.2} + \frac{(9 - 10.3)^2}{10.3} + \frac{(28 - 21.7)^2}{21.7} + \cdots + \frac{(6 - 9.1)^2}{9.1} = 6.39.$$

The number of degrees of freedom is equal to $6 - 1 - 1 = 4$, because there are six cells (after combining the first five), and one parameter θ is fitted. Because the calculated value of 6.39 is smaller than the tabled value of 9.24 (table A-3) for 4 degrees of freedom the conclusion may be drawn that the cobble data agree with a binomial model.

6.2   THE LOGNORMAL DISTRIBUTION

The lognormal distribution is a continuous distribution characterized by the property that the logarithms of the observations follow a normal distribution. In this section the formula for the distribution is given, estimation of parameters is discussed, calculation of confidence limits by four different methods is explained, and the negatively skewed lognormal distribution is introduced. Chapter 16 is about the interpretation of gold data through use of the lognormal distribution.

The lognormal distribution is discussed in detail in this book for several reasons. First, this distribution is followed more or less closely by many sets of geological data, especially by trace elements, that are characterized by the

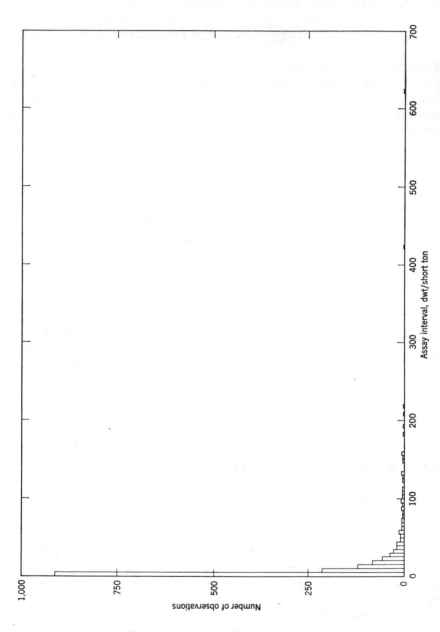

Fig. 6.5. Histogram to show approximately lognormal distribution of 1536 dwt gold assay values from the City Deep

fact that, compared with the mean value, most of the observations are small but a few are very large. Such data are encountered in diverse forms, for instance, in the incomes of individuals, magnitudes of earthquakes, gold assay values from a mine, heights of floods in a river, metal analyses from a survey in exploration geochemistry, and heights of buildings in New York City. In every one of these and many more instances most of the values are small relative to the mean, but a few are very large. The large values represent the millionaires, the catastrophic earthquakes (the Richter scale being logarithmic), the bonanza pockets of gold, the disastrous floods, the geochemical anomalies, and the skyscrapers.

Second, the detailed discussion points out that several statistical methods are available for analysis of the same set of data, that pitfalls may be found in using these methods, and that the geologist must choose among methods according to his exact objectives.

Figure 6.5 and table 6.5 present a typical lognormal distribution of gold assays from the City Deep mine, Central Witwatersrand, South Africa. The formula for the lognormal frequency distribution is

$$f(w) = \frac{1}{w\beta\sqrt{2\pi}} \exp\left[-\frac{1}{2\beta^2}(\ln w - \alpha)^2\right],$$

where the two parameters α and β^2 are the mean and variance, respectively, of the natural logarithm of the observation w. The mean of w is given by the formula

$$\mu = e^{\alpha + \frac{1}{2}\beta^2}$$

and the variance of w, by the formula

$$\sigma^2 = \mu^2(e^{\beta^2} - 1).$$

The lognormal distribution has been described in a book by Aitchison and Brown (1957), whose definition of the parameters α, β^2, μ, and σ^2 is the opposite of ours (table 6.6).

Just as there are many normal distributions depending on the values of the parameters μ and σ^2 so are there many lognormal distributions depending on the values of these parameters or on the values of the corresponding parameters α and β^2. The lognormal distribution is always skewed to the right, the amount of skewness depending only on the value of β^2, the variance of the logarithms of the observations. If the value of β^2 is small, so is the skewness, and the frequency distribution is nearly normal. In figure 6.6 are graphs of three lognormal distributions for α equal to 0, with different values of β^2. Although the distribution with β^2 equal to 0.1 is already noticeably skewed, it is more nearly normal than the other distributions with larger values of β^2.

TABLE 6.5. APPROXIMATELY LOGNORMAL FREQUENCY DISTRIBUTION OF 1536 DWT GOLD VALUES FROM THE CITY DEEP MINE, SOUTH AFRICA

Class interval (dwt/short ton)	Frequency	Cumulative frequency	Relative cumulative frequency (%)
0–5	910	910	59.24
5–10	208	1118	72.79
10–15	118	1236	80.47
15–20	80	1316	85.68
20–25	54	1370	89.19
25–30	33	1403	91.34
30–35	24	1427	92.90
35–40	13	1440	93.75
40–45	14	1454	94.66
45–50	8	1462	95.18
50–55	8	1470	95.71
55–60	10	1480	96.36
60–65	4	1484	96.62
65–70	4	1488	96.88
70–75	3	1491	97.07
75–80	1	1492	97.14
80–85	1	1493	97.20
85–90	4	1497	97.46
90–95	1	1498	97.53
95–100	7	1505	97.99
100–105	3	1508	98.18
105–110	2	1510	98.31
110–115	3	1513	98.51
120–125	2	1515	98.63
125–130	1	1516	98.70
130–135	5	1521	99.03
145–150	1	1522	99.09
150–155	1	1523	99.16
155–160	3	1526	99.35
180–185	1	1527	99.42
190–195	2	1529	99.56
205–210	2	1531	99.68
215–220	1	1532	99.72
245–250	1	1533	99.81
305–310	1	1534	99.87
420–425	1	1535	99.93
620–625	1	1536	100.00

TABLE 6.6. COMPARISON OF NOTATION FOR LOGNORMAL
DISTRIBUTION IN THIS BOOK WITH THAT OF AITCHISON AND
BROWN. (WHERE TWO SYMBOLS ARE GIVEN, THE FIRST IS
THE PARAMETER AND THE SECOND IS THE STATISTIC.)

	This book	Aitchison and Brown
Observation	w	x
Mean of observation	μ, m	α, a_1
Variance of observation	σ^2, V^2	β^2, b^2
Logarithm of observation	u	y
Mean of logarithms	$\alpha, \bar{u}$	$\mu, \bar{y}$
Variance of logarithms	β^2, s_μ^2	σ^2, v_y^2
Coefficient of variation	γ, C	η
Multiplying factor		
For geometric mean	ψ_n	ψ_n
For variance	Φ_n	χ_n

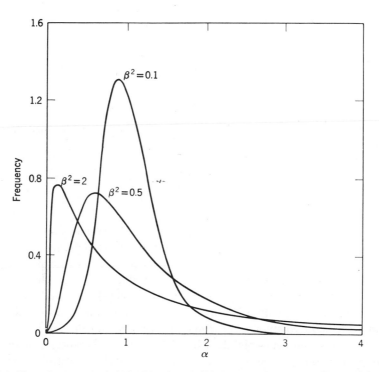

Fig. 6.6. Frequency curves of the lognormal distributions for three values of β^2 (after Aitchison and Brown, 1957, p. 10).

Parameter Estimation

The first problem is to estimate the parameters μ and σ^2. Because, for any distribution, the mean of sample means is equal to the population mean, the sample mean $\bar{w}$ is an unbiased estimate of μ. Similarly, because the mean of sample variances is equal to the population variance, the sample variance s^2 is an unbiased estimate of σ^2. Unfortunately, however, these estimates are not the most efficient ones. The most efficient estimates are obtained by first estimating α and β^2 and then estimating μ and σ^2. The estimate of α is the average value of the natural logarithms of the observations,

$$\hat{\alpha} = \bar{u} = \frac{\sum \ln{(w)}}{n},$$

which is efficient because the logarithms of the observations are normally distributed, and the sample mean is an efficient estimate of the population mean for observations so distributed. The estimate of β^2 is the variance of the logarithms of the observations,

$$s_u^2 = \frac{\sum (u - \bar{u})^2}{n - 1},$$

where u is the natural logarithm of w. Now, the unbiased efficient estimate m of μ may be obtained from the formula

$$m = e^u \psi_n(\tfrac{1}{2} s_u^2),$$

and the unbiased efficient estimate V^2 of σ^2 may be obtained from the formula

$$V^2 = e^{2u} \left[\psi_n(2 s_u^2) - \psi_n \left(\frac{n - 2}{n - 1} s_u^2 \right) \right] = e^{2u} \Phi_n(s_u^2).$$

Values of ψ are given in table A-7.

To illustrate parameter estimation for the lognormal distribution, we make use of 13 gold assay values from boreholes in the Welkom mine, South Africa (Krige, 1961, p. 18). A discussion of the geological significance of the calculations is postponed to Chapter 16. The first illustration is the use of the formulas for m and V^2. In table 6.7 column 1 lists the original 13 values in inch-pennyweights, column 2 lists the natural logarithms of these values, and column 3 lists the common logarithms. The familiar calculations for mean and variance are given below the data list. The note at the bottom of the table verifies that the calculations can be done in either natural or common logarithms because the natural logarithm of a number is 2.302585 times the

TABLE 6.7. CALCULATION OF MEANS AND VARIANCES OF OBSERVATIONS AND LOGARITHMS OF OBSERVATIONS, FOR 13 GOLD ASSAYS FROM THE WELKOM MINE, SOUTH AFRICA *

	w, in.-dwt	$u = \ln(w)$	$v = \log(w)$
	154	5.03695	2.18752
	525	6.26340	2.72016
	1560	7.35245	3.19312
	1252	7.13247	3.09760
	377	5.93225	2.57634
	70	4.24850	1.84510
	308	5.73010	2.48855
	109	4.69135	2.03743
	1221	7.10743	3.08672
	15	2.70806	1.17609
	48	3.87121	1.68124
	237	5.46806	2.37475
	68	4.21951	1.83251

$\sum w$	5,944	$\sum u$	69.76174	$\sum v$	30.29713
n	13	n	13	n	13
$\overline{w}$	457	$\overline{u}$	5.36629	$\overline{v}$	2.33055
$(\sum w)^2$	35,331,136	$(\sum u)^2$	4866.70036783	$(\sum v)^2$	917.91608624
$(\sum w)^2/n$	2,717,780	$(\sum u)^2/n$	374.361567	$(\sum v)^2/n$	70.608930
$\sum w^2$	6,108,382	$\sum u^2$	398.155209	$\sum v^2$	75.096706
SS_w	3,390,602	SS_u	23.793642	SS_v	4.487776
s_w^2	282,550	s_u^2	1.98280	s_v^2	0.37398
s_w	531	s_u	1.4081	s_v	0.6115

* Data from Krige, 1961, p. 18.

Note: $(2.302585)\overline{v} = \overline{u}$
$(2.302585)^2 s_v^2 = s_u^2$

common logarithm. In table 6.8 $\overline{u}$ and s_u^2 are repeated from table 6.7. Next m is calculated. The value of $e^{\overline{u}}$ is the natural antilogarithm of $\overline{u}$ obtained from a table available in any engineering handbook. The next line gives the value of 13 for n, and the next gives the value for the argument $\frac{1}{2}s_u^2$, used to determine the value of ψ from table A-7, using linear interpolation if necessary. By a similar calculation, the value of V^2 is obtained.

The preceding estimates of the mean and variance of observations that follow a lognormal distribution are quite complicated. Moreover, if the observations do not exactly or closely follow a lognormal distribution, these estimates may lead to biased estimates. Finney (1941) showed that the

TABLE 6.8. CALCULATION BY LOGNORMAL THEORY OF m, A STATISTIC TO ESTIMATE THE MEAN μ, AND V^2, A STATISTIC TO ESTIMATE THE VARIANCE σ^2, FOR THE DATA OF TABLE 6.7

	Calculation of m		Calculation of V^2
$\bar{u}$	5.36629	$\bar{u}$	5.36629
s_u^2	1.98280	s_u^2	1.98280
$e^{\bar{u}}$	214	$e^{2\bar{u}}$	45,825
n	13	n	13
$\frac{1}{2}s_u^2$	0.99	$\Phi_n(s_u^2)$	16.27
$\psi_n(\frac{1}{2}s_u^2)$	2.370	V^2	745,573
m	507	V	863

ordinary sample mean $\bar{w}$ is more than 90-percent efficient as an estimate of μ, provided that the coefficient of variation is less than 1.2, corresponding to a variance of the logarithms β^2 of about 0.9. Figure 6.7, based on Finney's work, gives a plot of the efficiency of the ordinary sample mean $\bar{w}$ as a function of the coefficient of variation. If the coefficient of variation is less than 1.2, we recommend using the ordinary sample mean $\bar{w}$ as the estimate of μ because little efficiency can be lost, considerable arithmetic is saved, and the question of bias does not arise. Because of the importance of the coefficient of variation, several ways of estimating it are now given.

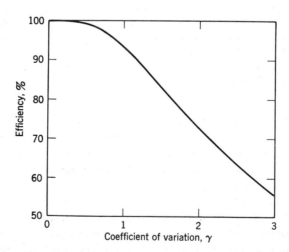

Fig. 6.7. Efficiency of the untransformed mean $\bar{w}$ as a function of the coefficient of variation for lognormally distributed observations (data from Finney, 1941).

Table 6.9. Methods to calculate C, the coefficient of variation, for lognormally distributed observations

	Calculations			
Item	(1)	(2)	(3)	(4)
Source of data	table 6.7	table 6.7	table 6.8	Krige (1960)
Sample size	13	13	13	33,031
Method of calculation	$\gamma = \sqrt{e^{\beta^2} - 1}$	$\gamma = \sigma/\mu$	$\gamma = \sigma/\mu$	$\gamma = \sqrt{e^{\beta^2} - 1}$
Calculations	$s_u^2 = 1.9828$	$s_w = 531$	$V = 863$	$s_u^2 = 1.05$
	$e^{s_u{}^2} = 7.26305$	$\overline{w} = 457$	$m = 507$	$e^{s_u{}^2} = 2.85765$
	$e^{s_u{}^2} - 1 = 6.26305$			$e^{s_u{}^2} - 1 = 1.85765$
C	2.50	1.16	1.70	1.36

In order to obtain the estimate C of the coefficient of variation γ, several methods can be used, as illustrated in table 6.9 for the Welkom data. The first method, presented in column 1, is to use the estimate of the variance of the logarithms s_u^2 to find C. The relation between these two quantities is as follows: the coefficient of variation is found by the formula

$$\gamma = \sigma/\mu,$$

which, if the expression for σ from the first part of this section is substituted, becomes

$$\gamma = \frac{\sqrt{\mu^2(e^{\beta^2} - 1)}}{\mu},$$

which reduces to

$$\gamma = \sqrt{e^{\beta^2} - 1}.$$

When the calculations in column 1 are performed by using the value of s_u^2 from table 6.7, C is found to equal 2.50. [Alternatively, the value of C may be obtained from table A-1 of Aitchison and Brown (1957, pp. 154–155).] Another estimate of s_u^2, based on a large number of assays from the mine as a whole, may be used in place of the estimate based on only 13 assays. Such an estimate, based on 33,031 assay values (Krige, 1960, annexures 1 and 2), is 1.05, which yields a much smaller C of 1.36 when the calculations of column 4 are performed. A second way to estimate C is to divide the estimated standard deviation of the original observations s_w by the estimated mean of the original observation $\overline{w}$; the result (column 2) is 1.16. A third way is to divide m by V (column 3) to yield the result 1.70.

All of the values for C in table 6.9 are different. Because it is based on a large number of mine samples, the value in column 4 should be reliable for the mine as a whole; whether it is applicable to the 13 assays depends on whether they are of typical ore, or, in statistical terms, on whether they are a random sample from the same population. Because they are based on only 13 observations with a large standard deviation, the first three values are different from one another, as well as from the value in column 4. The conclusion is that the coefficient of variation cannot be reliably estimated from so few lognormally distributed observations.

Calculation of Confidence Limits for the Population Mean

The next step is to calculate confidence limits for the population mean by one of several methods. Choice among the methods depends on the number of observations available, background information on consistency and nature of the mineralization, whether both upper and lower confidence limits or only lower confidence limits are required, and the exact purpose. The Welkom data are again utilized to explain the various methods. Although calculation of confidence limits, as summarized in tables 6.10 to 6.17, may appear rather involved, once the calculations are organized, they can be performed without too much trouble.

If a population variance of logarithms β^2 has been established from a large number of measurements and if the sample in question is known to come from this population, confidence limits can be calculated for the logarithms and then transformed into confidence limits for the original observations. As illustrated in table 6.10 for the Welkom assays, the initial data (lines 1 and 2) are a value of β^2 based on some 33,000 observations (table 6.9) and a value of $\bar{u}$ calculated from the 13 observations. In lines 3 to 8 of table 6.10, confidence intervals are calculated for the logarithmic mean α by using the familiar method (sec. 4.2) for normally distributed observations with the population variance known. Next, in lines 9 to 15, these values are transformed to confidence limits for the untransformed mean μ. Finally, in lines 16 to 17, the point estimate m of μ is calculated, based on the β^2 value.

If the population variance of logarithms β^2 is unknown, it must be estimated. Then several methods of calculation can be applied. The first one, the calculation of a confidence region for μ and σ^2 by a method explained by Mood (1950, p. 229), is applied to the Welkom data in table 6.11. Mood's method depends on the fact that, because the distributions of the sample mean and the sum of squares are independent for samples from a normal distribution, a joint confidence region for the parameters α and β^2 can be obtained by finding the values of α and β^2 that satisfy the joint inequalities

$$| \bar{u} - \alpha | < \frac{1.645\beta}{\sqrt{n}},$$

Table 6.10. Calculation of confidence limits for lognormally distributed observations with β^2 (the population variance of logarithms) assumed to be known

Line	Notation	Value	Calculations	Comments
1	β^2	1.05	From table 6.9	
2	$\bar{u}$	5.36629	From table 6.7	
3	$\beta^2/13$	0.0808		
4	$\sqrt{\beta^2/13}$	0.2842	$\sqrt{0.0808}$	Standard error of mean
5		1.645	From table A-2	5% point of standardized normal distribution
6		0.4675	1.645×0.2842	
7	α_L	4.8988	$5.3663 - 0.4675$	Lower confidence limit for α
8	α_U	5.8338	$5.3663 + 0.4675$	Upper confidence limit for α
9	e^{α_L}	134	From exponential table	
10	e^{α_U}	341	From exponential table	
11	$(n-1)/2n$	0.4615	$(13-1)/26$	
12	$[(n-1)/2n]\beta^2$	0.4846		
13	$\exp\left(\dfrac{n-1}{2n}\right)\beta^2$	1.6235	From exponential table	
14	μ_L	218	134×1.6235	
15	μ_U	554	341×1.6235	
16	$e^{\bar{u}}$	214	From table 6.8	
17	m	347	214×1.6235	

and

$$\chi^2_{95\%} < \frac{SS}{\beta^2} < \chi^2_{5\%}.$$

The solution will produce an 81 percent (90 percent squared) confidence region. The first step, illustrated in table 6.11, is to find upper and lower limits for β^2. The second step is to apply these limits to find upper and lower limits for desired values of α between the upper and lower limits for β^2. In table 6.12, four particular values for α have been calculated. Then, in figure 6.8, these four particular α, β^2 values have been plotted in α, β^2 coordinates.

Table 6.11. Calculation of parameter estimates and the joint confidence region for lognormally distributed observations by Mood's method: Part 1, calculation of upper and lower confidence limits for β^2

Line	Notation	Value	Calculation	Comments
1	$\bar{u}$	5.3663	From table 6.7	
2	SS	23.793642	From table 6.7	
3	n	13	From table 6.7	
4	$\sqrt{n}$	3.6056		
5		1.645	From table A-2	5% point of standardized normal distribution
6		5.23	From table A-3	95% point of χ^2, d.f. = 12
7		21.03	From table A-3	5% point of χ^2, d.f. = 12
8	$SS/\chi^2_{5\%}$	1.1314	23.793642/21.03	β^2_L
9	$SS/\chi^2_{95\%}$	4.5495	23.793642/5.23	β^2_U
10		0.4562	$1.645/\sqrt{13}$	

Additional values to complete the graph were plotted but are not listed in table 6.12.

Because each α, β^2 point has a corresponding μ, σ point, the region in the α, β^2 coordinates can be transformed into a region in the μ, σ coordinates. In table 6.13 the required transformation is calculated for 8 points. In columns 1 and 2 are given the α, β^2 values at the 8 points to be transformed. In columns 3, 5, 6, and 7 are given intermediate terms combined according to

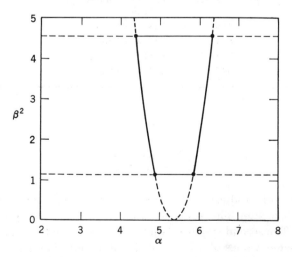

Fig. 6.8. Confidence region in α, β^2 space for lognormally distributed gold observations from the Welkom mine, South Africa (data from Krige, 1961, p. 18).

TABLE 6.12. CALCULATION OF PARAMETER ESTIMATES AND THE JOINT CONFIDENCE REGION FOR LOGNORMALLY DISTRIBUTED OBSERVATIONS BY MOOD'S METHOD: PART 2, CALCULATION OF UPPER AND LOWER CONFIDENCE LIMITS FOR VARIOUS VALUES OF α

β^2	β	$\dfrac{1.645\beta}{\sqrt{13}}$	$\alpha_L = \bar{u} - \dfrac{1.645\beta}{\sqrt{13}}$	$\alpha_U = \bar{u} + \dfrac{1.645\beta}{\sqrt{13}}$	$\dfrac{\beta^2}{2}$
1.1314	1.064	0.485	4.881	5.851	0.5657
4.5495	2.133	0.973	4.393	6.339	2.2748
2.0000	1.414	0.645	4.721	6.011	1.0000
3.0000	1.732	0.790	4.576	6.156	1.5000

the formulas at the beginning of this section to give the μ, σ values at the 8 points in columns 4 and 8. These points define the 81-percent confidence region graphed in figure 6.9. For comparison the 64-percent confidence region and the points $\bar{w}$, s_w and m, V are also graphed in this figure. The interpretation of the 81-percent confidence region is as follows. At a confidence level of about 90-percent the expected value of μ is between 232 and 5508; these are the μ values plotted as solid circles on the figure. Similarly, with a confidence of about 90 percent, the value of σ is between 336 and 53,824. With a confidence of exactly 81 percent the point μ, σ is expected to lie within the confidence region.

That the point $\bar{w}$, s_w lies outside the confidence *region* is not surprising because this point is calculated only from a sample; however, the individual $\bar{w}$ and s_w values lie within their respective confidence *intervals*. Notably the upper part of the confidence region extends to extremely high values, reflecting the uncertainty about this part of the region.

TABLE 6.13. TRANSFORMATION OF α, β^2 POINTS TO CORRESPONDING μ, σ POINTS

(1)	(2)	(3)	(4)	(5)	(6)	(7)	(8)
α	β^2	$\alpha + \frac{1}{2}\beta^2$	μ	μ^2	$e^{\beta^2} - 1$	σ^2	σ
4.881	1.1314	5.447	232	53,852	2.09999	113,100	336
5.851	1.1314	6.417	612	374,741	2.09999	786,900	887
4.393	4.5495	6.668	787	619,369	93.58511	57,963,700	7,613
6.339	4.5495	8.614	5508	30,338,064	93.58511	2,839,191,000	53,284
4.721	2.0000	5.721	305	93,153	6.38906	595,200	771
6.011	2.0000	7.011	1109	1,229,350	6.38906	7,854,000	2,802
4.576	3.0000	6.076	435	189,473	19.08554	3,616,000	1.902
6.156	3.0000	7.656	2113	4,466,000	19.08554	85,240,000	9,233

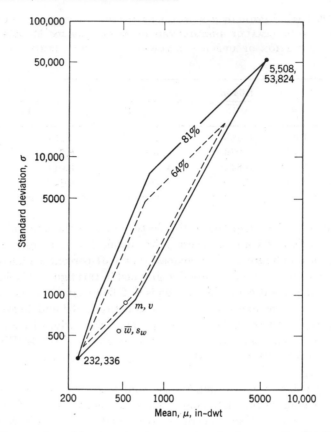

Fig. 6.9. 81-percent confidence region in μ, σ space for lognormally distributed gold observations from the Welkom mine, South Africa.

In both of the next two methods confidence limits are calculated by use of the t-distribution (sec. 4.4), proceeding as though the observations were normally distributed. The value of t for the 5-percent level with 12 degrees of freedom is obtained from table A-4. In the first of these two methods, given in table 6.14, the mean and variance of the untransformed observations from table 6.7 are used, and the calculations follow the familiar form of table 4.2.

In the second of these two methods, given in table 6.15, the mean and variance of the transformed observations from table 6.8 are used. In the first eight lines the 90-percent confidence limits for α are found, following the form in table 6.14. In lines 9 and 10 the powers to the base e of the values in lines 7 and 8 are calculated. Finally, the multiplier for the geometric mean, line 11, is applied to the values in lines 9 and 10 in order to obtain the confidence limits in lines 12 and 13.

Table 6.14. Calculation of confidence limits for log-normally distributed observations by use of the t-distribution (method 1 is based on mean and variance of untransformed observations from table 6.7)

Line	Notation	Value	Calculation
1	$\overline{w}$	457	From table 6.7
2	s_w	531	From table 6.7
3	$\sqrt{n}$	3.6056	$\sqrt{13}$
4	$t_{5\%}$	1.782	From table A-4 with 12 d.f.
5	$s_w/\sqrt{n}$	147	531/3.6055
6	$t_{5\%}s_w/\sqrt{n}$	262	1.782 × 147
7	μ_L	195	457 − 262
8	μ_U	719	457 + 262

The last method for calculating confidence limits was devised by Sichel (1952, 1966), one of the pioneer workers in the application of statistical methods to the valuation of lognormally distributed gold assays from South African mines. In Sichel's method, calculated in table 6.16, the value of SS_u for the transformed variable u is obtained from table 6.7 and the value of m is obtained from table 6.8. When SS_u is divided by n (line 3 of table 6.16), a

Table 6.15. Calculation of confidence limits for lognormally distributed observations by use of the t-distribution (method 2, based on mean and variance of transformed observations from table 6.8)

Line	Notation	Value	Calculation
1	$\overline{u}$	5.36629	From table 6.7
2	s_u	1.4081	From table 6.7
3	$\sqrt{n}$	3.6056	From table 6.11
4	$s_u/\sqrt{n}$	0.3905	
5	$t_{5\%}$	1.782	From table A-4 with 12 d.f.
6	$t_{5\%}(s_u/\sqrt{n})$	0.6959	
7	$\overline{u} - t_{5\%}(s_u/\sqrt{n})$	4.6704	
8	$\overline{u} + t_{5\%}(s_u/\sqrt{n})$	6.0622	
9	$e[\overline{u} - t_{5\%}(s_u/\sqrt{n})]$	106.7	
10	$e[\overline{u} + t_{5\%}(s_u/\sqrt{n})]$	429.3	
11	$\psi n(\tfrac{1}{2}s_u^2)$	2.370	From table 6.8
12	μ_L	253	
13	μ_U	1017	

TABLE 6.16. CALCULATION OF CONFIDENCE LIMITS FOR LOGNORMALLY DISTRIBUTED OBSERVATIONS BY A METHOD DEVISED BY SICHEL

Line	Notation	Value	Calculation	Comments
1	m	507	From table 6.8	
2	SS_u	23.793642	From table 6.7	
3	SS_u/n	1.83	23.793642/13	V (in Sichel's notation) = SS_u/n
4		0.5124	From Sichel's table B†	Lower multiplying factor
5		4.015	From Sichel's table B†	Upper multiplying factor
6	μ_L	260	507 × 0.5124	
7	μ_U	2036	507 × 4.015	

† Sichel (1966).

value of 1.83 is obtained. The next step is to enter Sichel's table B (1966, p. 14) by using n and V from line 3 to obtain, by interpolation, two multiplying factors (lines 4 and 5), which are used to obtain the confidence limits in lines 6 and 7 of table 6.16.

The estimated means and confidence limits obtained by the five methods are summarized in table 6.17, which shows above all that wide differences are obtained depending upon which method is used. However, the lower bounds agree better than the upper ones, a fortunate circumstance for practical mining application where the main concern is that the minimum expected

TABLE 6.17. COMPARISON OF CONFIDENCE LIMITS FOR LOGNORMALLY DISTRIBUTED OBSERVATIONS CALCULATED BY APPLYING FIVE DIFFERENT METHODS TO EXAMPLE DATA FROM THE WELKOM MINE, SOUTH AFRICA

Source table	Method	Estimated mean	Confidence limits	
			Lower	Upper
6.10	Apply normal theory; assume β^2 known	347	218	554
6.13	Calculate confidence region	507	232	5508
6.14	Assume $\overline{w}$ normally distributed; use t based on $\overline{w}$ and s_w^2	457	195	719
6.15	Assume m normally distributed; use t based on m and V^2	507	253	1017
6.16	Use Sichel's approximation	507	260	2036

grade of ore be estimated correctly. Of the five methods the best is the application of normal theory under the assumption that β^2 is known (table 6.10), provided that (a) a close estimate of β^2 is available from many observations, and (b) a convincing case can be made that the new observations in question are from the same population. Otherwise, we favor Mood's method of calculating a confidence region (tables 6.11 to 6.13). If the observations follow or seem to follow a lognormal distribution with β^2 larger than 0.9, the two methods that apply the t-distribution, under the assumption that the observations follow a normal distribution (tables 6.14 and 6.15), are inefficient. However, if the observations in fact depart from a lognormal distribution, the use of one of the t-distribution methods will yield conservative results because these methods are robust. Sichel's approximation (table 6.16) is convenient to use, particularly if an electronic computer is not available to implement Mood's method, although problems too involved to describe here can arise (Link, Koch, and Schuenemeyer, 1970).

The Negatively Skewed Lognormal Distribution

The negatively skewed lognormal distribution is one whose frequency curve is the translated mirror image of a lognormal distribution reflected about its natural lower limit, as discussed by Aitchison and Brown (1957, p. 16). Figure 6.10 gives the frequency curve for a negatively skewed lognormal distribution of a variable, $3 - x$, where 3 is the natural upper limit, and x has a lognormal distribution with parameters $\alpha = 0$, $\beta^2 = 0.25$.

Many negatively skewed distributions are found in geology, although not all of these are lognormal. Table 6.18 and figure 6.11 present an example of 40 potassium analyses of granites from South Africa, whose natural upper limit

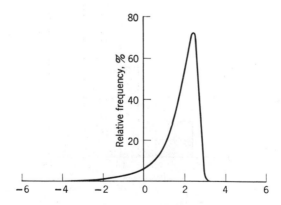

Fig. 6.10. Frequency curve of a negatively skewed lognormal distribution (after Aitchison and Brown, 1957, p. 16).

Table 6.18. Frequency distribution of potassium in granites from South Africa. (After Ahrens, 1963, p. 935.)

Class interval	Frequency	Cumulative frequency	Relative cumulative frequency (%)
1.0–1.6	2	2	5.0
1.6–2.2	1	3	7.5
2.2–2.8	3	6	15.0
2.8–3.4	5	11	27.5
3.4–4.0	9	20	50.0
4.0–4.6	18	38	95.0
4.6–5.2	2	40	100.0

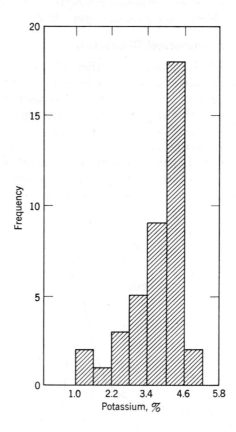

Fig. 6.11. Histogram of potassium in granites from South Africa (after Ahrens, 1963, p. 935).

might be taken as 5.4. In general, negatively skewed distributions are expected if many of the observations are at or near an upper limit: which may correspond to 100 percent, to the composition of a chemical element or other constituents in a pure mineral, or some other upper limit set by some natural process. Thus, quartz analyses in a glass-sand deposit, iron analyses in a hematite deposit, and chromium analyses in chromite may, but need not be, negatively skewed.

6.3   TRANSFORMATIONS

A *transformation* is a function of an observation that defines a new observation, that is,

$$u = f(w),$$

where w is the original observation, f is the transforming function, and u is the new observation. A familiar example is the linear transformation of inches to centimeters by multiplying by the constant 2.54. However, in this section the only kinds of transformation considered are nonlinear ones, that is, those for which a plot of u against w is not a straight line. An example is the logarithmic transformation whereby each observation is changed to its logarithm, according to the formula,

$$u = \ln (w).$$

The basic purpose of a nonlinear transformation is to change the shape of the frequency distribution. A linear transformation does not change the shape of the distribution, but only the scale.

There are three purposes for changing from one frequency distribution to another by a nonlinear transformation. The most important is to stabilize the variance. All transformations discussed in this book are primarily for this purpose, although, fortunately, they also tend to serve the other two purposes. The stabilization of variance may be explained as follows: An observation may be written (sec. 5.1) as its mean value plus a random fluctuation, that is,

$$w = \mu + e.$$

This notation implies that the mean value of e is 0, but its variance may or may not be equal to a constant, which is the requirement for stability. However, if the variance, σ_e^2, of e depends upon μ in some fashion, it may be possible to find a transformation,

$$u = f(w),$$

such that in the transformed equation,

$$u = \mu^* + e^*,$$

where the variance of e^* is a constant independent of the value of μ^*. Constant variance is also named *homoscedasticity*. As an example of nonconstant variance, consider the fluctuations associated with low-grade and with high-grade gold assays. It is known that the variability of gold assays from low-grade ore is much less than the variability of gold assays from high-grade ore. Therefore the average size of the fluctuation e depends upon the grade of the ore. For some gold ores a constant variance may be obtained by taking the logarithms of the observations.

The next most important purpose of a nonlinear transformation is to obtain additivity (sec. 5.11). For instance, if

$$w = e^y e^z e^e,$$

the factors y and z are clearly not additive because the effect of a change in z depends upon the value of y. However, through a transformation the factors can be made additive; in fact, through a logarithmic transformation the result is obtained that

$$u = y + z + e.$$

The third basic purpose is to transform the observations so that they follow a normal distribution. There are two advantages. The first is that, if the observations are normally distributed, the sample mean is an efficient estimate of the population mean (sec. 3.6); but if the observations follow a highly skewed lognormal distribution (sec. 6.2), the sample mean is a rather inefficient estimate of the population mean. The second is that all important statistical procedures for confidence intervals and hypothesis tests assume underlying normality. However, for these procedures departures of data from normality are seldom serious because many of the procedures are robust and work rather well even with observations that are not exactly normally distributed, and the mean is likely to be a quite efficient estimator except for frequency distributions with very long tails. The techniques that are rather robust are t, F as used in the analysis of variance, and any statistic based on sample means. Those statistics discussed so far that are less robust are chi-square and F as used to compare variances.

Transformations should never be made blindly, as a matter of course, because there are disadvantages that may be serious. The most important is that transformations may lead to biased estimates; for example, in estimating the mean of a lognormal distribution, greater efficiency may be obtained by transforming the observations to logarithms, but a bias is introduced because the antilogarithm of $\bar{u}$ is not on the average the population mean μ, as has been explained in detail in section 6.2. For lognormally distributed observations, the bias may be removed because a great deal is known about the lognormal and normal distributions. For observations following other distributions such bias may also be removable in principle, but practical methods to do so may not have been devised. A second disadvantage of transformations is that their use always requires more calculation. Even if an appropriate transformation is known, such additional work may not be economically justified because the efficiency of an estimate based on original untransformed data may be adequate.

In summary, specific advice on when to transform will be offered as the subject is developed. In general, it may be useful to transform observations when the conclusions based on the transformed scale can be understood, when biased estimates are acceptable, or when the amount of bias can be estimated and removed because the details of the distributions are known.

Some Common Transformations

For all of the distributions introduced in this chapter, transformations have been devised to make the distributions normal or nearly so. The only one widely used for geological data is the lognormal transformation whereby skewed distributions are more or less normalized by taking logarithms.

Transformations for observations from discrete distributions are not explained in this book because these data are uncommon in geology. Li (1964, I, pp. 505–526) explains the angular transformation for binomial data and the square-root transformation for Poisson data, but points out that transformations of such data are seldom helpful in practice.

However, one other transformation, the normal-score transformation for ranked data, has sufficient geological application to merit a brief account. Ranked data consist of items ordered according to preference, shades of color, intensity of weathering, or any other attribute not amenable to exact measurement. The procedure may be explained through a fictitious illustration. Suppose that four geologists rank six western states according to favorability for oil exploration and that someone wants to know if their judgments are mutually consistent. Table 6.19 lists the rankings, showing that geologist 1, for example, likes California the most and Arizona the least. Intuitively, one thinks it easier to rank the states at the top and bottom of the list than those in the middle; accordingly the normal-score transformation replaces the ranks by the standardized normal deviates obtained from table A-2. Table 6.20 shows that the differences between items near the center of the list are smaller with the normal-score transformation than those near the extremes.

TABLE 6.19. FICTITIOUS RANKINGS (BY FOUR GEOLOGISTS) OF
SIX STATES ACCORDING TO FAVORABILITY FOR OIL EXPLORATION

| | Geologist | | | |
State	1	2	3	4
A. Untransformed data				
Washington	3	2	1	1
Oregon	2	3	2	3
California	1	1	3	2
Idaho	4	5	6	6
Nevada	5	6	5	4
Arizona	6	4	4	5
B. Normal-score transformed data				
Washington	-0.20	-0.64	-1.27	-1.27
Oregon	-0.64	-0.20	-0.64	-0.20
California	-1.27	-1.27	-0.20	-0.64
Idaho	0.20	0.64	1.27	1.27
Nevada	0.64	1.27	0.64	0.20
Arizona	1.27	0.20	0.20	0.64

TABLE 6.20. COMPARISON OF RANKED AND NORMAL SCORE DATA
FOR FICTITIOUS ILLUSTRATION OF TABLE 6.19

Ranked data		Normal score data	
Rank	Difference between adjacent ranks	Normal score	Difference between adjacent normal scores
1		-1.27	
	1		0.63
2		-0.64	
	1		0.44
3		-0.20	
	1		0.40
4		0.20	
	1		0.44
5		0.64	
	1		0.63
6		1.27	

TABLE 6.21. ANALYSES OF VARIANCE FOR FICTITIOUS DATA OF TABLE 6.19

Source of variation	Sum of squares	Degrees of freedom	Mean square	F	$F_{10\%}$
A. Untransformed data					
States	5.60	5	11.20	12.00	2.27
Residual variability	14.00	15	0.93		
B. Normal-score transformed data					
States	12.50	5	2.50	9.37	2.27
Residual variability	4.00	15	0.27		

Statistical procedures previously discussed can be applied either to the original ranked data or to the transformed data. In table 6.21 the randomized-block design of the analysis of variance (sec. 5.7) is performed for both the original and the transformed fictitious data; the geologists correspond to four treatments and the states to six replications. Because the calculated F-values are larger than the tabled ones, for both kinds of data the judging is deemed inconsistent. However, the calculated F-value is smaller for the transformed than for the original data, a fact which reflects the concept that it is harder to rank the states in the middle of the list.

6.4 GRAPHICAL REPRESENTATION OF DATA

The old saw that a picture is worth a thousand words is as true for under-standing geological data as for anything else. Fortunately, geologists like graphs, so all that should be necessary is to point out some special graphs that statisticians have devised for the representation of data.

Graphs are easy to plot and easy to read. If the observations are numerous, computer plots (sec. 10.7 and 17.4) are convenient. Graphs illuminate various facets of data not readily displayed by other means of presentation, in particular extremely large and extremely small values, and, with enough observations, the general shape of a frequency distribution and whether there is one or more than one mode.

This section explains how to plot observations on some special graph papers (available from firms such as Keuffel and Esser, and Dietzgen, or from university bookstores). The techniques are purely for meaningful graphic represen-tation, and they are associated with informal inferences and indications rather than with formal statistical inference.

TABLE 6.22. PLOTTING PERCENTAGES FOR PHOSPHATE DATA OF TABLE 2.1

Class interval ($\% P_2O_5$)	Relative cumulative frequency ($\%$)	Plotting percentages
14–16	0.45	0.30
16–18	0.90	0.74
18–20	4.47	4.31
20–22	13.84	13.67
22–24	33.48	33.28
24–26	57.60	57.36
26–28	82.60	82.32
28–30	95.99	95.69
30–32	99.10	98.81
32–34	100.00	99.70

Normal Probability Graph Paper

Normal probability graph paper is scaled so that if observations are normally distributed their relative cumulative frequency in percent plots as a straight line. For fewer than 100 observations it is desirable but not essential to replace the relative cumulative frequency (r.c.f.) by a plotting percentage obtained by the formula

$$\text{plotting percentage} = 100 \times \frac{3(\text{r.c.f.}) - 1}{3n + 1},$$

where n is the total number of observations. In table 6.22 the plotting percentages are calculated for the phosphate data from table 2.1. For the first class interval the value is

$$100 \times \frac{3(1) - 1}{3(224) + 1} = 0.30.$$

In figure 6.12 the upper bound of the class interval is graphed against the plotting percentage.

If observations are normally distributed, their plot on normal probability graph paper defines a straight line. The straight line fitted by eye to the phosphate data does not exactly fit the points, but it is close enough to them to indicate that they are distributed nearly normally. The mean of the observations is estimated by reading the 50-percent point on the straight line; the value obtained of 25.2 agrees closely with the computed mean of 25.13 (sec. 3.3). The standard deviation of the observations is estimated by sub-

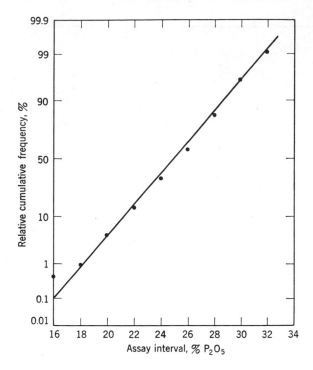

Fig. 6.12. Plot on normal probability graph paper of phosphate data from table 2.1.

tracting the 16-percent point from the 84-percent point and dividing the result by two. For the phosphate data the 16-percent point is 22.2, the 84-percent point is 28.1, the difference is 5.9, and the estimated standard deviation is 2.95, the same as the computed value (sec. 3.3).

Lognormal Probability Graph Paper

Lognormal probability graph paper is like normal probability graph paper except that it is scaled so that, for lognormally distributed observations, their relative cumulative frequency in percent plots as a straight line. In figure 6.13 the 13 Welkom observations from table 6.5 are plotted on this paper. The straight line fitted by eye shows that the points are more or less lognormally distributed, although not all of them are close to the line. The advantage of using the lognormal probability paper is that, by making a quick plot, it is easy to see if the observations are more or less lognormal. The mean and standard deviation cannot be obtained from this graph, however, as they can be from graphs on normal probability paper.

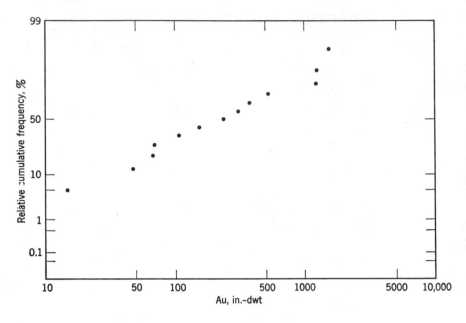

Fig. 6.13. Plot on lognormal probability graph paper of 13 gold observations from the Welkom mine, South Africa.

The Full Normal Plot (FUNOP)

The *FU*ll *NO*rmal *P*lot was devised by Tukey (1962, p. 22) to investigate a set of data for the occurrence of unusually large or small observations to determine which are likely to be outliers that belong in the data and which are probably mistakes. The procedure is similar to making an ordinary plot on normal probability paper, but a new quantity, the Judd, is calculated to make interpretation of the pattern of observations easier. The Judd is equal to the deviation of the observations from the median divided by the standard normal deviate corresponding to the plotting-point percentage. If the observations were distributed perfectly normally, the Judd would be nearly equal to their standard deviation. By inspecting an ordered list of the two-thirds of the Judds associated with the largest one-third and smallest one-third of the original observations one can see if any of the observations seem to be unusually large or small.

The calculations are explained with the 13 Welkom gold observations as an example. In table 6.23 the observations are ordered, and a plotting percentage is calculated for each. For each plotting percentage corresponding to the upper and lower one-third of the observations the standard normal deviate,

TABLE 6.23. CALCULATIONS FOR FULL NORMAL PLOT (FIG. 6.14) FOR 13 GOLD OBSERVATIONS FROM THE WELKOM MINE, SOUTH AFRICA

(1)	(2)	(3)	(4)	(5)	(6)	(7)	(8)
	Observation						
Rank	Raw, w	Logarithm, $u = \ln(w)$	Plotting percentage	s.n.d.	$u - $ median	Judd $= (6)/(5)$	Ordered Judds
1	15	2.70806	0.050	− 1.6449	− 2.76000	1.678	1.146
2	48	3.87121	0.125	− 1.1503	− 1.59685	1.388	1.330
3	68	4.21951	0.200	− 0.8416	− 1.24855	1.484	1.388
4	70	4.24850	0.275	− 0.5978	− 1.21956	2.040	1.447
5	109	4.69135	0.350				
6	154	5.03695	0.425				
7 (median)	237	5.46806	0.500				
8	308	5.73010	0.575				
9	377	5.93225	0.650				
10	525	6.26340	0.725	0.5978	0.79534	1.330	1.484
11	1221	7.10743	0.800	0.8416	1.63937	1.948	1.678
12	1252	7.13647	0.875	1.1503	1.66441	1.447	1.948
13	1560	7.35245	0.950	1.6449	1.88439	1.146	2.040

s.n.d., corresponding to 100 minus the plotting percentage, is obtained from table A-2. In column 6 the deviation of each observation from the median, 5.46806, is calculated, and in column 7 the Judd is calculated by the formula

$$\text{Judd} = \frac{u - \text{median}}{\text{s.n.d.}}.$$

The median, 1.466 (halfway between 1.447 and 1.484), of the Judds is approximately the standard deviation of the observations. If any Judd is more than twice the median, the corresponding observation is suspect, as well as those whose distance from the median is greater than that observation. None of the logarithms of the Welkom observations is suspiciously large or small. In figure 6.14 the Judds are plotted; the plot shows graphically that none of the observations is more than two Judds above the median.

Extreme-Value Graph Paper

Suppose that the daily rainfall at a station is recorded for a year. If the maximum rainfall for the year is taken as an observation and the process is repeated for 50 years, 50 observations, which may reasonably be expected to follow an extreme-value distribution, are collected. An essential part of the concept is that the observations in the distribution are ones *selected* from the

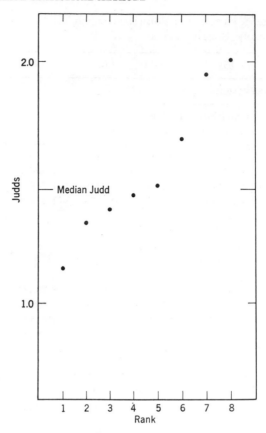

Fig. 6.14. Full Normal Plot (FUNOP) of Judds calculated for the 13 gold observations from the Welkom mine, South Africa.

total number of daily rain-gauge measurements as the largest for each year. Similarly, a radio-tube manufacturer might make life tests of tubes in samples of size 25. If the longest life in each batch of tubes is recorded, this value might also be expected to follow an extreme-value distribution.

Perhaps the extreme-value data of most interest in geology are annual flood records because so many geological processes are associated with unusually high floods. The observations in the extreme-value distribution may be the highest annual discharges or the highest annual floods. Note that these values are *selected* from observations made daily (or at some other period).

A book on extreme-value distributions by Gumbel (1958) explores the subject in great detail; we give only a brief introduction. One practical application is in bridge design, to predict from flood records how often a destructive flood is liable to occur. Another practical application is in rock mechanics,

TABLE 6.24. CALCULATIONS FOR PLOT ON EX-
TREME VALUE GRAPH PAPER OF MAXIMUM DAILY
DISCHARGE YEARLY FOR LAUREL HILL CREEK,
AT URSINA, PENNSYLVANIA*

Year	Maximum daily discharge (ft^3/sec)	Plotting percentages
1923	2070	1.79
1929	2170	4.46
1932	2370	7.14
1931	2780	9.82
1950	2900	12.50
1944	3100	15.18
1919	3110	17.86
1921	3110	20.54
1914	3220	23.21
1921	3330	25.89
1942	3580	28.57
1934	3660	31.25
1925	3780	33.93
1935	4020	36.61
1915	4140	39.29
1929	4270	41.96
1925	4300	44.64
1949	4360	47.32
1920	4410	50.00
1946	4470	52.68
1933	4860	55.36
1937	4860	58.04
1939	5010	60.71
1937	5010	63.39
1948	5150	66.07
1916	5630	68.75
1918	5630	71.43
1927	5630	74.11
1942 ˌ	5660	76.79
1947	5830	79.36
1917	5970	82.14
1927	5970	84.82
1945	6000	87.50
1940	6510	90.18
1924	8090	92.86
1941	9400	95.54
1936	10300	98.21

* Data from U.S. Geological Survey Circ. 204, p. 21.

to predict from train gauge measurements when rock failure is liable to occur.

For very large samples the distribution of extreme-value observations depends only upon the tails of the frequency distribution from which the observations are drawn. Thus, for the two illustrations, the extreme-value distributions are independent of the distributions of *all* rainfall and radio

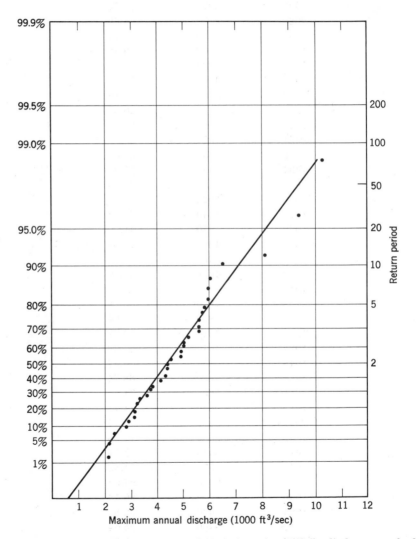

Fig. 6.15. Plot on extreme value graph paper of maximum daily discharge yearly for Laurel Hill Creek, Ursina, Pennsylvania (data from U.S. Geol. Survey Circ. 204, p. 21).

tube measurements. The distribution of extreme-value observations is named the *asymptotic extreme-value distribution* and has the formula

$$F(w) = \exp\{-\exp[-\alpha(w - a_n)]\} \qquad \alpha > 0,$$

where $F(a_n) = 0.3678$, $F(a_n - 1/\alpha) = 0.066$, and w is an observation. (Actually there are three similar distributions, but we work only with the one most pertinent for geological data.)

The use of extreme-value graph paper (available from Technical and Engineering Aids for Management, 104 Belrose Avenue, Lowell, Mass.) may be illustrated by plotting maximum daily discharge yearly from 1914 to 1950 for Laurel Hill Creek at Ursina, Pennsylvania (*U.S. Geological Survey Circular* 204, p. 21). In table 6.24 the maximum flood discharges are ordered in size from smallest to largest. For each observation a plotting percentage is calculated and plotted in figure 6.15. Inspection shows that the points lie near to a straight line, and a line was fitted to them by eye.

From the graph the return period can be read—that is, how long one expects it to be before a maximum annual discharge of a certain magnitude reoccurs; for instance, on Laurel Hill Creek, one expects an annual maximum discharge of 8500 cubic feet per second to occur about every 20 years. For bridge design purposes the data could be extrapolated to yield a return period of say, 1000 years; although highly uncertain, the prediction is better than nothing.

Other Graphical Methods

Many other graphical methods, not considered in this book, have been devised for representing data. A good source of information is an article by Mosteller and Tukey (1949).

6.5 MIXED-FREQUENCY DISTRIBUTIONS

Up to this point the geological data reviewed in this book could be related to one of the common theoretical frequency distributions; for instance, the phosphate observations (sec. 2.1) could be referred to the normal distribution and some gold data (sec. 6.2) to the lognormal distribution. However, many sets of geological data follow an empirical distribution that may be represented as a mixture of two or more of the common theoretical distributions. This mixture leads to the complication of having to estimate not only the parameters of the component theoretical distributions but also their proportions.

Mixed-frequency distributions may arise in nature in many ways; for instance, stream-carried sediments, whose particle-size distribution was approximately normal, could be introduced into a basin of distribution. Then, later, volcanic ash may be deposited in the same basin with another particle-size distribution, perhaps also normal but with different parameters. If the two kinds of sediments were mixed by waves and currents, the particle-size distribution of the resulting mixed sediment could be appropriately described by a mixed-frequency distribution, composed of two component distributions, each of them normal.

If the component distributions do not differ very much in mean and variance, if there is no theoretical basis for separating them, and if no other variables or variable to measure are available, a mixed distribution may be difficult to recognize and virtually impossible to decompose into its component distributions. For instance, if an investigator is presented with a set of unidentified measurements, he may not know whether he is dealing with a mixed distribution. However, if he learns that the distribution is one of human heights, he has a basis for expecting it to be a normal distribution or a mixture of normal distributions. If he discovers further that it is a distribution of heights of Caucasians and Pygmies, he will likely assume that the distribution is mixed; and if he is then supplied with data on another variable, for instance, skin color, he may be able to separate the two distributions without too much trouble.

These notions are clarified through a numerical illustration and some real examples. Some concluding remarks assert that most mixed geological distributions are difficult if not impossible to recognize, much less to separate, because the statistical techniques that might be used have such large variances of estimate for the parameters that little reliable information can be obtained about the parameters of the component distributions or about their proportions in the mixture.

An Illustration of Mixed Distributions

As an illustration of mixed distributions, consider five populations, named A to E, each composed of 500 observations with mean μ and standard deviation σ as specified in table 6.25. Each population may be thought of as the silica content of a sedimentary rock measured at various places. From these five populations, five distributions are defined, each consisting of 1000 observations, obtained by mixing these five populations. Distribution 1, which is population A taken twice, has the frequency distribution of table 6.26 and the histogram of figure 6.16-a. Distribution 2 consists of population A plus the 500 observations from population B. The other three distributions are defined similarly, as shown in table 6.26, with the resulting frequency distributions and histograms given in the table and the figure.

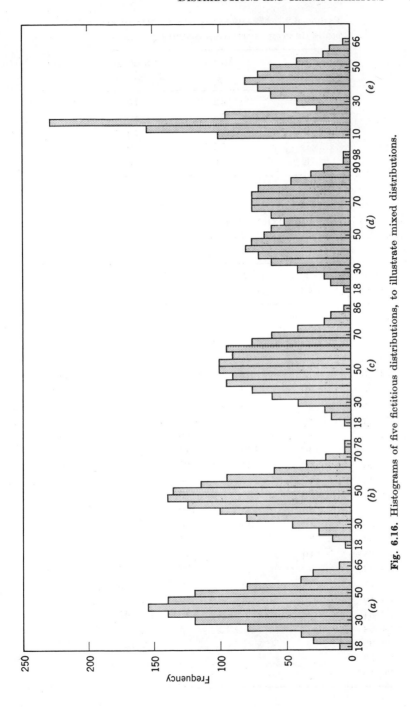

Fig. 6.16. Histograms of five fictitious distributions, to illustrate mixed distributions.

TABLE 6.25. MEANS AND STANDARD DEVIA-
TIONS OF FIVE FICTITIOUS DISTRIBUTIONS

Population	Mean, μ	Standard deviation, σ
A	42	10
B	52	10
C	62	10
D	72	10
E	42	3.16

TABLE 6.26. A FREQUENCY DISTRIBUTION OF FIVE FICTITIOUS DISTRIBU-
TIONS, TO ILLUSTRATE MIXED DISTRIBUTIONS

Class interval	Distribution				
	1 $(2A)$	2 $(A + B)$	3 $(A + C)$	4 $(A + D)$	5 $(A + E)$
0–4					
4–8					
8–12					30
12–16					160
16–20	10	5	5	5	230
20–24	30	15	15	15	95
24–28	40	25	20	20	25
28–32	80	45	40	40	40
32–36	120	80	60	60	60
36–40	140	100	75	70	70
40–44	160	125	95	80	80
44–48	140	140	90	75	70
48–52	120	135	100	65	60
52–56	80	115	100	60	40
56–60	40	90	90	50	20
60–64	30	60	95	60	15
64–68	10	35	75	75	5
68–72		20	60	75	
72–76		5	40	75	
76–80		5	20	70	
80–84			15	45	
84–88			5	30	
88–92				20	
92–96				5	
96–100				5	

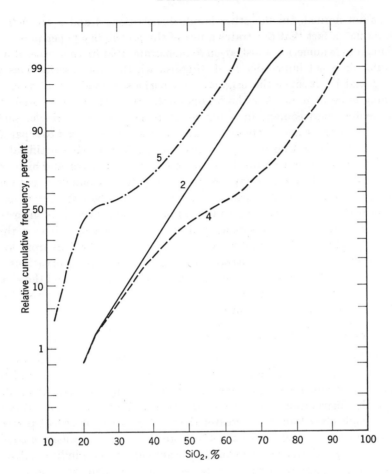

Fig. 6.17. Plot on normal probability paper of fictitious mixed distributions 2, 4, and 5.

The frequency distributions and histograms have the following interesting properties. Distribution 2 is only slightly skewed and is difficult to distinguish from a normal distribution. Distribution 3 is clearly nonnormal and displays three modes with the particular class interval chosen, although if these were real data, one would be hard pressed to decide if there were three modes present or merely one poorly defined mode. Distribution 4 displays two distinct modes, corresponding to the two population means at 42 and 72 and clearly is bimodal. Distribution 5 is skewed. Distributions 2, 4, and 5 are plotted on normal probability paper in figure 6.17. Distribution 2 plots as nearly a straight line, so it would be difficult to determine from the plot that it was not normal. On the other hand, distributions 4 and 5 are definitely nonnormal, a fact that illustrates a use of the probability paper plot.

This simple numerical illustration demonstrates that in some cases it would be difficult if not impossible to distinguish whether the distributions were mixed and to discover the original distributions, provided that there was nothing more to go on than the original data. If, however, the distributions represented silica content in sedimentary rocks, another variable, such as grain size, shape, color, or etching, might be measured as an aid to separation. The purpose for setting up the illustration in this somewhat artificial way, rather than by generating the distributions, frequency tables, and histograms by mathematical formulas implemented by electronic computer, is to point the way for the geologist to perform this work with tables of random normal numbers (Li, 1964, I, appendix table 1) or with punched cards in order to obtain the frequencies by sorting and counting. For a distribution that is evidently mixed, it is helpful to construct mixed distributions from known populations by varying the parameters and numbers of observations. One then grasps unforgettably the fact that many mixtures can yield the same or nearly the same distributions and, therefore, how difficult any statistical procedures for separating them must be.

Examples of Mixed Distributions

The Fresnillo mine affords an example of a mixture of two distinct distributions of silver/lead ratio present through a broad interface (Koch and Link, 1967, pp. 49–60). Although some of each distribution is present in all the veins, the distributions can best be delineated in the 2137 vein. There the area of high silver/lead ratio (greater than 400 grams of silver to 1 percent of lead) extends downward from the 470-meter level, and a transition zone lies in between. These areas were defined by scanning ratios for individual sample points on each level after the basis for selection had been established.

For each of the three areas of different silver/lead ratio we prepared a histogram of the logarithms of the different ratios (fig. 6.18). These fairly symmetrical distributions indicate how silver/lead ratios shift from one zone to the next. They overlap because defining three distinct regions is an over-simplification. The histogram for all three distributions combined is skewed with only one mode.

These high and low ratio types of mineralization could be separated because Stone and McCarthy (1948) have recognized and mapped the two kinds of mineralization designated as high and low sulfide and because there is also a theoretical basis. When the silver/lead ratio is less than 100, the correlation (sec. 9.3) of silver and lead is very high; the slope of the linear regression (sec. 9.2) is typically an increase of about 22 meter-grams of silver for a 1 meter-percent increase in lead; and the intercept is about 96 meter-grams of silver

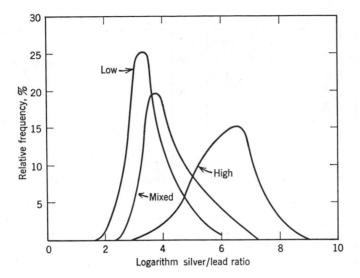

Fig. 6.18. Frequency curves of silver/lead ratios from three areas of the 2137 vein, Fresnillo mine.

per metric ton. These facts are consistent with the geological theory that the silver may be present in solid solution in galena. Elsewhere, where the silver/lead ratio exceeds 400, correlation of silver and lead is low; the slope of the linear regression relating silver to lead is typically about 154; and the intercept is about 631. These facts suggest that most of the silver is independent of lead; and, indeed, galena is sparse, and ruby-silver minerals are plentiful.

One geological field in which mixed distributions have been studied for many years and are reasonably well understood is that of size distributions in sediments. Pettijohn (1957, pp. 45–51) explains mixed-size distributions in sediments and gives references to the literature; for example, some size analyses of California alluvial gravels yield histograms (fig. 6.19) (Conkling and others, 1934, p. 86), which show that sand sizes are unimodal, whereas gravel sizes are bimodal and represent mixed distributions. Conkling writes that "the two maxima of these curves are characteristic of the alluvial gravel samples. . . . Probably the coarsest maximum represents the materials rolled and bounded along the channel bottom, and the finer percentage maximum represents the material deposited in the interstices between the boulders and cobbles from suspension." Again, this example indicates that geologic processes must be invoked and theory is needed to untangle mixed distributions.

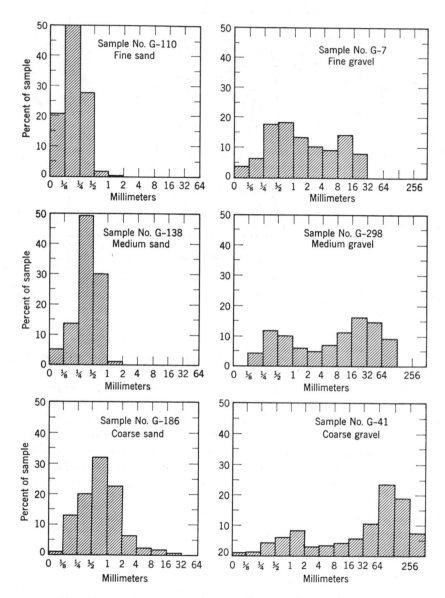

Fig. 6.19. Histograms of size distributions for some California alluvial gravels (after Conkling and others, 1934, p. 86).

Writing about particle-size distributions, Wilks (1963, p. 116) made the following remarks that are also pertinent to mixed distributions in general:

Cumulative particle-size distributions have been determined for many kinds of geological material. . . . The curve obtained in each case is thus characteristic of the sample of particles. As one might expect, these curves vary widely from sample to sample not only from one kind of material to another but from one sample to another of the same material. . . . If the sample consists of a mixture of particles produced by a mixture of two (or more) kinds of geological disintegrating action, each of which would produce . . . normal distributions, the overall particle size distribution . . . would be the sum of the two (or more) normal distributions and would, in general, be non-normal. There is a strong temptation (which occurs not only in geology but in biology and other fields) to make the converse statement, namely, that if a non-normal distribution exists it can be inferred that it is the sum of two or more normal distributions. This is a hazardous inference, since there could be homogeneous grinding, crushing, or sorting actions that do not produce normal particle-size . . . distributions. To make any headway in verifying such a converse statement, one would need reasonably good information or strong grounds for some hypothesis as to how many different major grinding, crushing, or sorting actions were involved in producing the resulting sample of particles, and one would also need to know that each action would by itself produce a normal distribution. . . . One could then decompose the composite sample distribution into its normal components and estimate the fraction of particles involved in each component distribution. This, however, is a rather involved mathematical exercise, even for a mixture of two or three populations, which should be undertaken only when one has good information or a strong hypothesis that a small number of different distributions have been mixed. Such an analysis for a mixture of more than three or four distributions should be made, if at all, only with great caution.

Concluding Remarks

Many frequency distributions encountered in geology are mixed. Sometimes they can be sorted out because the parameters of the component distributions are quite different from one another or because some other variable can be measured. Thus for the California sediments the means and standard deviations, as well as the lithologies, are very different. Other times the mixed distributions are difficult if not impossible to sort out because the parameters of the component distributions are not very different from one another or because some other variable cannot be measured. Thus for the Fresnillo data a large number of ratios had to be classed as mixed mainly because the silver/lead assays were for total silver and lead without differentiating the silver that is in solid solution in galena. If assays were available for distinguishing the two kinds of silver in the mixed-ratio zone, a much cleaner separation would have been possible.

Thus, if mixed distributions are to be separated effectively, one or more of these elements must be present:

1. A theory for the functional forms of the component distributions and preferably also for their parameters.
2. Additional variables, such as lithologies and geographic positions.
3. Rather different parameters for each of the component distributions.
4. Only two, or at most a few, rather than many, component distributions.

REFERENCES

Aitchison, John, and Brown, J. A. C., 1957, The lognormal distribution, with special reference to its uses in economics: Cambridge Univ. Press, 176 p.

Conkling, H., Eckis, R., and Gross, P. L. K., 1934, Ground water storage capacity of valley fill: California Div. Water Resources, Bull. 45, 279 p.

Dixon, W. J., and Massey, F. J., 1969, Introduction to statistical analysis: New York, McGraw-Hill, 638 p.

Feller, William, 1968, An introduction to probability theory and its applications, v. 1: New York, John Wiley & Sons, 509 p.

Finney, D. J., 1941, On the distribution of a variate whose logarithm is normally distributed: Jour. Royal Statistical Soc., Supp. 7, no. 2, p. 155–161.

Griffiths, J. C., 1966, Exploration for natural resources: Jour. Operations Research Soc. Am., v. 14, p. 189–209.

Gumbel, E. J., 1958, Statistics of extremes: New York, Columbia Univ. Press, 375 p.

Haldane, J. B. S., 1941, The fitting of binomial distributions: Annals of Eugenics, v. 11, p. 179–181.

Hoel, P. G., 1962, Introduction to mathematical statistics, 3rd ed.: New York, John Wiley & Sons, 427 p.

Koch, G. S., Jr., and Link, R. F., 1967, Geometry of metal distribution in five veins of the Fresnillo mine, Zacatecas, Mexico: U.S. Bur. Mines Rept. Inv. 6919, 64 p.

Krige, D. G., 1960, On the departure of ore value distributions from the lognormal model in South African gold mines: Jour. South African Inst. Mining Metall., v. 61, p. 231–244.

———, Developments in the valuation of gold mining properties from borehole results: 7th Commonwealth Mining Metall. Cong., v. 1, p. 18.

Li, J. C. R., 1964, Statistical inference, v. 1, Ann Arbor, Mich., Edwards Bros., 658 p.

Link, R. F., Koch, G. S., Jr., and Schuenemeyer, J. H., 1970, The lognormal frequency distribution in relation to gold assay data: U.S. Bur. Mines Rept. Inv., in press.

Mason, Brian, 1962, Meteorites: New York, John Wiley & Sons, 274 p.

Mood, A. M., 1950, Introduction to the theory of statistics: New York, McGraw-Hill, 433 p.

Mosteller, Frederick, and Tukey, J. W., 1949, The uses and usefulness of binomial probability paper: Jour. Am. Statistical Assoc., v. 44, p. 174–212.

Pettijohn, F. J., 1957, Sedimentary rocks: New York, Harper & Bros., 690 p.

Sichel, H. S., and Rowland, R. S., 1961, Recent advances in mine sampling and underground valuation practice in South African gold fields: Trans. 7th Commonwealth Mining Metall. Cong., p. 1–21.

Sichel, H. S., 1966, The estimation of means and associated confidence limits for small samples from lognormal populations, *in* Symposium on mathematical statistics and computer applications in ore valuation: South African Inst. Mining Metall., p. 106–122.

Slichter, L. B., 1960, The need of a new philosophy of prospecting: Mining Eng., v. 12, p. 570–576.

Stone, J. B., and McCarthy, J. C., 1948, Mineral and metal variations in the veins of Fresnillo, Zacatecas, Mexico: Am. Inst. Mining Metall. Engineers Trans., v. 178, p. 91–106.

Tukey, J. W., 1962, The future of data analysis: Annals of Math. Statistics, v. 33, p. p. 1–67.

U.S. Geological Survey, 1952, Floods in Youghiogheny and Kiskminetas river basins, Pennsylvania and Maryland, frequency and magnitude: U.S. Geol. Survey Circ. 204, 22 p.

Wilks, S. S., 1963, Statistical inference in geology, *in* Donnelly, T. W., ed., The earth sciences, problems and progress in current research: Univ. of Chicago Press, p. 105–136.

SAMPLING AND VARIABILITY
IN GEOLOGY

In Chapters 7 and 8, making up Part III of this book, the univariate statistical methods explained in Part II are applied to geological data.

In Chapter 7 general principles of geological sampling are discussed; and ways of interpreting data derived from several different techniques of sampling, such as diamond-drill-hole and hand-specimen sampling, are examined. The sampling techniques are those most commonly used by geologists; some more specialized methods are discussed in Chapter 15. In the final section of Chapter 7, the range of geological variability of various substances in different rocks is reviewed.

Variability in geological data is the subject of Chapter 8. Different kinds of variability—including that introduced by sampling, sample preparation, and chemical analysis, as well as the inherent natural variability—are appraised.

Chapter 7

Geological Sampling

Having introduced fundamental statistical procedures for data analysis in Chapters 2 to 6, we can now focus our attention on geological problems to a greater extent. The discussion in this chapter is on several methods of geological sampling, starting with collection of hand specimens, and relates them to the statistical concepts, especially that of variability, discussed previously. That the sampling of rocks is not easy is abundantly illustrated by mines that never paid because the grade of ore estimated by sampling was higher than the grade actually mined.

It is convenient to distinguish four kinds of variability in geological data: (a) natural variability, that inherent in the rock body or other geological object being sampled; (b) sampling variability, that introduced by the physical sampling process; (c) preparation variability, that introduced in readying the geological sample for chemical analysis by crushing, splitting, etc.; and (d) analytical variability, that introduced by the chemical or physical determination of substances in the geological samples. This chapter is about the first two kinds of variability, and Chapter 8 is about the other two kinds.

7.1 INTRODUCTION

A geological sample (sec. 3.1) is taken to obtain information about a rock body for a particular purpose. The rock body, whether it be a formation, outcrop, cylinder of rock removed by drill core or cuttings, or whatever else, should first be *classified* (*stratified*) in all the geologically meaningful ways that one can think of. Then, one or more geological samples of the rock body should be obtained to satisfy the following requirements:

1. The geological samples are selected with at least one element of randomness introduced in the sampling process.
2. None, or as little as possible, of any sample is lost during collection.
3. The most inexpensive feasible methods are used to sample.
4. Some way is available to remove the sample from the field to the laboratory or other place in which the sample is to be studied.

Section 7.2 is a discussion of geological factors that affect sampling, and sections 7.3 to 7.7 are about some specific methods of geological sampling. The purpose is not to treat the technology of sampling but rather to discuss statistical analysis of sampling data. Therefore, representative real data are considered, but no attempt is made at an exhaustive examination. Finally, in section 7.8, distributions of coefficients of variation according to types of geology and classes of constituents are discussed. First, some sampling problems that may be even more familiar to the reader than those of rocks are sketched to set the stage for the geological discussion.

A Few Sampling Problems from Disciplines Other Than Geology

In this section we sketch four statistical problems from disciplines other than geology in order to convey some understanding of how statisticians attack real problems. Just as in field mapping a contact between two formations, dating a fossil assemblage, or classifying a rock, the textbook rules are not always followed, so in statistics one must choose, improvise, and be arbitrary. For further reading, a book by Wallis and Roberts (1956), who give several hundred examples, is excellent.

Clearly defined purposes, necessary in order to obtain valid results, are present in these four problems. In the first two, prediction of elections and evaluation of biomedical research, the target populations are unmistakable, and the relations between the target and the sampled populations can be clearly specified. In the other two problems, the relation of smoking to lung cancer and the cause of polio, the target populations are hazy, and their relations to the sampled populations are obscure. The first two problems allow prospective studies to be made because most of the data are gathered after the problems are conceived; the second two problems have retrospective elements because some or all of the data are already at hand. Geological problems parallel to these different combinations are pointed out in the discussion.

Prediction of Elections

For the last few years, the American television networks have competed energetically during nights of the general elections to predict the winning candidates and their shares of the total vote. This frantic activity, live on prime time, is intended to entertain the public and to impress them with the mastery of network men over electronic computers. Their achievements and their mistakes are particularly entertaining to those familiar with statistics and computing. A brief examination of election predicting demonstrates the judicious blending of many elements that is required whenever statisticians work with real data. Ingenuity is required, as well as a knowledge of statistics and the subject matter.

In election prediction both the purpose and the target population are well defined. The purpose is to make an accurate prediction as early as possible, and preferably earlier than another network. The target population consists of the votes cast in that particular election, and the sampled population is those votes, less any that are lost by voting-machine malfunctions, that are stolen by dishonest election judges, etc. Alternatively, the target and sampled populations may be said to coincide if one takes the viewpoint that only the votes that are counted matter. The sample, which consists of early returns, clearly is not a random one, as some precincts consistently provide early returns election after election; and some voters, such as retired people, consistently vote early. But one can pretend that the sample behaves essentially as though it were random, as must be done time and again, more often than not, in geological sampling.

Three kinds of information go into an election prediction: (a) prior information from opinion polls, straws in the wind, conjectures of knowledgeable people, etc.; (b) current information on returns from "key" precincts, which are those chosen because past results indicate that their voting patterns can be more readily interpreted than those of other precincts; and (c) current information derived from the total vote, broken down by states and sometimes into finer subdivisions, usually by county. These three categories of information are combined in a mathematical model, and once the returns begin to supply the two latter kinds of information, decision rules are used either to predict the winner or to conclude that not enough returns are yet in to do so. As the returns come in, these rules are continually applied until all the races being considered are predicted.

The most important element in the mathematical model is the returns from the key precincts, which are chosen judiciously to reflect the electorate. In key precincts such factors are desirable as stable boundaries and populations rather than the changing boundaries and itinerant populations that are found in areas being changed from single-family to multiple-family dwellings.

Once chosen, the statistical properties of the key precincts, obtained from their performance in past elections, are estimated to forecast their behavior in the current election. Two characteristics of the voters must be established, their average behavior and their variability; that is, how they ordinarily vote and how consistently they vote. Once these characteristics are determined by statistical analysis of the historical data, rules can be devised for announcing a winner. The third element, the total vote, enters into the mathematical model mainly in close elections, where the fragmentary results from the early returns in the key precincts are insufficient on which to base a decision.

In summary, election predicting depends upon combining a knowledge of both subject matter and statistics. In order to apply the statistics, many formal assumptions must be ignored, including those about random samples and normal distributions, but experience shows that the applications are successful. The situation is like that in geological sampling where statistical assumptions seldom are fully satisfied. The purposeful selection of key precincts is similar to the geologist's intuitive decisions based on knowledge of the science, perhaps to look for fossils from certain localities where he thinks that they may be found. The continual data analysis is similar to the plan of collecting fossils until enough are found to date a formation. The main difference is that time on election night is crucial, as a prediction is valuable only if made within minutes, but even here one may draw an analogy to the exploration of a prospect by a company that wants to acquire the land as soon as, but no sooner than, it learns that the prospect will make a valuable mine.

Financial Support of Health Programs

In 1964 the National Institutes of Health (NIH) spent about one billion dollars in direct financial support of the health research of the United States. Most of this money was spent on twenty thousand individual projects at more than one thousand universities and medical schools. This system was studied to learn if the American people were getting their money's worth and to see if any changes would increase the program's effectiveness (National Institutes of Health, 1965).

Teams of scientists and administrators visited projects to evaluate the quality of research and the effectiveness of the administrative support. The key to scheduling these visits was an effective sampling scheme to allow the teams to inspect enough projects without spending an exorbitant amount of money, but also to assure that the structure of the NIH was thoroughly evaluated. Because simple random sampling would have required too much time and travel, the statisticians had to devise a sampling scheme to keep time and travel down to a reasonable level.

The statisticians determined that the requirements would be met if teams of six to eight experts spent two or three days each at 25 or 30 institutions and inspected 8 to 10 projects per institution. The institutions were selected as a stratified random sample (Cochran, 1963), stratified first by geography, and second by the amount of NIH dollars contracted to that institution. The number of institutions to be visited in any area was set proportional to NIH dollars for the area, so that those with $2 in support had twice the chance of being visited as those with $1 of support. Because small institutions might have only one or two NIH projects, geographically contiguous small institutions were grouped for sampling into a "single" institution. After an institution was selected the projects to be visited at it were determined in a similar fashion.

Applying this plan led to a sampling scheme with many subjective and arbitrary elements. However, the basic procedure was a random sample taken within constraints to yield an effective sample. The nonrandom elements permitted sampling efficiency, and the random elements ensured the basic integrity of the sample. This solution demonstrates that simple random sampling is seldom the most economical or efficient way to sample, especially if the target population, the objectives, and the resources for sampling are well known.

Smoking and Lung Cancer

One of the controversial topics of our time is whether smoking causes lung cancer. For the last few years both the death rate from lung cancer and cigarette consumption have risen dramatically. Lung cancer has changed from a relatively rare disease to a common one, and the fatality rate from lung cancer is now about equal to that from automobile accidents, which is about 40,000 deaths a year in the United States. There is no doubt that both cigarette smoking and the death rate from lung cancer have increased. The controversy arises from the relationship, if any, between these events. The death rate from automobile accidents has also risen spectacularly since the turn of the century, but no one would blame cigarette smoking for this. Also, other chemical alterations to the environment of our industrial society have risen during the same period.

The studies that bear on this issue have been of two types. Retrospective studies try to match people who have contracted or died from lung cancer with otherwise-alike people with no history of the disease. These studies show that people with lung cancer smoke or smoked more than those without the disease. Prospective studies have been devised to compare the incidence of cancer and the death rates of smokers and nonsmokers. These studies also show higher death rates from lung cancer for smokers than for nonsmokers. Both types of study have been attacked as lacking the proper controls to define properly the populations under study; for example, in one study the health of both the tested and the control groups was much better than that of the general population even though the smokers did have more lung cancer.

Of the many difficulties in studies of smoking, four may be stressed. First, target and sampled populations with which to work are difficult to define. Second, whatever happens takes a long time; lung cancer takes decades to develop, whether or not it is caused by smoking. Third, the effects are confused by other events: a potential cancer victim can be killed accidentally, and most smokers are unhealthier and have higher death rates than other people. Fourth, the cause of lung cancer may be the same factor as whatever induces people to smoke, perhaps a common genetic characteristic that tends to make people smoke and also raises their chance of getting lung cancer.

For this problem and for many others like it an answer is wanted today. Much is at stake socially and economically, but, unfortunately, the unequivocal results will require years to obtain; hence the controversy over the evidence and the conclusions that have so far been reached. The example is an excellent one of how economic interest and rationalizations can sway opinion in both directions. Useful conclusions can already be drawn, but until careful experiments are completed, airtight results cannot be expected.

1954 Polio Vaccine Trials

The 1954 polio vaccine trials were one of the largest medical experiments ever conducted. A brief account may be of interest because many geological experiments are of similar size—comprehensive petrological and paleontological studies on a continent-wide or world scale, analyses of data from one or several oil fields, or sampling of large ore deposits. And these experiments may have the further similarity that a rare occurrence is investigated: polio in the medical experiment, and perhaps oil or ore in the geological experiment. Our discussion and a full account of the vaccine trials by Brownlee (1955) show how complex the application of statistics to a real problem usually is.

Participants in the vaccine trials included 312 state and local health officials, 54 physical therapists, 22 epidemiological intelligence officers, scientists in 28 laboratories, a 17-member advisory committee, countless teachers and physicians, and 1,829,916 children. The purely administrative problems were themselves immense and probably contributed to the imperfection of the scientific work.

Much of the experiment consisted of giving polio vaccine to second-grade children, with first- and third-grade children serving as control groups. This procedure was clearly unsatisfactory because there is no reason to believe that the risk of getting polio was equal for each of the three groups. However, for about 40 percent of the trial, the children of the first three grades were combined. One-half of them, selected at random, received vaccine and the other half, a solution that has no effect on immunity (a placebo). Only the national evaluation center knew which children received the vaccine and which received the placebo.

Some statisticians would say that this part of the trial was the only part with any scientific validity. Out of the total number of 400,000 randomly selected children of mixed grade levels, there were 33 cases of paralytic polio in the group which received the vaccine and 115 cases in the group which did not receive it. Assuming a Poisson distribution (sec. 6.1), it is clear that the incidence of paralytic polio was less in the group which received the vaccine. Unfortunately, there is some evidence that even with the randomized sample the group which received the vaccine had less propensity toward contracting polio (Brownlee, 1955).

Today the effectiveness of polio vaccine is well established, although the type of vaccine now used is much safer and its method of administration more simple than the vaccine used in the trials. It is interesting to note that, despite an intensive effort to conduct scientifically valid tests, only a small part of the work would stand up to criticism. How this state of affairs came about is easy to surmise. The administrative difficulties were so large, the pressure to get on with the trials so great, and the responsibility for the work so diffuse that there cannot have been time enough to think through the experiment before it was begun.

An interesting aftermath to these trials is that once they had concluded and the vaccine was released for public use, it became clear that some of the vaccine was itself causing polio rather than preventing it. Evidently at least some of the vaccine produced commercially was different from that used in the trials. The point is that the trials did not establish that the vaccine could be made commercially in large quantities or establish the safety standards that should be used in its manufacture and in quality control.

The complications that arose in testing polio vaccine were caused primarily by the rarity of polio as a disease—its incidence being about one in a thousand —and the fact that the number of cases varies widely from season to season. Moreover, the vaccine was imperfect and evidently effective only about three-fourths of the time. The combined consequences of the rarity of the disease and the imperfectness of the vaccine meant that an extremely large trial was needed, although the enormousness of the trial actually made was probably not necessary.

For instance—assuming an incidence of only 0.5 per thousand and a 50 percent efficiency of the vaccine—if a total of only 200,000 children had participated in the trial, 100,000 of whom had received the vaccine, the standard deviation between the mean assuming no effect and a 50-percent effect would be about 3. Fifty children on the average would get polio in the untreated group and 25 children would get polio in the treated group. The standard deviation of the difference would be $\sqrt{75} = 8.5$ and $(50 - 25)/8.5 = 2.9$. On the other hand, if it were argued that the vaccine would be worthwhile even if it were only 10 percent effective, a large trial *would* have been necessary.

In summary, this and the other examples show above all that in sampling problems, as in most things, it pays to plan ahead.

7.2   GEOLOGICAL FACTORS THAT AFFECT SAMPLING

The geological factors that affect sampling depend on the goal of the investigator, who must ask first what is the purpose of sampling and then what area should be sampled, what items should be sampled, and finally, what attributes of these items should be sampled. From the answer to these questions can be formulated a *sampling plan*, which consists of the procedures to be followed.

What is the purpose of sampling? It may simply be discovery (Chap. 12), perhaps hunting for a needle in a haystack—the aim being the most efficient possible coverage of an area with the resources available. Another objective is to estimate variability, for instance, comparing rocks to learn if their variability is the same, a circumstance which might suggest that their geology has something in common. This subject is raised by a comparison of coefficients of variation in section 7.8 and is further considered for gold deposits in Chapter 16. However, perhaps the most usual aim of sampling in geology is to estimate the mean value of one or more constituents of a rock.

Unless a geologist carefully thinks out and specifies his goal in sampling, he should expect nothing but trouble. In economic problems, for instance, mine sampling, specifying an objective is much easier than in academic problems. Once a purpose is defined a sampling plan should be devised and followed, at least at first. If it turns out to be unsatisfactory, it can be revised, but each change introduces new subjective elements, difficult to evaluate statistically.

What area should be sampled? Although this question is liable to receive an arbitrary answer, influenced by travel costs, state and national boundaries, or location of an interesting rock body too high in the mountains or too deep in the earth, it is one that deserves careful thought. To promptly find small bodies of mineable chromite, one would be inclined to search in ultramafic rocks; but to play a long shot, one might decide to search for less well-known sources of chromite in the hope of finding the mineral in another environment. On the other hand, for estimating chromite abundance in rocks of all types, enough information on its abundance in ultramafic rocks may be at hand, and all sampling might be concentrated in other lithologies. For other problems similar questions arise; one will suffice to make the point: In a study of granite tectonics, is more learned from a reconnaissance of the Sierra Nevada or from a detailed map of a Maryland quarry? Statistics can aid in thinking through such questions from the stage of sampling to the final data analysis.

What items and what attributes should be sampled? These questions are simpler to answer than the others. Often the entire rock is to be sampled to determine rock chemistry or rock physics. To determine rock chemistry, whether expressed in chemical elements, oxides, minerals, or otherwise, and whether completely or for only one or a few constituents, is the most usual purpose for sampling and the only one discussed in detail in this chapter. In order to determine rock physics; properties such as strength, hardness, rate of transmission of seismic and other waves, and paleomagnetism, may be studied. Besides examining chemical and physical properties of rocks, one samples many other items—fossils and structural properties such as strike and dip and current directions.

We can think of no better way to summarize these introductory remarks than by quoting from Griffiths (1962, p. 604), who writes the following remarks about sedimentary petrology that apply to the whole science of geology:

> Any scientific investigation is no better than its sampling plan; inadequate sampling cannot be subsequently offset by any procedure, experimental or statistical. The problem of sampling arises in the initial stages of an investigation when setting up the most efficient means of achieving the main objective of the experimental program, and it crops up again at various stages throughout the experiment in attaining required levels of precision of estimates from different measuring techniques. Because of its fundamental role in experimentation, the sampling pattern should be decided upon at the same time as the overall strategy of the program, i.e., at the beginning; generally in sedimentary petrography it is resolved as the experimenter becomes aware of it, a certainly inefficient and, possibly, disastrous practice.

Estimation of the Mean

Because estimation of the mean is the principal purpose of most geological sampling, the basic statistical concepts introduced previously are reviewed herewith. Precision of an estimate based on independent observations depends upon the fundamental law about the standard error of the mean, $\sqrt{\sigma^2/n}$, where the variance σ^2 is the *total* variance from all of the sources differentiated in the introduction to this chapter. In order to make the standard error of the mean small, only two ways are possible, as explained in the discussion of confidence intervals (sec. 4.4): either reduce the variance σ^2 or increase the number of observations n.

The largest reduction in the standard error of the mean almost always is afforded by reducing variability through taking advantage of the natural stratification of the rock body sampled, as is discussed throughout this chapter. The importance of deciphering the stratification, if it is not obvious, and then devising an appropriate sampling plan cannot be over-emphasized. The standard error of the mean can also be reduced by better sampling, better sample preparation (sec. 8.4), or better physical or chemical analysis (sec. 8.5).

The second way to reduce the standard error of the mean is to increase n, although the advantage is not commensurate with the increase in effort after a few tens of observations (sec. 4.4). Even though σ^2 may be larger, several observations by an inexpensive method may provide a more accurate estimate than one or a few expensive observations; for instance, several inexpensive rotary-drill holes may yield more information than one expensive diamond-drill hole. This concept, discussed in sections 7.3 to 7.8 and in Chapter 15, demands emphasis because many geologists do not recognize that there may be merit in deliberately accepting an increase in variability (or sloppiness, in plain words) if then more observations can be afforded.

Origin of Natural Variability

The natural variability of substances in rocks plays an important part in planning rock sampling, and this variability depends fundamentally on the geologic processes that formed the rocks. However, although a knowledge of rock genesis may someday enable the variability of a rock to be predicted, this time has not yet come.

One can speculate that, as a general rule, rocks formed at high temperatures should be less variable than those formed at low temperatures, provided that equilibrium has been attained. This generalization is supported by two excellent studies of major and minor elements in basalts by Manson (1967) and Prinz (1967). In rocks formed at lower temperatures, such as sedimentary ones, variability increases, as a rule. However, rocks formed at high temperatures are less likely to have attained equilibrium, as shown by hydrothermal veins and by most igneous extrusive rocks other than basalt flows.

Barth (1962) made the same point when he observed that sedimentary rocks are differentiated into extreme compositional types, thus being, *as a group*, more variable than igneous rocks. However, individual sedimentary rock types, such as mature sandstones, may be extremely uniform in composition. Therefore, for sampling, the most that can be said is that, as a class, igneous rocks formed at high temperatures should have a better understood variability than sedimentary rocks about which information has not yet been obtained. Mason (1962) also discusses this topic.

The variability of rocks also depends on variability in the composition of the constituent minerals. We may cite some examples, but they do not suggest general rules to us. Chemical composition of garnet is highly variable, but in a given rock the limited evidence suggests that composition is rather uniform or varies gradually from place to place (Engel and Engel, 1960). Silver content of native gold in different placer deposits is also highly variable, but in a given placer it is uniform or changes only gradually and regularly (sec. 15.2). Chemical composition of quartz is practically invariant, all quartz being composed of nearly pure SiO_2.

TABLE 7.1. TENTATIVE CLASSIFICATION OF SOME ROCK TYPES ACCORDING TO THEIR UNIFORMITY FOR SAMPLING

Uniform rocks	Nonuniform rocks
Igneous rocks	
Basalt	Andesite
Diabase	Diorite
Gabbro	Rhyolite
Obsidian	Granodiorite
Anorthosite	Granite
	Most pyroclastic rocks
Metamorphic rocks	
Marble	Hornfels
Quartzite	Mylonites
Eclogite	Slate
	Schist
	Gneiss
Sedimentary rocks	
Limestone	Gravel and conglomerate
Chert	Shale
Diatomite	Marl
Arenite	Wacke
	Dolomite
	Glacial till
	Rocks of intermediate composition, e.g., calcareous sandstone, arenaceous limestone

Thus table 7.1 classifies some rock types according to their uniformity. This table suggests that process of formation is related to sampling, but not in a clear way. A comprehensive petrologic study should generate insights into this subject.

Configuration of the Rock Body to be Sampled

Besides natural variability, the main geologic factor affecting sampling is the configuration of the rock body to be sampled, that is, its shape and position in the ground. It is convenient to distinguish large-scale structure or macrostructure, from small-scale structure, or microstructure, on the scale of a hand specimen; these are discussed in turn with the aid of a few familiar geologic examples. More important for developing a sampling plan than rock history or genesis is rock geometry. Sampling methods explained in this chapter and sampling designs taken up in section 8.6 should be planned to take account of it.

Macrostructures may be tabular or linear in shape and may be concordant or discordant to regional geologic structures and stratigraphy. Tabular and concordant rock bodies include sedimentary beds, pyroclastic beds, lava flows, sills, and the metamorphic equivalents of all of them. The entire rock body may be sampled, as in a petrological study; or only part may be of interest, as in an oil trap or a manto ore deposit.

Most tabular but discordant rock bodies are igneous or clastic dikes and veins. Again, the entire discordant rock body may be of interest; but, especially for veins, sampling may be confined to or concentrated in particular parts of a vein where it intersects another vein or a favorable bed or where it changes in attitude.

Linear or lathlike rock bodies formed in sedimentary rocks at the time of deposition—bodies that are concordant to regional structure and stratigraphy —include shoestring sands and organic reefs, which localize petroleum in many places, and in some places localize ore bodies such as uranium on the Colorado Plateau and lead in southeast Missouri. Formed secondarily, but also essentially concordant, are crests of synclines, which also localize petroleum and ore bodies such as the gold at Bendigo, Australia; troughs of synclines, which contain zinc ore at Franklin, New Jersey; and sheared-off limbs of folds which localize iron ore at Dover, New Jersey.

A few linear rock bodies are discordant to regional structure—the principal types being volcanic pipes, such as that at Cripple Creek, Colorado, containing gold ore; that at Braden, Chile, containing copper ore; salt domes, which may localize salt and oil; and chimneys of limestone, such as those localizing lead, zinc ores at Santa Eulalia, Mexico.

The position of either tabular or linear bodies relative to the surface of the earth or underground workings affects the physical method of sampling used and also the design of a sampling plan. Tabular bodies may be planar or folded; linear bodies may be straight or curved; and either may be horizontal or inclined. One illustration of the kind of problem encountered is enough. To sample a horizontal coal bed deep below the surface, expensive drilling may be the only feasible way. On the other hand, if the bed is inclined and crops out in the area of interest, sampling the outcrop may be inexpensive. Then a good sampling plan will evaluate the information from the outcrop so that as few holes as possible need be drilled where the inclination takes the bed far below the surface. This scheme depends on the coal bed being uniform laterally. These comments on macrostructures emphasize the importance of making a specific definition of the target population (sec. 3.2).

As used here, microstructure consists of particular size and shape, and changes in these variables from place to place. In at least two instances, microstructure is such that large volumes of rock must be sampled. First, if a rock is coarse-grained, perhaps a conglomerate or a porphyry, large volumes are needed (sec. 7.6), an extreme case being the sampling of a pegmatite (Norton and Page, 1956). Second, if the constituent of interest is sporadically distributed, large volumes of rock must be taken, as in diamond sampling. Thus the variability on the scale of the microstructure may be the dominant source of the natural variability, or it may be small compared with the variability on the scale of the macrostructure.

Sampling Methods to Take Account of Natural Variability

When the geological factors that affect sampling are known, sampling methods to take account of them can be devised. The sampling methods depend on previous factual knowledge about the variability, on factual knowledge gained from preliminary sampling, on the variability expected from geological experience in similar rocks, or on an interpretation of the pattern of variability to be expected from geological theory. Once something is known about the pattern of natural variability, a sampling plan can be devised incorporating a sampling design (sec. 8.6) that specifies the number and pattern of observations and the statistical methods for analyzing the observations.

Two important kinds of sampling are *stratified sampling* and *pattern sampling*. Stratified sampling is always desirable if a meaningful way to do it can be found because the larger the body of rock, the larger the variance, other things being equal. A few examples of stratified sampling follow. Sampling may be stratified based on the bedding of bedded rocks; for example, in sampling gold in the Homestake mine, one takes advantage of the fact that most gold is known to be confined to the Homestake Formation. Therefore, even if one's interest is in the gold in the several rock formations for the purpose of an academic study rather than for mining, sampling should be concentrated in the Homestake Formation because the larger amount of gold there is associated with a higher variability that is more difficult to estimate. Similarly, in sampling for oil, most attention may be concentrated in known reservoir rocks and little in other rocks. In sampling intrusive igneous rocks, stratification might be based on rock type to avoid such a problem as collecting a mixture of zircons from different rock types and thereby producing a multimodal distribution that would be difficult if not impossible to interpret. In sampling metamorphic rocks, stratification should be based on rock types defined by metamorphic zones as well as by original lithologies—for instance, quartz formed by hydrothermal alteration should be distinguished from primary quartz.

The second kind of sampling, pattern sampling, which is discussed in detail in section 8.6, is touched on here because stratified and pattern sampling are almost always combined in practice. As an example, if a well-defined linear structure, such as a shoestring sand, a reef, or a linear outcrop belt, is present, a desirable pattern often would be to traverse across the structure, with sampling within each traverse relatively closely spaced compared with the distances from one traverse to another.

Finally, if natural variability displays a trend recognizable by trend-surface analysis (Chap. 9), the sampling may be stratified to take advantage of the trend.

7.3   HAND-SPECIMEN SAMPLING OF OUTCROPS

Undoubtedly most geological samples, excluding those taken by mining and petroleum geologists, are hand specimens from outcrops. Many fundamental geological theories and conclusions rest on information gained from study of these specimens, and it is therefore surprising that little attention has been paid to the validity of hand-specimen sampling of outcrops. In this section some representative studies of outcrop sampling are first reviewed; then, important work by a group of Pomona College geologists (who compare sampling of outcrops by hand specimens with sampling by the best feasible alternative so far proposed, a hand-carried portable drill) is considered in some detail.

The properties of rock outcrops are seldom like those of the unexposed part of the same rock body. It is well known that, except for accidents such as glaciation, the outcrops of ore bodies are unlike the rocks below the zone of weathering, and a large amount of literature (McKinstry, 1948, pp. 242–276) explains how to determine from the outcrop the mineralogy of the ore body at depth. Studies of rocks other than ore bodies are seldom detailed enough to provide general principles about the relation between outcrops and unexposed rocks near the surface and at greater depth. For rocks formed at surface temperatures and pressures, the changes should not be too severe, at least not as much as those of ore bodies. Because of this unsatisfactory state of knowledge, everything that follows in this section refers to a population of outcrops rather than of rock bodies, as would be preferred if the information were available.

In hand-specimen sampling of outcrops a great amount of bias can be introduced. The human tendency to collect oddities is revealed by the usual rock collection, which contains more strange than common rocks. Figure 7.1 is an example, which Ager (1963, p. 187) believes to be more the rule than the exception, of two histograms picturing proportions of mollusks obtained by bulk sampling and by hand picking from the surface. Ager writes that "the species overrepresented in the hand-picked samples are, in every case, forms

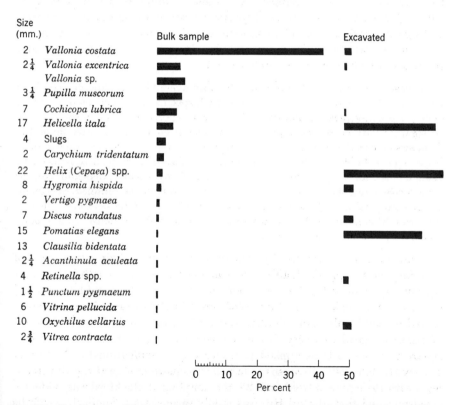

Fig. 7.1. Comparative histograms for quaternary mollusks obtained by bulk sampling (*left*) and by hand picking from the surface (*right*); Arreton Down, Isle of Wight, England (after Ager, 1963, p. 187).

which are conspicuous by reason of size or color . . . paleontologists themselves add to all their natural difficulties by their subjective and unscientific modes of collecting, and generally speaking, do not seem to realize the huge subjective errors which come into these normal collecting methods." His remarks apply all too truly to geology in general as well as to paleontology.

Not only do exposed rocks differ from unexposed ones, but an outcrop itself is not uniform. Thus, an outcrop may have only a few places where it is easy to break out a specimen; for instance, most rocks break along joints, so that the tendency to get specimens with more than the true proportion of joint surfaces may be disastrous if the specimens are coated with calcite and the lime content of the rock is under study.

Some Representative Works on Hand-Specimen Sampling

Books on field geology contain instructions on how to take conventional hand specimens from outcrops; for instance, Compton (1962) gives good advice. Miesch believes that hand specimens can be taken much better with a sledge hammer and chisel than with a geologic pick. Relying on as yet unpublished data from an extensive study of Cambrian outcrops throughout the western United States, he writes (1967, p. A-3):

In most geochemical field problems the only available sampling devices are a hammer and chisel or similar tools. (Most rock specimens are collected by means of only a geologic pick and hammer. However, the portion of an outcrop that can be sampled with an ordinary pick and hammer is commonly greatly limited in comparison with that which can be sampled using a 2-pound sledge hammer and a heavy steel cold chisel. Thus, bias in sampling may be significantly reduced by using a chisel for sampling many types of rock.) Consequently, depending on the type of rock being studied, not all parts of the outcrop can be sampled. Specimens are generally taken only where the rock protrudes, as along bedding planes or fractures. This is especially true where the rock is hard and dense, as are many quartzites, dolomites and granites.

If an outcrop is sampled by two or more people, their consistency of sampling can be appraised, although clearly it will matter whether they visit the outcrop alone or together, whether they are a professor and his students, etc. Thus Ager and Guber (Ager, 1963, pp. 240–241) counted species represented by 850 fossils at one locality and 1000 at another in Iowa for a study of paleontological diversity. Table 7.2 compares their results by the chi-square test (sec. 6.1) as applied to a two-way classification by Dixon and Massey (1969, p. 240). Because the tabled chi-squared value at the 10-percent level with 16 degrees of freedom is 25, the sampling at Rockford was evidently consistent and that at Bird Hill was mildly inconsistent. Similarly, Griffiths (1967, Chap. 2) and his students have studied consistency in sampling sandstones.

TABLE 7.2. COUNTS OF FOSSIL SPECIES AT THE ROCKFORD AND BIRD HILL LOCALITIES, IOWA*

Rockford†			Bird Hill‡		
Name	Number of specimens		Name	Number of specimens	
	Ager	Guber		Ager	Guber
Spirifer hungerfordi Hall	156	83	Productella walcotti Fenton and Fenton	146	33
Atrypa devoniana Webster	139	103	Douvillina arcuata (Hall)	135	50
Schizophoria iowaensis (Hall)	57	31	Atrypa rockfordensis Fenton and Fenton	103	35
Spirifer whitneyi Hall	23	20	Spirifer whitneyi Hall	73	23
Spirifer whitneyi subsidus Fenton and Fenton	22	15	Schizophoria iowaensis (Hall)	67	29
Atrypa rockfordensis Fenton and Fenton	21	11	Heliophyllum solidum (Hall and Whitfield)	43	20
Douvillina arcuata (Hall)	15	14	Spirifer hungerfordi Hall	38	10
Floydia gigantea (Hall and Whitfield)	15	3	Strophonella hybrida (Hall and Whitfield)	27	4
Heliophyllum solidum (Hall and Whitfield)	13	11	Spirifer whitneyi subsidus Fenton and Fenton	25	8
Productella walcotti Fenton and Fenton	12	8	Undetermined stick bryozoans	22	8
Diaphorostoma antiquuam Webster	8	4	Cranaenella navicalla (Hall)	13	2
Petalotrypa formosa Fenton and Fenton	5	10	Charactophyllum nanum (Hall and Whitfield)	12	2
Strophonella reversa (Hall)	5	2	Strophonella reversa (Hall)	10	13
Lioclema occidens (Hall and Whitfield)	4	7	Atrypa planosulcata Webster	8	4
Paracyclas sabini White	4	1	Crinoid fragments	6	2
Worm tubes	4	1	Platyrachella cyrtinaformis (Hall and Whitfield)	5	0
Miscellaneous	13	10	Miscellaneous	19	15
TOTAL	516	334	TOTAL	752	248

* After Ager, 1963, p. 240–241.

† χ^2 (16 d.f.) = 27.

‡ χ^2 (16 d.f.) = 20.

Grout (1932) studied replicate sampling of granite, banded gneiss, and a large differentiated diabase sill. Some sampling was done by taking hand specimens judged as representative by geologists experienced in the particular rock bodies and some by taking chips from scattered points on the outcrops. Grout's tables are interesting, although not enough data are given for a statistical analysis; they reveal the state of thinking that prevailed among geologists until recently: one of worry and fretfulness—but not enough worry to do anything about it.

Outcrop Sampling by Pomona College Geologists

For several years McIntyre, Baird, Welday, and Richmond, working at Pomona College, have been intensively studying igneous rocks in the San Bernardino Mountains in Southern California, primarily to investigate chemical variation in igneous rocks in much the same way as we study chemical variation in veins of the Frisco and Fresnillo mines. Their work, all of which to date is referenced by Baird and others (1967), provides a fascinating account of the sources of variability in igneous rocks; several points are discussed here.

Richmond (1965) compared outcrop sampling by hand specimens and by diamond drill. Cactus Flat, the area investigated, is part of a larger area studied by the Pomona group in the San Bernardino Mountains and is the type locality for the Cactus quartz monzonite, the rock unit sampled. Outcrops cover about two-thirds of the area.

Although any drill could be used in principle, in practice, when samples are being taken away from roads, a readily portable one is the only one likely to compete with hand sampling. The drill used by Richmond weighs 26 pounds, not counting drill rod, core barrel, water, and tools (manufactured by Packsack Diamond Drills, Ltd., 1385 Hammond St., North Bay, Ontario, Canada). Compared with a conventional diamond drill requiring one or two trucks for transportation, this drill is truly portable; drills of intermediate sizes that can be carried by animals or helicopters are also available.

Richmond (1965, p. 54) describes his sampling plan as follows:

Most of Cactus Flat lies between 5750- and 5900-foot elevation contours, and a tracing with boundaries defined by these conditions was made on graph paper. . . . The tracing included 360 squares, each covering a map area of 200 × 200 feet. Each square was assigned a number, and numbers were drawn at random to determine the order of sampling. The center of each 200-foot square designated for sampling was occupied. Where no outcrops were seen to occur, another 200-foot square was occupied, and it was necessary to examine 46 squares in order to complete the planned sampling of 30 squares.

Sample selection was further randomized within 200-foot squares where outcrops were observed. The location of two 20-foot squares was determined by random numbers drawn from a box of tags numbered 1 to 100, and each small square was then sampled, A, by a 10-inch core (1-inch diameter) obtained by diamond drilling, and B, by a hand specimen collected by hammer and chisel within 5 feet of the core hole. In all, 120 specimens were collected, made up of two samples of 60 specimens each, collected two from each of thirty randomly selected 200-foot squares, and paired by collecting method within a 5-foot radius within squares.

Richmond kindly supplied his original data so that his results could be recomputed in one of the analysis-of-variance formats explained in this book. The results for magnesia, the only one of the six constituents for which a difference was found between the two methods of sampling, are presented here. The analysis of variance, table 7.3, partitions the variability into that associated with the different locations and sampling methods and that which is residual. The single degree of freedom comparing hammer and drill sampling was also broken out. The analysis of variance is a randomized-block design with the 30 replications due to location and sampling methods as treatments. Because the computed F-value of 6.57 comparing the two kinds of sampling with the residual variability is greater than the tabled value, the two methods of sampling are evidently different on the average for magnesium. However, with this being the only constituent for which a significant difference is obtained and with the computed F-value being rather low, one can feel reasonably happy about hammer sampling for this rock.

TABLE 7.3. ANALYSIS OF VARIANCE IN MAGNESIA DATA FROM CACTUS FLAT, SAN BERNARDINO MOUNTAINS, CALIFORNIA*

Source of variation	Sum of squares	Degrees of freedom	Mean square	F	$F_{10\%}$
Sample location	2.5013	29	0.0863	13.96	1.41
Sampling method	0.0544	3	0.0181	2.94	2.15
Hammer versus drill	0.0407	1	0.0407	6.57	2.77
Residual variability	0.5375	87	0.0062		

* Data from Richmond, personal communication, 1967.

Since this section was written studies similar to those of the Pomona College geologists have been started by several investigators, and some results have been published. Before too long much better information on how to sample outcrops should be available. Meanwhile, outcrop sampling by diamond drill rather than by hand tools is the surest way to obtain reliable samples, because of the following advantages:

1. Lack of the bias that is introduced in hammer sampling through collecting specimens that break off easily, for example, along joints.

2. Avoidance of the bias introduced in hammer sampling because the collector sees the rock to be sampled and therefore avoids rocks deemed to be "peculiar" or "atypical," rocks which may actually be part of the population to be sampled. In drill sampling, the first interval of core, say 6 inches or a foot, can be rejected to ensure that the rock type in the second interval of the sample will be initially unknown to the collector.

3. Uniform sample orientation—for example, vertical cores to make consistent the relation of the samples to any layering or lineation in the rock.

4. Uniform size of the geologic samples, which makes it unnecessary to consider sample-volume variance (sec. 7.4), even though evidently this source of variance is seldom large.

5. Uniform sample shape. Although of no known advantage, this factor may have future value. It is at present fanciful, but if someone, someday, learns that the shape of geological specimens is significant, those who collected samples of uniform shape will have data from which information can still be extracted.

7.4   DRILL SAMPLING

The proof of the pudding is in the eating. The best way to investigate the reliability of the sampling of a rock body by drilling was never put better than by I. B. Joralemon some forty years ago, when he wrote (1925, p. 614):

... in the development of the New Cornelia Copper Co., at Ajo, Ariz., over 1000 ft. (304 m.) of test pitting was done to check diamond drilling. The average of diamond-drill samples and of large channel samples checked to within 0.005 percent copper. The bulk samples, consisting of every tenth bucketful crushed and quartered mechanically, averaged 0.15 percent higher than the channel and drill samples. The drill and channel samples were accepted as correct. Several million tons of this ore have now been mined and treated. All of this ore to date has averaged 1.51 percent copper, compared with the estimate of 1.54 percent. The grade has been a little higher than the estimate in one part of the orebody and lower in another part. This property has proved that, in a disseminated orebody, it is possible to sample and estimate the ore correctly without reducing the grade by a factor of safety.

To evaluate drill sampling, mining companies generally follow a method like that explained by Joralemon, although the analyses are seldom published. When the actual mining of the rock is impracticable, other methods to investigate consistency within and among drill holes must be followed; it is these methods that are emphasized in this section. Unfortunately, for most of the example data, mining results for empirical evaluation of reliability are unavailable.

Analysis of data is explained in the following subsections. Technology of drill sampling is not discussed because sufficient information and bibliographies are given in standard works, for instance, those by McKinstry (1948), Truscott (1962), and Jackson and Knaebel (1932).

Diamond-Drill Sampling

Because it can bore a hole for a long distance in any direction, a diamond drill is a versatile sampling machine that can yield a reliable sample under the proper conditions. Required are a suitable machine, an experienced operator, appropriate operating rules, and suitable rock conditions. The hole is drilled to a usual diameter of $\frac{3}{4}$ to 3 inches (table 7.4), with an annular bit impregnated with diamonds, while water is being circulated in the hole as a lubricant. Usually a core is obtained, which provides a visual record of the geology as well as a solid-rock sample whose location is certain. Also sludge, the cuttings from the hole, is often collected in settling tanks at the collar of the hole, particularly when core recovery may be impossible or too expensive. Sometimes, composition of a rock body can be more closely determined from diamond-drill cores than from other kinds of samples that are more numerous and larger in volume, as at Naica, Chihuahua, Mexico (G. K. Lowther, personal communication), for lead, zinc ore and at the Giant Yellowknife mine, Northwest Territories, Canada (Dadson and Emery, 1968), for gold ore.

TABLE 7.4. DIAMOND-DRILL HOLE AND CORE DIAMETERS

Name	Core diameter		Hole diameter	
	in.	cm	in.	cm
XRT	0.75	1.905	1.1875	3.01625
EXT	0.9375	2.38125	1.5	3.81
AXT	1.28125	3.254375	1.875	4.7625
BX	1.625	4.1275	2.3125	5.87375
NX	2.125	5.3975	2.9375	7.46125

Besides in the general references cited at the beginning of this section, diamond-drilling technology is fully described in a book by Cumming (1956), who includes references to the periodical literature. Technology is rapidly changing because more and more diamond-drill holes are being bored; demands are made to drill increasingly difficult rock types; and wages, a large part of the total cost, are increasing more rapidly than the prices of equipment and diamond bits. Diamond drilling is increasingly used to sample rocks for purposes other than the directly economic ones such as discovering or estimating grade of ore and evaluating dam sites; for instance, in 1967 the U.S. Geological Survey bored several thousand feet of holes in one area of Nevada alone to investigate regional stratigraphy and structure.

The geologist embarking on diamond-drill sampling must answer a number of questions.

1. Should sludge as well as core be collected?
2. What size core should be taken?
3. At what intervals should the core and sludge be sampled, and, for each interval, how many replicate increments of rock material should be taken for chemical or other analysis?
4. How should the reliability of the sampling results be appraised?

Statistics can aid in making these decisions. Those decisions concerned with relations within a single drill hole of the core and sludge are considered here first; then those that relate the samples to the rock body are discussed.

Sludge and core assays seldom agree, partly because sludge is mixed before coming out of the hole, but mostly because losses of the valuable mineral or minerals are almost certain to be different for the two; for instance, McKinstry (1948, p. 95) records that at the Colquiri tin mine, Bolivia:

> In twelve drill hole intersections, core recovery averaged 84.3 percent, most of the missing core representing sections of friable vein-matter. Core assays averaged 1.44 percent tin; core and sludge assays, combined in proportion to theoretical volume, averaged 1.87 percent tin as compared with channel samples for the same veins, though not in the identical places, which averaged 1.89 percent tin.

Obviously, tin that was lost in the core turned up in the sludge. Because of such biases, sludge should be taken unless (a) the diameter of the hole is large enough so that the amount of sludge proportional to that of core is small, (b) core recovery is excellent, (c) although core recovery is less than excellent, reliable weighting factors have been developed empirically, or (d) cost of obtaining sludge is prohibitive. Systematic core and sludge data from many drilling programs would be highly informative but do not seem to have been published.

An ideal geological sample from a diamond-drill hole would consist of 100 percent of the rock displaced recovered in core and sludge, the chemical analyses of which could then be averaged by weighting them according to the cross-sectional areas of the core and of the annular opening between the core and the wall of the hole. McKinstry (1948) and Cumming (1956) supply full details. However, core recovery is never 100 percent because pieces of rock break up in the hole and are ground up by the drill bit. Sludge recovery is usually worse, since some sludge is not recovered for one or more of many causes: because of insufficient water pressure to wash it out of the hole, because it is washed into cracks in the rock, or because, although washed out of the hole, it does not settle in the tanks provided at the collar of the hole. For these reasons percentages of recovery must be estimated and the weighting factors adjusted.

The weighting factors must be determined empirically because the observations for core and sludge constitute two statistical samples, each with its own mean and variance. For instance, at the Chuquicamata copper mine, Chile, Waterman (1955) published in a careful study the example data of table 7.5.

TABLE 7.5. SLUDGE AND CORE ASSAYS FROM DIAMOND-DRILL HOLES IN THE CHUQUICAMATA COPPER MINE, CHILE *

Item	Core recovered (%)	Core assay (% copper)		Sludge recovered (%)	Sludge assay (% copper)	
		Soluble	Insoluble		Soluble	Insoluble
Observation	44	0.46	0.39	81	0.72	0.13
	50	1.12	0.35	67	0.72	0.10
	58	2.42	0.21	95	1.35	0.10
	57	0.95	0.22	102	1.31	0.10
	61	0.30	0.23	73	0.95	0.11
	80	1.51	0.23	48	0.98	0.12
	64	1.18	0.13	115	1.05	0.12
	83	1.24	0.28	91	1.07	0.10
Mean	62.13	1.14	0.26	84.0	1.02	0.11
Standard deviation	13.52	0.65	0.08	21.3	0.23	0.01
Coefficient of variation	0.22	0.57	0.32	0.25	0.23	0.11

* After Waterman, 1955, p. 60.

Taking for simplicity only the soluble copper, one could theoretically combine these assays according to the formulas

$$\bar{w} = \frac{k_1 \bar{w}_c + k_2 \bar{w}_s}{k_1 + k_2}$$

and

$$s_{\bar{w}}^2 = \left(\frac{k_1}{k_1 + k_2}\right)^2 \frac{s_c^2}{n} + \left(\frac{k_2}{k_1 + k_2}\right)^2 \frac{s_s^2}{n} + 2 \frac{k_1 k_2}{(k_1 + k_2)^2} \frac{s_c s_s}{n} r_{cs},$$

where k_1 and k_2 are percentage recoveries of core and sludge, r is the correlation coefficient (sec. 9.3), and the subscripts c and s stand for core and sludge. As Waterman points out, it is impossible to determine if this estimate has any validity elsewhere in the mine. Therefore Waterman's empirical solution is satisfactory.

Because geological interpretation can be made of the standard deviations and coefficients of variation of table 7.5, a good reason for calculating them can be provided. In the core the coefficient of variation, which is lower for the insoluble than for the soluble copper, presumably reflects a more uniform mineralization. In the sludge the same relation holds; moreover, the coefficients of variation for both soluble and insoluble copper are lower than those for the core, a fact which shows that the sludge must be mixed from one interval to another. More insoluble copper was lost in the sludge than in the core.

Sludge and core observations can be compared by a paired t-test to learn if they are evidently the same or different. Applied to the Chuquicamata data in table 7.5 to compare sludge with core assays, paired t-values are -0.66 for soluble copper and -4.98 for insoluble copper. The first of these values is outside and the second is inside the critical region defined by the tabled t-value of 1.895 for 7 degrees of freedom at the 10-percent level. The conclusion is that core and sludge assays agree for the soluble copper but not for the insoluble copper.

Sample-Volume Variance

For drilling in most geological environments, as core diameter increases, so does cost per foot. Although larger holes are easier and quicker to drill than smaller ones, the higher diamond consumption and cost of the larger equipment outweigh the savings in labor. However, recovery of a large-diameter core is better than that of a small core and should yield a geological sample that is better because its volume is larger. Although the mean of observations based on large volumes of rock should not be expected to be different on the average from those based on small volumes of rock, the corresponding variances might be expected to be smaller. The theory of sample-volume variance that has been advanced states that variance of observations is inversely proportional to sample volume. However, the limited published data suggest that decrease in variance is not related to increase in volume in any simple way, and for some constituents and some rocks the decrease is insignificant within the range of practical sampling.

Some calculations for sample-volume variance are demonstrated for a small set of data published by Krige (1966, p. 74), consisting of gold values from 12 drill holes ($1\frac{3}{8}$ inch diameter) bored through a narrow gold-bearing conglomerate bed. The cores were sawn in half longitudinally, and the two halves assayed. The results are in table 7.6. The in.-dwt values for whole core are the averages of the two halves; the means and variances in the last two lines of the table are nearly the same. If the values from the two halves of the core

TABLE 7.6. TEST OF SAMPLE-VOLUME VARIANCE IN DRILL CORE FROM A SOUTH AFRICAN GOLD MINE *

Borehole number	In.-dwt value		
	First half core	Second half core	Whole core
1	180	235	207.5
2	131	123	127.0
3	182	173	177.5
4	135	117	126.0
5	91	113	102.0
6	117	112	114.5
7	109	90	99.5
8	179	178	178.5
9	100	58	79.0
10	227	204	215.5
11	304	226	265.0
12	279	285	282.0
$\bar{w}$	169.5	159.5	164.5
s^2	4893.18	4642.45	4508.68

* Data from Krige, 1966, p. 74.

were statistically independent observations from a single population, the ratio of the variance of the values for the whole core to the average of the variance of the values for the two halves should be one-half. But, because the calculated ratio of 0.94 is only slightly less than 1, little is gained by assaying the whole core rather than only one-half of it. Table 7.7, which gives the results of a randomized-block analysis of variance of these data, shows that practically all of the variability is among the bore-holes rather than between the core halves. Thus, the 12 boreholes must be considered to have sampled different statistical populations with different means.

TABLE 7.7. RANDOMIZED-BLOCK ANALYSIS OF VARIANCE FOR THE DATA OF TABLE 7.6

Source of variation	Sum of squares	Degrees of freedom	Mean square	F	$F_{10\%}$
Among bore holes	99,103.00	11	9009.36	17.12	2.23
Between core halves	600.00	1	600.00	1.14	3.23
Residual variability	5,789.00	11	526.27		

TABLE 7.8. SAMPLE-VOLUME VARIANCE IN DRILL CORE FROM TWO HOLES IN THE
CLIMAX MOLYBDENUM MINE, COLORADO*

Line	Item	Volumes and variance ratios		
1	Relative volume	1	2	3
2	Volume (cc)	218	436	654
3	Theoretical ratio of variances	1	1/2	1/3
	Observed ratios of variances			
4	Hole 1	1	0.89	0.83
5	Hole 2	1	0.76	0.67
	Observed ratios of variances with location effects removed			
6	Hole 1	1	0.64	0.72
7	Hole 2	1	0.98	0.78

* Data from Hazen, 1962, p. 21.

The results of two other investigations agree with those detailed for Krige's gold data. In the first, conducted by Hazen (1962), two NX size drill holes were bored at the Climax molybdenum mine in Colorado. The 101 feet of core from the first hole and the 52 feet from the second hole were sawn into 2-foot-long intervals, and then longitudinally at angles of 60, 120, and 180 degrees, to yield relative volumes of 1, 2, and 3. In lines 4 and 5 of table 7.8 are the observed variance ratios for the two holes compared with the theoretical ratios; the observed variance is not inversely proportional to the sample volume. Because within each hole there is a slight systematic change in ore grade which might affect the variance ratios, this trend was removed (sec. 9.5), and the data were analyzed again. The results are in lines 6 and 7 of the table, for which, as before, the variance is not inversely proportional to the volume, between-location variability being large compared with within-location variability.

Another study is of gold assays from nine hundred 1-foot intervals of EX drill core obtained from several drill holes bored in the course of regular company operations in the Homestake mine. We combined (table 7.9) the gold assays for the 1-foot intervals to simulate the assays that would have been obtained, assuming perfect sample preparation and assaying, from 2-, 3-, 4-, and 5-foot intervals. As is true for the previous examples, the observed variances are not inversely proportional to the sample volume.

For these three examples variance does not decrease linearly as sample volume increases, although there is a tendency for a small decline. In general, variance probably declines as volume increases, but we do not know of any data for which sample-volume variance holds, although doubtless there is some. The reason is that, for real data, the among-location variability is large

TABLE 7.9. SAMPLE-VOLUME VARIANCE IN DRILL CORE FROM THE HOMESTAKE GOLD MINE, SOUTH DAKOTA

Line	Item	Volumes and variance ratios				
1	Interval length (ft)	1	2	3	4	5
2	Volume (cc)	129	258	387	516	645
3	Observed variances (ppm)	327	218	166	141	125
4	Theoretical variances (ppm)	320	160	107	80	64

compared with the within-location variability; and increasing the sample volume reduces the effect of within-location variability but has no effect on the among-location variability. Because the total variability is a weighted sum of these two kinds of variability, reducing the smaller source does not appreciably reduce the sum.

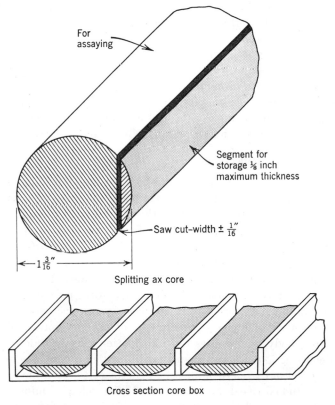

Fig. 7.2. Core sawn longitudinally with a diamond saw (after Cumming, 1956, p. 238).

If sample-volume variance were important, it would be desirable to assay all or most of the core from a drill hole; and the common industry practice of splitting the core longitudinally, saving one half for a record and assaying the other half, would be bad. A method increasingly used, which is conservative if little is known about the variance relations, is to save a thin segment sawn longitudinally out of the core, as illustrated by figure 7.2. This method is desirable if rock lithology is of particular interest and if the valuable constituent has an erratic distribution, as in sampling precious metal ore from a geologically unknown environment.

Relation of Drill-Hole Samples to the Rock Body

The relation of the drill-hole samples of core and sludge to the rock body sampled can be compared by taking a bulk geological sample of the rock body, by considering the variance of multiple drill holes in a rock body on various scales, and by comparing observations in deflected drill holes. Unfortunately, no generalizations can be made because practically all of the studies extant have been made by individual companies in particular environments, and the information has not been published.

Mining companies commonly check results of drill-hole samples by taking bulk samples, as explained in the quotation from I. B. Joralemon at the beginning of this section. Sometimes the means calculated from the two kinds of sampling agree; sometimes they do not. When they do agree, the variability of the drill-hole samples is generally a little higher than that of the bulk samples, but not as much as would be expected if sample-volume variance applied. Comparisons may be made by an analysis of variance (Chap. 5).

Diamond-Drill-Hole Deflections

The final subject to be discussed about diamond drilling is drill-hole deflections. When diamond-drill holes are bored to obtain information about the grade of an ore body, quite often the reliability of the information obtained from the small-diameter cylinder of rock and/or from the accompanying sludge is in question. Then, the diamond-drill bit may be deflected so that a second hole is bored through the ore body or other geological entity of interest. The bit is deflected by placing a wedge-shaped piece of metal in the original hole to force the diamond-drill bit out of the hole at an angle. Even though one does not expect several samples close to one another to provide as good information as the same number of samples spread more evenly across a rock body, nonetheless the information may be valuable, and the cost of deflecting is much less than that of drilling a new hole, particularly when many feet of barren rock must be penetrated to reach the rock body.

We have studied (Koch and Link, 1969) the value of information from deflected diamond-drill holes by analysis of 795 assays from original intersections and deflections of boreholes in South African gold mines, information supplied us by D. G. Krige (1967, personal communication). When these were arranged in order of size and a cumulative sum tabulated, we found that 21 percent of the value was in only three observations with the extremely high values of 12,528, 15,317, and 23,037 in.-dwt. Although these observations cannot be considered exceptional, they verge on being extreme even for this highly skewed lognormal distribution, with a coefficient of variation of 3.79. The remaining calculations are based on 792 observations—the three highest ones are omitted because their inclusion obscures meaningful comparisons. Had, say 100,000 observations been available, the percentage of extremely high values could have been estimated, and disregarding the three highest ones would have been unnecessary. The problem is one often met in geology where highly skewed distributions are common (secs. 6.2 and 6.3).

Table 7.10, the upper half of the correlation matrix (sec. 9.2) for boreholes with deflections, shows that the correlations are of moderate size. Were they perfect (equal to 1), the deflections would provide no new information about the mean grades of the gold-ore bodies.

The correlations in table 7.10 can be used to calculate the relative variances of the mean for holes with no deflections, one deflection, two deflections, etc. The relative variances for from zero to three deflections are presented in table 7.11. If the observations from the original intersections and the deflections were uncorrelated, the relative variances in the column farthest to the right would be obtained, that is, $1/2 = 0.500$, $1/3 = 0.333$, and $1/4 = 0.250$, because the variance is inversely proportional to the number of observations per hole according to the formula $\sigma_{\bar{w}}^2 = \sigma^2/n$ (standard error of the mean law). However, because the original and deflection observations are correlated, the variances of the means based on combining them are larger than $1/2$, $1/3$, and $1/4$. Thus, an increase in precision is gained by making deflections, but not so large an increase as that obtained by drilling a new borehole, from which to get an independent observation.

TABLE 7.10. CORRELATION MATRIX FOR BOREHOLES WITH DEFLECTIONS IN SOUTH AFRICAN GOLD MINES

Item	P	(0)	(1)	(2)	(3)
Original intersections deflected	(0)	1.0	0.537	0.821	0.592
Deflection 1	(1)		1.0	0.439	0.213
Deflection 2	(2)			1.0	0.589
Deflection 3	(3)				1.0

TABLE 7.11. RELATIVE VARIANCES OF FROM ZERO TO THREE DEFLECTIONS OF OBSERVATIONS FROM BOREHOLES IN SOUTH AFRICAN GOLD MINES

Number of deflections per hole	Number of observations per hole	Number of boreholes	Observed relative variance of mean for hole	Relative variance if original intersections and deflections were entirely uncorrelated
0	1	241	1.0	1
1	2	241	0.768	$1/2 = 0.500$
2	3	101	0.733	$1/3 = 0.333$
3	4	27	0.649	$1/4 = 0.250$

Whether these results hold for other geological situations we do not know, but analysis of these data suggests that deflections are well worth the additional cost in these South African mines and presumably in other situations as well.

Drill Sampling with Other Than Diamond Drills

Drill sampling with other than diamond drills is not discussed in detail in this book for two reasons: (a) the general procedures for statistical analysis of diamond-drill-hole sample data should apply, and (b) few detailed recent data have been published although many reside in company files. However, a few differences in the geological characteristics of the samples taken by the various methods of drilling may be mentioned; the references cited at the beginning of this section supply full details on the engineering problems.

Cores are sometimes cut by methods other than diamond drilling. If the cores are of small diameter, obtained by drilling with bits other than diamond bits or by driving a split pipe into sediments, the statistical analysis should be the same as for data from diamond-drill holes. If the cores are of large diameter, as those obtained by calyx drilling, sample preparation must be carefully controlled to avoid introducing variability; in this way, enough core should be obtained so that sludge can be ignored.

Other methods of drilling yield only cuttings, rather than both cuttings and core. In churn drilling a hole usually 12 inches or smaller in diameter, although sometimes larger, is drilled in a wet hole by lifting a cable-hung bit and dropping it to impart a churning motion that breaks up the ground; the mud thus formed can be bailed out at intervals. Holes must, of course, be bored vertically downward. The large volume of sample may compensate for the mixing and for contamination from the walls, contamination that can be controlled by casing the hole. McKinstry (1948, pp. 71–81) gives full details on precautions in churn drilling and in sample handling. Applications to placer sampling are given in section 15.2.

In recent years cuttings have been increasingly obtained by rotary drilling with a noncoring bit, with the advantages over diamond drilling of greater speed and lower cost. Holes are drilled either wet, with the cuttings flushed out of the hole in the return flow of the water or mud, or dry, with the cuttings blown out with air. Studies have been made to compare sampling in rotary drilling with other kinds of sampling; some results, not published with detailed numerical information (Harding, 1956), suggest that mixing of cuttings from one interval to another and incomplete blowing out of particles in dry drilling may be serious problems.

In many underground mines samples are taken from holes bored with the air drills used to drill blast holes (McKinstry, 1948, p. 106). The samples have the advantage of being inexpensive, but the disadvantage is that, in cases of which we are aware, correspondence of results with those from diamond-drill-hole samples or from channel or chip samples (sec. 7.5) is poor. Another problem is that annular corrugations that sometimes form in the holes lead to the riffling out of heavy particles. Again the data necessary to evaluate this kind of sampling quantitatively are not available.

7.5 CHANNEL AND CHIP SAMPLING

In channel sampling a slot or channel is cut in a rock face. The rock fragments broken out of the slot constitute the geological sample. The dimensions of the channel may be maintained by cutting it so that a board of appropriate size, say 2 by 6 inches in cross section, fits into it. In chip sampling, a similar procedure is followed, except that small chips are collected across the face, and no attempt is made to cut a channel. McKinstry (1948, pp. 37–45) and Truscott (1962, pp. 10–12) give full details; Parks (1949, pp. 76–90) explains application in the copper mines at Butte, Montana.

Channel and chip samples are commonly taken in underground metal mines. These samples provide most of the experience and published data on these two methods; but there is no reason why the methods cannot be applied in academic geology to outcrop sampling. Formerly, it was a mark of a competent mining engineer to insist on channel sampling; but today, with increased labor costs, chip sampling or broken-rock sampling (sec. 15.1) is more common for routine control of ore grade. If more accurate results are required, the tendency is to use a much larger bulk sample derived from driving a mine working (sec. 15.4).

In metal mines channel or chip samples are taken systematically at regular intervals in the mine workings. They may be cut from the face, walls, back (roof), or, rarely, floor of tunnel-shaped openings and at various places in the other workings. The length of an individual channel or line of chips tradition-ally is no longer than 5 feet, but shorter intervals are taken when mineralogy changes obviously, particularly if the minerals differ in hardness or friability. In countries that use the metric system English- and American-trained engineers cut channels that are 2 meters long.

Better results can be expected from samplers who are entirely disinterested in the results of their work rather than from technically trained men who are liable to introduce prejudice. The educated man tends to include too much high-grade material because of optimism or alternatively, too much low-grade material for a "safety factor." In South Africa excellent results are obtained by native samplers, and in Latin America one of us has seen good sampling done by a carefree crew that sampled in the morning, slept in the afternoon, and played guitars all night in a local bar.

Channel or chip sampling of outcrops is done in the same way as of under-ground workings and chip sampling of outcrops should be inexpensive enough to do routinely in academic geology. Whether most academic geologists sample outcrops by single hand specimens or by chips is a mystery, because this interesting information is seldom revealed in their papers. The best argument for chips is that wider coverage and the advantages of averaging are obtained; the best argument for single hand specimens is that the chemical or modal analysis is related directly to the thin section, the mineralogy of which is studied.

Data Analysis

Common mine practice is to check the assays of channel or chip samples by cutting new samples from the same locations and comparing the assays obtained. Means and standard deviations computed from each set of observa-tions should be close to one another although individual observations seldom agree, the differences between original and check observations naturally being greater for the higher observations. In figure 7.3 are graphed typical data of 267 pairs of check chip samples from an unidentified South African gold mine (Rowland and Sichel, 1960, p. 256). .

Therefore, although in channel and chip sampling individual check observa-tions are not expected to be the same, agreement on the average can be sought and tested for by the paired t-test. The procedure is demonstrated in table 7.12, which lists 20 check-sample gold values from another unidentified South African gold mine (Rowland and Sichel, 1960). Because these data were taken by two different samplers, the effect of natural variability is con-founded to an unknown extent by the difference in technique between the two men. The African chipper obtains a mean grade that is 1.57 ppm higher than that of the European sampler, but significantly, except for the one

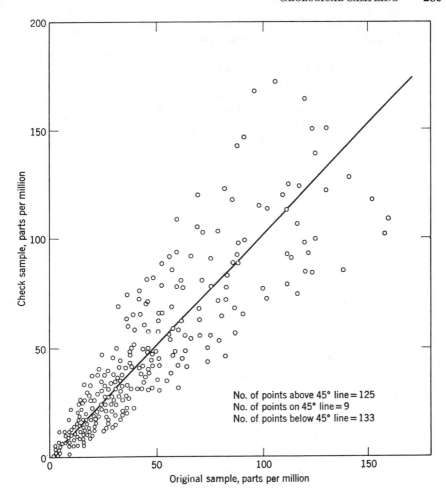

Fig. 7.3. Assays of paired-chip samples from a South African gold mine (after Rowland and Sichel, 1960, p. 256).

extremely high-value location whose paired values are 78.00, 22.29, the African in most places cut lower value samples and most of the discrepancy stems from this one location.

As often happens with real data, the precise statistical analysis to use is obscure. A paired t-test applied to all the observations yields a t-value of 1.54, which is outside the critical region defined by the tabled value of 1.729 with 19 degrees of freedom at the 10-percent significance level. However, arguing that the difference of 55.71 is an outlier (sec. 6.4) that should be removed, one obtains a t-value of -3.92 with 18 degrees of freedom, which is inside the

Table 7.12. Check-sample gold values from an unidentified South African gold mine [*]

Check-sample gold values (ppm)		
African chipper	European sampler	Difference
5.83	8.91	−3.08
3.77	7.37	−3.60
1.54	1.37	0.17
2.91	3.43	−0.52
9.09	10.29	−1.20
4.63	6.17	−1.54
3.77	6.17	−2.40
18.00	21.43	−3.43
78.00	22.29	55.71
1.71	1.54	0.17
1.37	3.26	−1.89
2.23	4.63	−2.40
5.14	3.77	1.37
4.29	4.63	−0.34
4.11	4.46	−0.35
15.77	17.14	−1.37
0.86	1.89	−1.03
1.37	3.94	−2.57
2.74	3.77	−1.03
4.46	3.60	0.86
Sum 171.59	140.06	31.53
Mean 8.57	7.00	1.57

[*] Data from Rowland and Sichel, 1960.

critical region. Alternatively, the number of positive and negative differences may be compared by a sign test (Dixon and Massey, 1969, p. 335), which has the advantage of being distribution-free and which also leads to the conclusion that the European sampler is somewhat higher on the average than the African chipper. No definite conclusions can be reached from the small amount of published data, although the evidence points to some bias between the two men.

Channel and chip sampling have often been compared with one another, and the conclusion has usually been drawn that the additional expense of channel sampling is unjustified, provided that the chip sampling is carefully done. Likewise channel and chip sampling have been compared with diamond-drill-hole sampling at many mines, with varied results. Few numerical results of these investigations have been published.

The best way to evaluate channel and chip sampling usually is to compare them to bulk sampling from the same mine workings or with mine production. In section 15.4 some comparisons made at the Kilembe mine, Uganda, are given. Another interesting study was made by Bradley (1923) in the Alaska Juneau gold mine, where highly variable ore of a low grade required careful evaluation. Paired t-tests applied to Bradley's data show no inconsistency among channel samples, muck samples, and millheads.

7.6   MODAL ANALYSIS

Modal analysis is a method to estimate the mineralogical composition of rocks, either by superimposing a square grid on a thin section and recording the mineral species present at each grid intersection or, alternatively, by measuring the lengths of mineral intersections on traverses evenly spaced across the grains. Modifications are possible, such as making plane sections other than thin sections. Modal analysis has been discussed in a book by Chayes (1956) and also in other works, of which some of the more significant ones later than Chayes' book are reviewed by Griffiths (1967, Chap. 9).

In this section we review some statistical problems in modal analysis and reinterpret the results of a key experiment performed by Chayes and Fairbairn (1951) to evaluate operator variability. Only point counting is discussed because it yields results equally as good as traversing and is easier to do.

The principles of modal analysis are simple to state. Suppose that a rock is composed of minerals that can be visually distinguished. If n points could be taken at random in the rock and the mineral species present at each point recorded, then the ratio of the number of points for each mineral to the total number of points is an estimate of the volumetric proportion of the particular mineral in the rock. Moreover, if the points were randomly distributed over a plane that randomly cut the rock, the same argument holds.

Although it is impracticable to take points at random throughout a rock and inconvenient to take points at random on a plane (thin section) passing through a rock, taking points on a plane systematically but with a random starting point yields statistics having the same properties. The statistical analyses for the counts are associated with the binomial distribution (sec. 6.1) under the assumptions that no difficulties arise in identifying the mineral at each point and that the thin section is chosen from the rock at random. However, if the grain size of the rock is large, binomial theory will underestimate the counting error unless more than one thin section is taken (Chayes, 1956, p. 93).

The hardest problem in modal analysis is mineral identification. Skill is required to identify the mineral at each point, even if the minerals counted have been selected to minimize the difficulty (as by grouping together all opaque minerals, or both feldspars in a two-feldspar rock). It is even more difficult to measure the length of a traverse across a given mineral because to the mineral identification problem is added the necessity of recognizing accurately the boundaries between minerals.

Aside from the problems of excessive grain size and operator variability, the maximum accuracy to be expected from point counting may be specified. Table 7.13 gives the standard errors of point counts calculated from the binomial distribution; for instance, if a mineral makes up 20 percent of a rock, 100 points must be counted to reduce the standard error of the mean to an absolute accuracy of 4 percent or to 20 percent (4/20) of the content of the mineral in the rock. The table therefore provides a sound basis for calculating how many points should be counted to obtain a desired accuracy in modal analysis of a rock whose mineralogical composition is more or less known.

The crucial problem in modal analysis is operator variability, which is the variability introduced by different persons counting the points. Chayes and Fairbairn (1951) report on two experiments conducted at the Massachusetts Institute of Technology (MIT) in which five graduate students were trained in the point-counting technique. Each operator examined each of five thin sections cut from a single piece of granite from Westerly, Rhode Island, on a schedule devised as a Latin square experimental design (Dixon and Massey, 1957, p. 171). The entire experiment was repeated six months later. The original results are tabulated by Chayes and Fairbairn.

Table 7.14 is an analysis of variance of quartz observations from the second experiment; analyses of variance for the other data gave similar results. The

TABLE 7.13. STANDARD ERRORS OF POINT COUNTS CALCULATED FROM THE BINOMIAL DISTRIBUTION

Number of points counted	Percentage of mineral in rock			
	1	5	20	50
25	1.99	4.36	8.00	10.00
100	1.00	2.18	4.00	5.00
400	0.50	1.09	2.00	2.50
900	0.33	0.73	1.33	1.67
1600	0.25	0.55	1.00	1.25

TABLE 7.14. ANALYSIS OF VARIANCE OF QUARTZ OBSERVATIONS IN AN EXPERIMENT
TO DETERMINE OPERATOR VARIABILITY IN MODAL ANALYSIS *

Source of variation	Sum of squares	Degrees of freedom	Mean square	F	$F_{10\%}$
Replication	0.838	4	0.209	0.19	2.48
Among operators	11.154	4	2.788	2.48	2.48
Among slides	13.438	4	3.359	2.99	2.48
Residual variability	13.481	12	1.123		

* Data from Chayes and Fairbairn, 1951.

four sources of variability are the replications (because there were five tests,
perhaps on different days, or perhaps the operators were more tired for one
than for another), the operators, the slides, and the residual variability. The
table indicates slightly more variability among the analysts and definitely
more variability among slides than would occur by chance. The among-slide
variability is large enough to indicate that the rock was of distinctly different
composition at the five places where the slides were cut. The replication
variability is extremely small. [Chayes and Fairbairn (1951) pooled the
replication and residual mean squares to obtain analytic error, but because
the replication mean square is so small, they probably underestimated the
analytic error in these data.]

Table 7.15 compares the observed and the theoretical mean squares for all
the data. Except for biotite, the residual mean squares are smaller in the
second MIT test, an indication that the operators, after six months' experience, had improved their techniques. Most of the observed mean squares

TABLE 7.15. OBSERVED AND THEORETICAL MEAN SQUARES FROM AN EXPERIMENT
TO DETERMINE OPERATOR VARIABILITY IN MODAL ANALYSIS *

Mineral	Residual mean square		Theoretical binomial mean square
	First MIT test	Second MIT test	
Quartz	1.65	1.12	0.79
Microcline	2.36	1.40	0.91
Plagioclase	1.29	0.689	0.87
Biotite	0.157	0.403	0.12
Muscovite	0.160	0.082	0.047
Opaque minerals	0.045	0.013	0.030
Non-opaque minerals	0.020	0.012	0.016

* Data from Chayes and Fairbairn, 1953, p. 710.

were larger than those predicted by binomial theory. Therefore some variability remained in excess of that expected if the operators were perfect; however, this excess variability was only 50 percent at most. In the first test the results of operator 4 were distinctly different from the others, and in the second test his results were still different, although less so. We have not investigated these data in enough detail to learn whether operator 4 was evidently better or worse than the others. The different results presumably stem from identification problems.

In conclusion, if a competent operator makes the counts, point counting is an effective technique for finding volumetric mineralogy whose accuracy can be reasonably well predicted from binomial theory. The conversion, if required, to chemical analyses is a further step not considered in this book.

7.7   OTHER SAMPLING METHODS

In this section we mention some other methods for geological sampling that do not fit into the previous sections of this chapter. These methods are only a few of many; each specialized field has its particular problems, and we can mention only some.

For sampling in paleoecology and also in paleontology in general, Ager (1963, especially pp. 220 ff.) offers good advice. Krumbein (1965) has also written an article on sampling in paleontology, with a good list of references. Mosimann (1965) has written a long article on statistical methods for pollen analysis, based on binomial and multinomial models (sec. 6.1).

Sampling of uranium ores and statistical analysis of the data present special problems because the composition of the samples changes rather rapidly with time and because chemical analyses must be reconciled with gamma-ray data. These problems have been studied by Grundy and Meehan (1963), by Schottler (1965), and by others; but no very complete investigation has yet been published.

7.8   RANGE OF GEOLOGICAL VARIABILITY

Above all, devising a good plan for geological sampling depends on knowing or estimating the variability in the geological entity to be sampled. Thus in their studies of the Southern California batholith (sec. 7.3) Baird and others (1967, p. 11) write that "evaluation of geochemical distributions over heterogeneous plutonic assemblages . . . requires considerable knowledge of the chemical variability within the individual rock bodies." We have found this true in sampling gold deposits (Chap. 16).

In this section we summarize a few facts and introduce some concepts about the range of geological variability. Variability depends on the quantity of the substance studied in the rock body, on the chemical form of the substance, and on the geological process or processes that formed the rock body and the substance. The dependence is illustrated for the element carbon.

A few data for the element carbon in coal, graphite, and diamond demonstrate the great range of variability and point up the difficulty in obtaining the information needed to evaluate variability. The carbon content of coal ranges from more than 95 percent in anthracite to less than 70 percent in lignite; with such high amounts of carbon present, coefficients of variation are low; a typical one, calculated from 30 observations made on the Pittsburgh coal bed (Pennsylvanian) in 30 counties of Pennsylvania is 0.0087, with a mean of 84.58 percent. For graphite mines, carbon content ranges from a few percent up to 30 percent; a typical coefficient of variation from 188 trench samples in the Benjamin Franklin and Just graphite mines in Pennsylvania is 0.61, with a mean of 1.66 percent.

The variability of carbon content in the form of diamond in diamond deposits must be extreme. Giles (1961, p. 840) writes that typical South African diamond mines that work kimberlite in place have grades of 0.079 ppm at the Premier mine and of 0.018 ppm at the Jagersfontein mine; these grades are only about one-fourth to one-tenth as large as those at the lowest economic limit of gold dredging. Although we have been unable to find size distributions for a complete spectrum of diamond sizes, those for large diamonds are highly skewed. As some mines, such as Jagersfontein, owe their profitability to the larger stones, we may surmise coefficients of variation of the order of magnitude of 1000. In summary, variability in carbon depends mainly on the amount of carbon in the rock body, but also on the geologic processes that formed the carbon-bearing mineral and the rock, which clearly are very different for coal, graphite, and diamond.

A complete study of the range of geological variability would treat all minerals and chemical elements, or at least the common ones, and would be most worthwhile.

In studying geological variability, investigators have used one or another of the methods explained in this book. In a series of papers Ahrens (1963, with references to his earlier papers) confines himself to frequency distributions and histograms, which are not satisfactory for this purpose. A better measure is sample variance, s^2, even if the distribution under study is not normal. However, for most purposes the coefficient of variation, C, is most informative because it gives a relative measure of variability which takes into

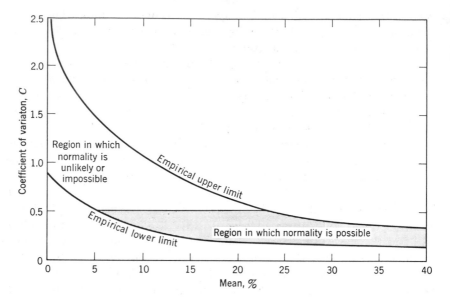

Fig. 7.4. General relation between the mean and the coefficient of variation.

account both the mean and the variance; therefore, the comparisons in this section are made with it.

Figure 7.4 graphs the general relation between the mean and the coefficient of variation. As the mean, expressed in weight percent, increases, the coefficient of variation must tend to decrease until, when the mean reaches 100 percent (a point not shown on the figure), the coefficient of variation must be zero. On the figure are sketched schematically the upper and lower limits for coefficients of variation, based on empirical data discussed in the next subsection; the data are insufficient to define these limits closely. Notably, extremely high coefficients of variation, above say 2 or 2.5, are found only for substances present in minute proportions, such as trace elements or precious metals in ores. On the other hand, extremely low coefficients of variation, below say 0.2, are generally found only for geological substances present in amounts measured in the range of a few tens of percent.

On figure 7.5 the region of expected data, which lies between the empirical upper and lower limits of the coefficient of variation, is divided into two subregions—one in which observations may, but need not, follow a normal frequency distribution and the other in which normality is unlikely or impossible. The dividing line is arbitrarily defined by the fact that, for a distribution of nonnegative observations to be approximately normal, its mean must be a few standard deviation units above zero. If the not very rigorous criterion is adopted that the mean be two standard deviation units above zero, then the

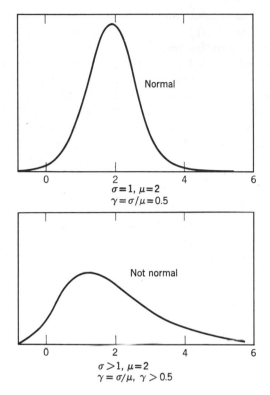

Fig. 7.5. Contrast between normal and non-normal distributions in relation to the coefficient of variation.

coefficient of variation must be less than 0.5 for normality to be possible although not necessary (fig. 7.5). On the graph the plotted horizontal line corresponding to a coefficient of variation of 0.5 shows that substances whose means are between about 5 and 25 percent and whose coefficients of variation are less than 0.5 may be normally distributed; that is, an observed coefficient of variation of less than 0.5 does not guarantee normality, but it does make it a reasonable assumption. As pointed out in section 6.2, the coefficient of variation must be greater than 1.2 before the efficiency of the normal assumption drops enough to make it worthwhile to consider making a transformation.

In summary, figure 7.4 shows that if a substance is present in a certain amount, say 10 percent, the coefficient of variation may be expected to be between about 0.3 and 1.1; and, if it is below 0.5, the observations may be normally distributed. The size of the coefficient of variation must depend fundamentally on the geological variability, provided that the variability introduced by the sampling process is well controlled (Chap. 8). Thus this graph points up once again that postulated "laws" about distribution of elements have little if any meaning. Quantification of the empirical and arbitrary boundaries on this graph would be a fruitful study.

Geological variability depends on the processes that formed the rock. The specific sources of variability include the chemistry and amount of the substance, the geologic processes that formed the rock, and the geologic processes that formed or redistributed the substance in the rock. Thus for gold (Chap. 16), coefficients of variation are, as expected, high for deposits in metamorphic rocks and low for deposits of the Carlin, Nevada, type, where the size of the gold grains is minute. For all substances, data both existing and potential could be organized for the improvement of sampling practice.

Observed Coefficients of Variation

Ideally, observed coefficients of variation for a great variety of geologic substances and environments would be discussed in this subsection, but this ideal cannot be realized because data are lacking. Therefore figure 7.4 had to be based on the data listed in table 7.16 and graphed in figure 7.6—data which are coefficients of variation calculated from 50,057 observations made on various substances in 484 mineral deposits. These data (Hazen and Meyer, 1966, p. 5) are a veritable rag bag of miscellaneous observations derived from samples from churn-drill holes, diamond-drill holes, cut channels, and test pits; the data sources are detailed in the original paper. As these data were taken for many different reasons and were statistically analyzed by Hazen

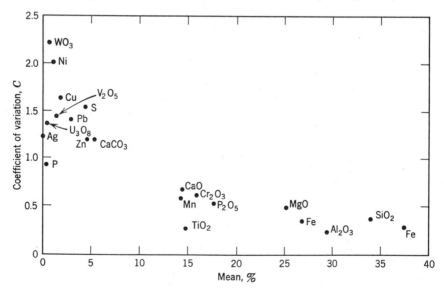

Fig. 7.6. Coefficients of variation from table 7.16 (after Hazen and Meyer, 1966)

TABLE 7.16. SUMMARY STATISTICS FOR 50,057 OBSERVATIONS FROM 484 MINERAL DEPOSITS *

Substance and chemical formula in which analyses are reported	Mean (wt %)	Variance	Coefficient of variation	Number of deposits	Number of observations
Silver, Ag	0.02002	0.000595	1.218	18	971
Alumina, Al_2O_3	29.40	41.16	0.218	39	8888
Lime, CaO	14.53	91.83	0.659	7	1099
Calcite, $CaCO_3$	5.28	39.14	1.185	24	1059
Chromite, Cr_2O_3	15.92	90.66	0.598	40	3592
Copper, Cu	1.66	7.33	1.626	16	843
Magnetite, Fe	26.81	80.58	0.335	30	2810
Iron, miscellaneous deposits, Fe	37.44	105.56	0.274	14	1449
Magnesia, MgO	25.20	146.41	0.480	2	347
Manganese, Mn	14.25	67.63	0.577	63	7711
Nickel, Ni	1.02	4.12	1.981	24	2091
Phosphorus, miscellaneous deposits, P	0.277	0.0668	0.934	18	1307
Phosphate rock, P_2O_5	17.60	83.68	0.520	32	1569
Lead, Pb	2.81	16.05	1.427	8	342
Sulfur, S	4.32	43.84	1.532	5	494
Silica, SiO_2	33.89	143.35	0.353	42	6044
Titanium, TiO_2	14.72	14.74	0.261	4	621
Uranium, U_3O_8	0.402	0.299	1.359	3	519
Vanadium, V_2O_5	1.27	3.36	1.444	25	1286
Tungsten, WO_3	0.587	1.685	2.21	34	4019
Zinc, Zn	4.47	28.17	1.19	36	2976

* From Hazen and Meyer, 1966.

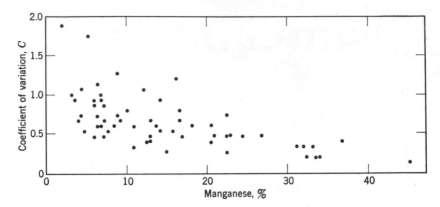

Fig. 7.7. Coefficients of variation for 7711 manganese assays from 63 deposits (after Hazen and Meyer, 1966).

and Meyer for another purpose, they at least are haphazard rather than selected to support our ideas. Other recorded data that we have examined behave similarly.

For individual chemical elements and other substances present in greater than trace amounts a relationship similar to that in figure 7.6 holds; for instance, figure 7.7 graphs the coefficients of variation for 7711 manganese assays from 63 deposits, as tabulated by Hazen and Meyer.

REFERENCES

Ager, D. V., 1963, Principles of paleoecology: New York, McGraw-Hill, 371 p.

Ahrens, L. H., 1963, Element distributions in igneous rocks, pt. 6, Negative skewness of SiO_2 and K: Geochimica et Cosmochimica Acta, v. 27, p. 929–938.

Baird, A. K., McIntyre, D. B., and Welday, E. E., 1967, A test of chemical variability and field sampling methods, Lakeview Mountain tonalite, Lakeview Mountains, Southern California batholith: California Div. of Mines and Geol., Spec. Rept. 92, p. 11–19.

Barth, T. F. W., 1962, Theoretical petrology: New York, John Wiley & Sons, 2nd ed., 416 p.

Bradley, P. R., 1925, Estimation of ore reserves and mining methods in Alaska Juneau Mine: Am. Inst. Mining Metall. Engineers Trans., v. 72, p. 100–120.

Brownlee, K. A., 1955, Statistics of the 1954 polio vaccine trials: Jour. Am. Statistical Assoc., v. 50, no. 272, p. 1005–1013.

———, 1965, A review of "Smoking and Health:" Am. Statistical Assoc. Jour., v. 60, no. 311, p. 722–739.

Chayes, Felix, and Fairbairn, H. W., 1951, A test of the precision of thin-section analysis by point counter: Am. Mineralogist, v. 36, p. 704–712.

Chayes, Felix, 1956, Petrographic modal analysis: an elementary statistical appraisal: New York, John Wiley & Sons, 113 p.

Cochran, W. G., 1963, Sampling techniques: New York, John Wiley & Sons, 413 p.

Compton, R. R., 1962, Manual of field geology: New York, John Wiley & Sons, 362 p.

Cumming, J. D., 1956, Diamond drill handbook: Toronto, Canada, J. K. Smit and Sons, 655 p.

Dadson, A. S., and Emery, D. J., 1968, Ore estimation and grade control at the Giant Yellowknife Mine, *in* Ore reserve estimation and grade control: Canadian Inst. of Mining and Metall., spec. v. 9, p. 215–226.

Dixon, W. J., and Massey, F. J., Jr., 1969, Introduction to statistical analysis: New York, McGraw-Hill, 638 p.

Engel, A. E. J., and Engel, C. G., 1960, Progressive metamorphism and granitization of the major paragneiss, northwest Adirondack mountains, New York: Bull. Geol. Soc. Am., v. 71, p. 1–58.

Giles, G. S., 1961, Diamond mining practice in South Africa: *in* 7th Commonwealth Mining and Metall. Cong., South African Inst. of Mining and Metall., v. 2, p. 839–850.

Griffiths, J. C., 1962, Statistical methods in sedimentary petrography, *in* Sedimentary petrography, v. 1: New York, Macmillan, 609 p., p. 565–617.

———, 1967, Scientific method in analysis of sediments: New York, McGraw-Hill, 508 p.

Grout, F. F., 1932, Rock sampling for chemical analysis: Am. Jour. Sci., v. 24, p. 394–404.

Grundy, W. D., and Meehan, R. J., 1963, Estimation of uranium ore reserves by statistical methods and a digital computer *in* Geology and technology of the Grants uranium region: Soc. Econ. Geologists, Mem. 15, p. 234–246.

Harding, B. W. H., 1956, Prospect sampling by air-flush drill: Inst. Mining Metall. Trans., v. 66, p. 79–87.

Hazen, S. W., Jr., and Berkenkotter, R. D., 1962, An experimental mine-sampling project designed for statistical analysis: U.S. Bur. Mines Rept. Inv. 6019, 111 p.

Hazen, S. W., Jr., and Meyer, W. L., 1966, Using probability models as a basis for making decisions during mineral deposit exploration: U.S. Bur. Mines Rept. Inv. 6778, 83 p.

Jackson, C. F., and Knaebel, J. B., 1932, Sampling and estimation of ore deposits: U.S. Bur. Mines Bull. 356, 155 p.

Joralemon, I. B., 1925, Sampling and estimating disseminated copper deposits: Trans. Am. Inst. Mining Eng., v. 72, p. 607–627.

Koch, G. S., Jr., and Link, R. F., 1969, A statistical analysis of some data from deflected diamond-drill holes: *in* Weiss, Alfred, ed., *A decade of digital computing in the mineral industry*, New York, Am. Inst. Mining Engineers, p. 497–504.

Krige, D. G., 1966, Two-dimensional weighted moving average trend surfaces for ore valuation, *in* Symposium on mathematical statistics and computer applications in ore valuation: South African Inst. of Mining and Metall., p. 13–79.

Krumbein, W. C., 1965, Sampling in paleontology, *in* Handbook of paleontological techniques: San Francisco, W. H. Freeman, p. 137–149.

Link, R. F., Koch, G. S., Jr., and Schuenemeyer, J. H., 1970, The lognormal frequency distribution in relation to gold assay data: U.S. Bur. Mines Rept. Inv., in press.

McKinstry, H. E., 1948, Mining geology: Englewood Cliffs, N.J., Prentice-Hall, 680 p.

Manson, Vincent, 1967, Geochemistry of basaltic rocks: major elements, *in* Hess, H. H. and Poldervaart, Arie, eds., Basalts: New York, John Wiley & Sons, p. 215–270.

Mason, Brian, 1962, Meteorites: New York, John Wiley & Sons, 274 p.

Miesch, A. T., 1967, Theory of error in geochemical data: U. S. Geol. Survey Prof. Paper 574-A. 17 p.

Mosimann, J. E., 1965, Statistical methods for the pollen analyst: multinomial and negative multinomial techniques, *in* Handbook of paleontological techniques: San Francisco, W. H. Freeman, p. 636–673.

Norton, J. J., and Page, L. R., 1956, Methods used to determine grade and reserves of pegmatites: Am. Inst. Mining Metall. Petroleum Eng. Trans., v. 205, p. 401–424.

Parks, R. D., 1949, Examination and valuation of mineral property: Cambridge, Mass., Addison-Wesley Press, 446 p.

Prinz, Martin, 1967, Geochemistry of basaltic rocks: trace elements, *in* Hess, H. H. and Poldervaart, Arie, eds., Basalts: New York, John Wiley & Sons, p. 271–324.

Richmond, J. F., 1965, Chemical variation in quartz monzonite from Cactus Flat, San Bernardino Mountains, California: Am. Jour. Sci., v. 263, p. 53–63.

Rowland, R. S., and Sichel, H. S., 1960, Statistical quality control of routine underground sampling: Jour. South African Inst. Mining Metall., v. 60, p. 251–284.

Schottler, G. R., 1965, Statistical analysis of gamma-ray log sample data from a uranium deposit, Ambrosia Lake area, McKinley County, New Mexico: U.S. Bur. Mines Rept. Inv. 6645, 49 p.

Truscott, S. J., 1962, Mine economics: London, Mining Pubs., Ltd., 471 p.

Waterman, G. C., 1955, Chuquicamata develops better method to evaluate core drill sludge samples: Mining Eng., January, 1955, p. 54–62.

Chapter 8

Variability in Geological Data

This chapter is closely related to Chapter 7 on geological sampling. Chapter 7 is mainly concerned with natural variability and sampling variability. Now, in Chapter 8, two other kinds of variability, those introduced by preparation and chemical analysis of the samples, are discussed.

For brevity, the term *chemist* is used in this chapter to distinguish workers in several disciplines (including analytical chemistry, spectroscopy, fire assay, and x-ray analysis) from geologists. The word *chemist* is used rather than *analyst*, which would be more appropriate, because the word *analyst* can be confused with statistical analyst or with a statistical analysis. Similarly, the phrase *chemical analysis* is used as a general expression for a variety of chemical and physical operations which are applied to geological materials to yield numerical results.

8.1 PROBLEMS OF GEOLOGICAL VARIABILITY

Geological variability is immense. The first step in an investigation is to decide on the purpose of the investigation, to choose a substance or substances of interest, and to decide in what detail the variability need be known. Total information cannot be extracted from a rock any more than a census enumerator can extract total information from a citizen. Too many purposes or too many substances will make a study fall of its own weight.

The four kinds of variability distinguished in this book may be expressed in a general statistical model, in which any univariate observation is written as

$$w = \mu + \xi_n + \xi_s + \xi_p + \xi_a + e$$

where ξ_n is the natural variability, ξ_s is the sampling variability, ξ_p is the preparation variability, ξ_a is the analytical variability, and e is the random fluctuation not accounted for by the other sources of variability. The coefficients of variation recorded in section 7.8 reflect an unknown mixture of these four variability sources and of the random error. Even in a designed experiment, it is difficult to separate the different sources of variability, but the aim is for the coefficient of variation to reflect natural variability closely. For unevaluated data, one can only hope that the size of the coefficient of variation expresses the natural variability.

In this chapter, natural, sampling, preparation, and analytical variability are taken up in turn in sections 8.2 to 8.5. Natural variability is a property of rock bodies; the other three kinds of variability stem from operations performed on single geological samples or specimens. Then in section 8.6, on experimental designs, variability in rock bodies is related to variability in single geological samples and to the decisions that must be made to determine what precision is required. This subject of decision making, introduced here for the first time, is explored at length in later chapters, especially Chapter 14.

8.2   NATURAL VARIABILITY

Natural variability is the variability inherent in a rock body (sec. 7.2). It is the variability that ideally would be discovered if only it could be separated from the other sources of variability. The concept may be explained by the expository device of considering some kinds of natural variability in the gold-bearing Pinyon Conglomerate of Paleocene age in northwestern Wyoming. Because little has yet been written (Antweiler and Love, 1967), the exposition need not be constrained by the facts.

The Pinyon Conglomerate crops out across an area of more than 100 square miles and has a volume of a few tens of cubic miles, according to Antweiler and Love (1967, fig. 2 and p. 2), who write that

... the Pinyon Conglomerate was deposited by rivers that flowed from the quartzite source area eastward and southwestward across Jackson Hole. . . .[In] the area of maximum conglomerate deposition . . . more than 5,000 feet of strata, largely conglomerate, was deposited.

Of the many geologic models that may be postulated for the Pinyon Conglomerate, one could be a model of the composition, and another could be a model of the process of formation. A composition model might be regarded as a tabular body that is bounded by undulating upper and lower surfaces and that thins out laterally. This tabular body is disrupted by stream valleys, folds, and sills; and it is cut by joints, faults, and dikes. Within the defined space are rock fragments ranging in size from grains a few microns across up to boulders. The proportions and arrangement of these rock fragments account for the variability in the composition model. They also provide the basic information for the model of the process of formation; from them the geologist may test hypotheses about geological processes, for instance, about stream channels within the formation which may be marked by changing porosity, permeability, grain size, composition, etc.

Yet at the present state of geological knowledge total variability is far too complicated a concept to be useful. Instead, specific kinds of variability, for instance, that of the gold particles or that of the pebbles, must be surveyed. If one had x-ray eyes, he could see into the Pinyon Conglomerate, learn exactly where the gold is, and measure its variability in terms of grain size, shape, composition, or other variables of interest. One could see the pebbles, identify the composition of each, and thus define the composition model. Still in the realm of make-believe, one could continue backward through time in a time machine to contemplate the process model and determine whence and how these gold grains and pebbles were carried into place. Such speculation is not idle daydreaming but is a fruitful way to contemplate the problem in order to devise a practical sampling scheme.

If the reader keeps in mind the geological model that we have set up for the Pinyon Conglomerate, he will understand why the distinction between target and sampled population concerns us. The reason is that the variability in the sampled population may not be the same as that in the target population. Even if the target population is potentially available for sampling through defining it as the uneroded Pinyon Conglomerate, only a small fraction of the formation is close enough to the surface to actually sample, and there is little hope that this portion is representative. If the target population is defined as the entire formation that was deposited in Paleocene time, no chance at all exists to sample it representatively because some of the formation has been eroded away.

Natural variability is different in the Pinyon Conglomerate for the many attributes that may be studied; for instance, the variability in the proportion of quartzite pebbles is likely to be low (sec. 6.1). On the other hand, the variability of the gold is likely to be high; from one typical locality, Antweiler and Love (1967, personal communication) took 44 geological samples whose mean was 0.2158 ppm and whose coefficient of variation was 3.28.

Natural variability may best be examined by stratifying the rock body. Thus the Pinyon Conglomerate may be stratified in the conventional geologic sense according to time or lithologic breaks, or it may be stratified according to geographic position, if, for instance, the gold increases in a certain direction. There may be a trend, usually a linear one in variability (sec. 9.4), or an area of influence, for instance, a well-sorted conglomerate in which the pebble variability is lower than elsewhere.

8.3   SAMPLING BIAS AND VARIABILITY

Sampling variability, which is that introduced by the process of physical sampling, is discussed for various kinds of sampling in Chapter 7, and in this section only the main conclusions are reviewed and summarized. Because the sampling variability obscures the natural variability, it should be kept small; however, its size should be appropriate to that of the other sources of variability, as there is little advantage to low sampling variability if the preparation and analytical variability are excessive. The main sources of sampling variability stem from losing material that should be in the sample and from contaminating the sample by adding extraneous material to it.

Some of the many sources of contamination are stated in Chapter 7. Sample sacks may contain material left over from a previous job; diamond-drill sludge may contain material from another interval than that being sampled; gravel may cave from the walls of test pits; and steel chips broken off picks and chisels may find their way into the sample. These are accidental sources of contamination; intentional contamination, named *salting*, is discussed entertainingly by McKinstry (1948, pp. 67–69) and by other authors that he references.

Losing material from the sample is serious because one can seldom determine if the loss is biased or not, and usually it is biased. Sample may be lost when rock fragments fly away from the outcrop rather than fall into the sample sack, when diamond-drill core is ground up, when sludge settles in cracks in the rock rather than being washed out of the hole, when dust from a rotary drill blows into the air, etc. These losses almost always introduce bias because the various sized fragments differ in grade (sec. 15.1). Even if bias is absent, variability is surely increased because the desired volume of sample is reduced to an unknown extent that is different from sample to sample.

The scale on which sampling variability is measured must be taken into account by prior information or estimation before sampling begins. For instance, in sampling the Pinyon Conglomerate in Wyoming to determine the proportion of quartz pebbles, geological samples should be larger than one pebble (sec. 6.1); but in sampling to determine the chemical composition of individual pebbles, no geological sample should be larger than one pebble.

8.4   PREPARATION BIAS AND VARIABILITY

Sampling bias and variability, discussed in the preceding section, are clearly the geologist's responsibility. Analytical bias and variability, discussed in the following section, are clearly the chemist's responsibility. Preparation bias and variability, the subject of this section, are not clearly in the province of either of these disciplines. Therefore they are often ignored by both professions, and the task is delegated to technicians whose work may be good, but more often it is bad because of lack of training or interest or both. Probably in most geological sampling more bias and variability are introduced in sample preparation than in any other stage, although we cannot prove this contention because the literature is sparse and lacking in numerical data.

This section is limited to discussing in turn the purposes of sample preparation, then some of the many problems, and finally some of the physical methods. The reader is warned that he should understand the many tiresome details involved because no one else is likely to take the time and care to prepare his samples competently for the chemist.

Purposes of Sample Preparation

Sample preparation has two principal purposes: (a) to reduce the amount of the geological sample and (b) to make it uniform. The first purpose is nearly always necessary because the earth is very large and the amount of material that the chemist can analyze is nearly always very small. The sample may be reduced, either in bulk by removing a certain weight of rock, or in number, perhaps by taking a certain number of sand grains or a certain number of microfossils.

The second purpose of sample preparation is to make a uniform sample by blending (or mixing), which is one of the most difficult processes in sample preparation. After this process a portion usually is removed.

Problems in Sample Preparation

The problems in sample preparation discussed in this book are systematic losses, contamination, blending, concentration, splitting, and alteration. In all of these problems, variability is bound to be introduced, and the goal is to keep it as low as possible. Moreover, serious bias can be introduced extremely easily but usually can be held down to a manageable level by planning. The problems can be tackled with the methods explained in the next subsection.

Systematic losses are those that occur in every sample that is prepared. If, every time a rock is pulverized, a certain weight or proportion of it is lost, trouble results, although it may not be serious unless one constituent of the rock is more liable to be lost than another. One way to determine systematic losses is to weigh the sample at various stages during its preparation, remembering that the sample will gain in weight through contamination as well as decrease in weight through loss. Another way is to prepare samples differently and compare the results. This work should be done on extra material specifically obtained for the test, not on the experimental material itself.

Contamination is the addition to the sample of foreign material; if done intentionally with intent to defraud, contamination amounts to salting (sec. 8.3). Whenever geological samples are crushed or pulverized, contamination is inevitable, as the equipment is abraded as well as the rock. Therefore, one must decide how much and what kind of contamination is acceptable, and then he can take appropriate action. For instance, pulverizers with ceramic rather than metal grinding plates can be used if contamination from iron or other metals is a problem. Because a penalty is paid in greater time and therefore cost, one should determine if the additional expense is justified.

Blending is difficult to do properly, as is well known to industries whose product depends on blending, for instance, the paint industry. It is best avoided in sample preparation. Many ingenious blending devices have been invented, but nearly all of them impart some kind of concentration when blending rock powder, the usual geological material to be blended, especially when the particles vary in shape or specific gravity, as with mica flakes, gold, platinum, and asbestos. Besides the concentration initially present, a concentration is often introduced in blending a sample. For instance, differential settling is well established for such substances as rock powders stored in buildings subject to vibration, molybdenum ore samples carried by airplane after they are pulverized and blended, and samples trucked from a sample preparation plant to a chemical laboratory a few miles away.

The best way to reduce the bulk of samples is by splitting. Although many problems arise, splitting is a better way to reduce the bulk of a sample than blending and scooping some out. Sample splitters of many different designs have been developed, and studies of their performances are discussed in the next subsection. The aim is to ensure that, when a particle goes through a splitter, it has a known chance of going into a particular compartment.

Finally, alteration of a sample during its preparation may be a problem. An example is found in preparing copper sulfide ores, when sulfide minerals can easily be changed to oxides by heat, which is introduced by drying sludges from drill holes or by crushing and grinding. Another example is found in preparing coal samples whose moisture, gas content, etc., are important and can be easily altered. Many other kinds of samples deteriorate with age, and the advantage of speedy analysis must be balanced against the drawbacks of doing chemistry under field conditions.

Methods of Sample Preparation

Of the many methods for preparation of geological samples, some of those most used are listed in table 8.1. The methods have been grouped according to the physical or chemical process used: these may be concentrating, leaching, pulverizing, splitting, or blending. The table also lists some geological materials for which the methods are suitable. Most of the methods are intended for samples weighing a few kilograms or more and are also usually

TABLE 8.1. SOME METHODS OF SAMPLE PREPARATION

Processes	Examples of geological materials
Concentrating processes:	
Washing Gravity concentration Heavy liquid concentration Flotation Elutriation Screening Precipitation Magnetic concentration Electrostatic concentration Hand picking	Unconsolidated or weakly consolidated sediments and sedimentary rocks, especially those with grains varying in size, shape, specific gravity, magnetic properties, or electrostatic properties
Leaching and other chemical processes:	
Solution in water, cyanide solutions, and other reagents, depending on specific nature of the material Amalgamation	Soluble rocks and minerals
Pulverizing:	
Drying Crushing Grinding Pulverizing	Rocks whose grains are relatively uniform in size or specific gravity Other rocks and minerals not amenable to preparation by concentration or leaching and other chemical processes
Splitting:	
Splitting	Any sample composed of discrete grains or fragments
Blending:	
Rolling Shaking	Any sample composed of discrete grains or fragments

used in industry for processing ores, although the machines are scaled down in size for sample preparation. Preparation of smaller samples is explained in specialized works on petrography, for instance, those by Johannsen (1918), Krumbein and Pettijohn (1938), and Milner (1962).

Whole volumes have been written about sample preparation, and to comment in detail on the many methods would expand this book far too much. A few of those most important to a geologist are mentioned, and some of the few published numerical results are cited. Of the works on sample preparation, the following are particularly pertinent. Peele (1941) and Taggert (1945, sec. 19, 208 pp.) explain the preparation of ores and large samples weighing as much as a few tons or even more; Lundell and others (1953) have written on inorganic chemical analysis and rock analysis in particular; and, Milner (1962), German Mueller (1967), and L. D. Muller (1967) discuss sample preparation in sedimentary petrology.

In practice, several of these methods are often combined, and then a flow chart is helpful. Figure 8.1 is a flow chart devised by Schottler (1968, personal communication) for processing gold ore from a saprolite near Kershaw, South Carolina.

We next turn to a discussion of some specific methods of sample preparation. Ideally, each method would be described in terms of experiments performed as follows for all geological materials of interest. Two or more samples would be prepared by each preparation method, with appropriate randomization. When the analytical results were returned, the sources of variability could be partitioned through a nested analysis of variance, as is done for the data from the Kilembe, Uganda, mine (sec. 15.4). However, it is quite impossible to offer any such description because the experiments have not been performed, and the data are not available. Coxon and Sichel's excellent work (1959) discussed below comes as close as any, although they did not quite achieve the overall partitioning of sources of variability.

Concentration

In the first group of methods, those for concentration of samples, quantitative concentration may or may not be required. If it is required, as for preparation of placer samples (sec. 15.2), three classes of material must be distinguished: the *heads*, which constitute the material to be concentrated, the *concentrate*, which is the material enriched in the substance of interest, and the *tails*, which is the remaining material impoverished in the substance of interest. If the concentration were ideal, the concentrate would contain all of the substance of interest and the tails would contain none. However, the tails always contain at least a trace of the valuable substance, often in an amount too small to analyze directly. Therefore, the only check on the efficiency of a quantitative mechanical concentration process is usually to

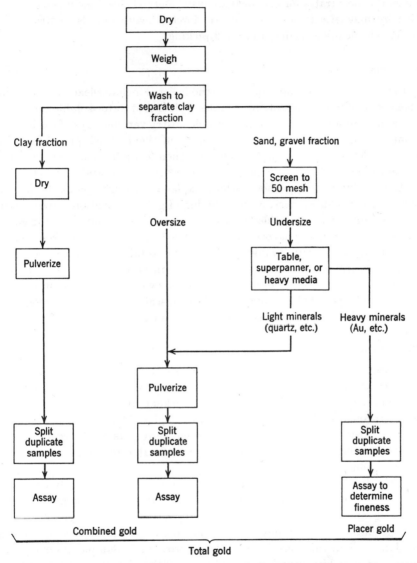

Fig. 8.1. A flow chart for sample preparation of gold in saprolite (after Schottler, 1968).

reconcentrate the tails or a part of them by the same or another method. Most of the recorded information on quantitative concentration is in the placer mining literature (sec. 15.2).

If quantitative concentration of samples is not required, the problem of sample preparation is simplified. Usually the aim is to obtain fairly pure mineral concentrates for chemical analysis. Methods used include concentration by magnetic forces; electrostatic forces; heavy liquids; elutriation in water, air, or other media; and hand picking.

Leaching

In table 8.1, the second group of methods listed includes leaching and other chemical methods for dissolving samples. Solution of water-soluble substances is seldom a problem. In exploration geochemistry various ways of leaching are commonly used, as explained by Hawkes and Webb (1962, pp. 35 ff.). Once the samples are in solution, mixing and then sampling from the solutions present a few problems, but they have been thoroughly studied by chemists.

Another interesting problem, whose explanation might point the way for sampling other substances, is the leaching of gold. In gold ores, where large volume samples are desirable because the size and distribution of the gold is extremely varied, it has been proposed to leach samples of from one to several kilograms with cyanide solutions for total extraction of the gold. Because nearly complete recoveries of gold can be made with cyanide solutions in production-scale mills, the problems, evidently not yet surmounted (Green, 1968, personal communication), lie in scaling down the operation and in devising laboratory-scale equipment that is easy to clean.

Pulverizing

Another group of methods comprises those that pulverize samples. If the samples are wet, the first step is drying, which should cause no particular problems provided that the temperatures are kept low enough so that nothing is lost through sublimation or is altered through oxidation or other chemical change. Also, no contamination from such sources as air currents blowing material about can be allowed. The amount of water evaporated is measured if its amount is important.

After drying, pulverization requires several (occasionally only one) steps to reduce the particle size of the sample. Although the names crushing, grinding, and pulverizing are conventionally given to three stages of reducing size gradationally, the physical actions involved may overlap, and the fundamental distinctions if any are controversial. Sample fragments are broken by both hitting and rubbing; in both of these processes, contamination by the apparatus is inescapable; also dust losses of the geological sample may be unavoidable. Pulverizing is probably the worst necessary part of sample preparation; the other bad part, blending, is a step that nearly always can and should be avoided.

While setting up a laboratory for pulverizing ore samples a few years ago we investigated some of the many machines for crushing and pulverizing rocks on a laboratory scale. Most of these machines are scaled-down models of ones used to crush ore for production. All that we have seen have some or all of these faults: excessive dust losses, difficult cleanup, excessive contamination, or shoddy construction (misaligned parts, defective bearings, broken castings, etc.). The crushers and grinders can quite readily be modified to reduce losses in receiving and discharging material; once this modification is done, these machines, which reduce material down to approximately 8 mesh per inch, introduce less contamination and have fewer problems than do the pulverizers. The modifications needed are to catch any entering material that is thrown back out and to collect without loss the material after it has been reduced in size. Pulverizers are of two general types. In the first the sample is fed between two plates, one or both of which rotate, and is collected below them; dust losses tend to be high; the plates frequently have to be refaced, and they become misaligned. In the second the sample is contained in a sealed chamber in which it is pulverized by the action of hard pucks or balls; there is little dust loss in this type, but much excessively fine powder is produced, as all of the rock stays in the chamber until the pulverizing action is completed.

Better machines for crushing and pulverizing may be made; we have not seen them all. For none have we seen any authentic records evaluating quality of the sample preparation, an evaluation which is very important as indicated by an interesting experiment that Coxon and Sichel (1959, p. 496) performed on high- and low-grade gold ore. Figure 8.2 shows that, for both high- and low-grade ore, reproducibility increases rapidly as the sample is crushed finer. In fact, one pulverized sample has about 100 times the precision (square of the ratio of the coefficients of variation) of one coarse-crushed sample. Moreover, Coxon and Sichel point out that "from the histograms it is also evident that ideally the total sample as received should be pulverized before assay . . . economically this is not possible, and hence sample splitting will be with us for some time to come." But an experiment like theirs provides a basis for establishing the necessary steps to obtain acceptable preparation variability and cost.

A sample preparation laboratory always has many samples coming into it; through Parkinson's law they increase to clog the capacity of the plant, and the aim will be to get rid of them as fast as possible. To do this, the tendency is to coarse-crush, split, and throw out the material rejected in the splitting, which is an acceptable procedure unless it leads to the trouble pictured in figure 8.2. Accordingly, tables of "safe" sizes at which to split have been constructed. Those by Taggert (1945) are the most widely used, but as Davis (1963, p. 259) points out in a study of sample preparation at the Kilembe, Uganda, mine, the generalities are of little value, and each case should be evaluated on its own.

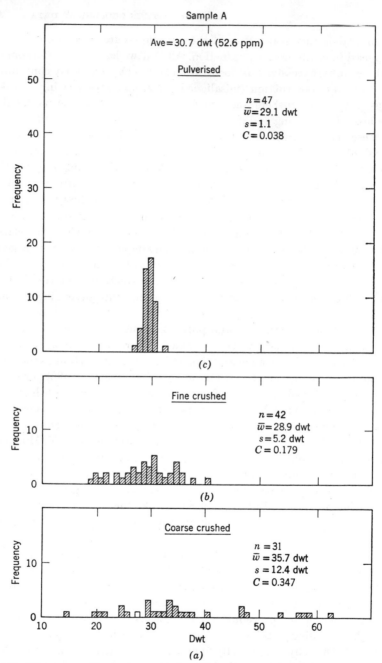

Fig. 8.2. Histograms illustrating reproducibility of mine assay values when identical sample is split after (*a*) coarse crushing, (*b*) fine crushing, and (*c*) pulverizing (after Coxon and Sichel, 1959, p. 496).

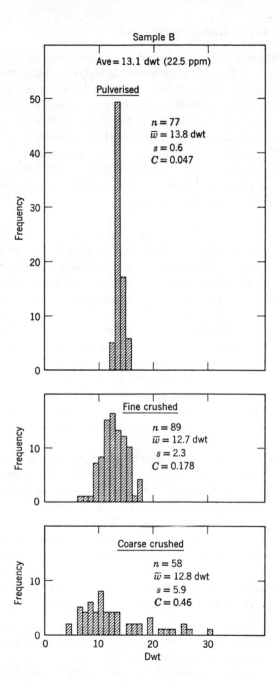

If contamination is a problem, several precautions can be taken. Lundell and others (1953, p. 809) state that breaking rocks by hand methods that emphasize hitting rather than rubbing reduces iron contamination. Also ceramic plates in machine pulverizers, or ceramic or agate mortars, can be used to substitute silica contamination for metal contamination if silica contamination is less undesirable. Klugman (1966, personal communication) reports that most pulverizer plates are cast from scrap metal that may contain many kinds of metal impurities different from one lot to the next, and presumably crusher plates are made in the same way. One could have these plates custom cast to control this problem.

Splitting

Splitting, the fourth method in table 8.1, is used to divide fragmented material—either originally disaggregated material such as gravel or soil, or pulverized rock—into two or more parts. One or more of several devices may be used; the most common is the Jones splitter, in which the material to be split is dumped onto a knife edge so that one-half falls on one side and the one-half on the other. Splitting is almost always needed to prepare samples for chemical analysis and is often needed to reduce samples for grain-count analysis. The requirements differ, depending on the precise purpose of the splitting and on the amount of material to be handled. For grain-count analyses a specified fraction of the total material, often one-half, must go to a given compartment of the splitter; whereas in order to get a standard amount of material for chemical analysis, a specified volume or weight may be collected.

The fairly extensive literature on sample splitting, which is more complete than for most other kinds of sample preparation, reflects the various requirements. The *Journal of Sedimentary Petrology* has published several papers over the years, among the best of which are those by Wentworth and others (1934), Otto (1937), Kellagher and Flanagan (1956), and Flanagan and others (1959). Milner (1962) and Griffiths (1967) discuss sampling for sedimentary petrology, and Boericke (1939) and Coxon and Sichel (1959) discuss splitting of mine samples. Manufacturers of splitters also have information.

Flanagan and others (1959, p. 108) clearly explain desirable qualities for a laboratory splitter that should also apply to larger ones:

1. The splitter should be able to reduce both small and large amounts of sample.
2. The splitter should be capable of sampling a wide variety of particle sizes.
3. The time required for the operation should be short.
4. All particles in the lot to be sampled should have an equal chance of being sampled.
5. The sampler should be easily cleaned and require a minimum of maintenance.
6. The operation of the sampler should be simple.

7. The materials of which the sampler is made should not contaminate the sample.

A review of the relative merits of different types of splitters would take too much space in this book. However, the reader is warned that bias and unnecessary variability can easily be introduced by the Jones splitter, which has been more or less the standard one. For an extensive research project, one is well advised to test a splitter on representative material of the type to be split. Experiments made with other materials, particularly artificial ones, are liable not to apply to the problem at hand; for instance, a splitter may satisfactorily split lead shot in sand but may not do at all well on flakes and flattened nuggets of gold in fibrous gangue minerals.

The most-used splitters are the Jones splitter and cone splitters, which divide the material outward from the apex of a cone. Another splitter that has been recommended for laboratory use was devised for the U.S. Atomic Energy Commission office in Grand Junction, Colorado. This splitter feeds powder down a chute by a vibrating action (as in a Frantz magnetic separator) to discharge onto a rotating wheel about the size of a roulette wheel, with compartments rather like those that catch the roulette ball. These compartments catch the powder; and by varying their sizes, one can collect a sample of a certain fixed volume or divide an initial volume into fractions of different proportions.

Blending

Finally, blending, the last of the methods in table 8.1, should be avoided in favor of splitting because it is difficult to do properly. We have found no experimental data of particular interest, although presumably something is known in industries like paint manufacturing that do a lot of blending. Materials may be blended wet or dry; dry blending is more often done in geology. It seems probable that blending usually segregates rather than blends particles differing from the average in specific gravity, size, or shape. If blending cannot be avoided, one should take samples for analysis from several parts of the blended pile—or from top and bottom of the blending container—to establish the blending variability. The literature shows that problems that arise are extremely specific for the substances that are blended.

8.5   ANALYTICAL BIAS AND VARIABILITY

If a geologist could have his wish, he would send replicate increments of perfectly prepared geological samples to several chemists, none of whom could communicate with each other, and each chemist would make determinations by several chemical methods. After the resulting observations were statistically analyzed, the geologist would know within definite limits the composition of his rock.

This ideal state of affairs does not seem likely to be realized in everyday analytical practice in the forseeable future, although Britten (1961, p. 1008) writes that in South Africa "under good conditions of application two assayers work independently using separate reagents and separate equipment including assay balances. At the Rand Refinery the obviously beneficial provisions of entirely separate and independent accommodation and equipment has been an unqualified success for many years." Until it becomes economically feasible to perform all chemical analyses of geological samples under such controlled conditions, geologists must realize that mistakes in chemical analysis can always happen and probably will.

As already mentioned in this chapter, the problems of analytical bias and variability lie primarily in the discipline of analytical chemistry, thought of in its broadest sense as including other methods of analysis, like spectroscopic or x-ray analysis. In this book, space permits only a brief treatment of a few of the many ways to evaluate analytical data statistically. Dalrymple and Lanphere (1969, Chap. 7) give an excellent example of the application of statistics to the evaluation of the reliability of data for potassium-argon dating.

That the accuracy of chemical analyses made in well-run laboratories has been increasing in recent years is illustrated by Tatlock (1966) in an excellent review of 55 older and 110 modern chemical analyses of Australasian tektites. Tatlock suggests that "more than half of these older alkali and titania determinations are decidedly inaccurate and misleading." For instance, in figure 8.3a K_2O is plotted against Na_2O for 110 modern tektite analyses made by five different chemists to define a joint range-of-variability region. Then, in figure 8.3b, 55 older analyses are plotted, one-half of which fall outside the joint region.

However, in some laboratories, the accuracy of routine chemical analyses of geological materials may have actually deteriorated over the years, probably because fewer geologists and mining engineers do their own analyses and are familiar with the analytical methods. Therefore the geologist must rely in part at least on consistency of the observations. Some ways to investigate consistency are explained in this section. One of the simplest is to determine if the analyst favors certain digits in reporting analyses, a subject fully developed by Hazen (1967, pp. 80–83). For instance, for 90 assays from a drill-sampling project, Hazen states that the frequency of occurrence of digits in the tenths position, from 0 to 9, was as follows: 8, 10, 7, 7, 10, 6, 10, 17, 2, 13, instead of the expected average frequency of 9 for each digit. The

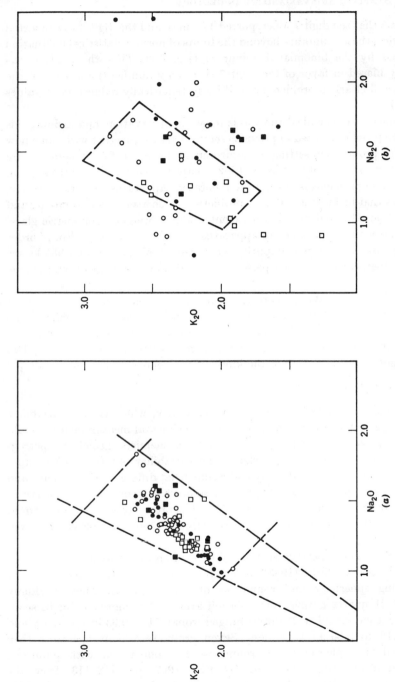

Fig. 8.3. Variation of K_2O with respect to Na_2O in analyses of Australian tektites (after Tatlock, 1966, p. 124). In (a) are plotted 110 precisely monitored analyses; in (b) are plotted 55 older analyses. Symbols identify analysts cited in the original paper.

list shows that the digit 7 was reported 17 times, and the digit 8 was reported only twice; these counts are beyond the limits of normal statistical fluctuation predicted by the binomial distribution (sec. 6.1). Thus the analyst was strongly biased in favor of the digit 7 (his lucky number?) and against the digit 8, and clearly a problem arose if he systematically reduced most values of 8 to 7.

The purpose of chemical analysis is to get an observation, and in doing this all of the problems stated in previous sections arise—together with some new ones. The problems in setting up environmental controls for a laboratory are detailed in many books; Wilson (1952, Chap. 5) gives a good introduction. One of these problems is getting the laboratory apparatus to work correctly, a process that includes calibration problems, which also arise in sampling and preparation, although to a lesser extent. Balance weights, volumetric glassware, and many other kinds of apparatus must be calibrated. Also, in order to compensate for reagent impurity when geological samples and blanks are run together, calculations are performed to subtract the blanks from the true samples.

In addition to these general problems, specific ones arise for particular substances, for instance, certain elements. Solving specific problems like these, which can be challenging and interesting, requires a knowledge not only of chemistry but also of geology and geochemistry. Fleischer and Chao (1960) give a good account of the kinds of problems that come up.

Operator Variability

The analytical problem of operator variability, which is that variability introduced by the people who make the experimental measurements, should always be evaluated. The operators may be chemists, geologists, petrographers, engineers, or other workers. The variability is introduced because different operators will naturally make somewhat different observations, and even the same operator will tend to be somewhat inconsistent from day to day and from hour to hour on a single day. Operator variability is introduced in section 7.6 on modal analysis; in this section, a single example is considered in more detail.

The example of operator variability is from an experiment performed by Rosenfeld and Griffiths (1953) to test the visual-comparison technique for estimating sphericity and roundness of quartz grains. They randomly selected 21 quartz grains from a beach sand and mounted them in semipermanent mounts. The 21 grains ranged from 0.71 to 0.93 in sphericity and from 0.10 to 0.80 in roundness. Seven graduate-student operators then estimated the sphericity and roundness by comparison with published standard charts (reproduced by Griffiths, 1967, pp. 112–113, from the original papers). In this book we discuss only the analysis for sphericity

TABLE 8.2. AVERAGE MEAN SQUARES IN AN EXPERIMENT TO TEST VISUAL-COMPARISON TECHNIQUES FOR ESTIMATING SPHERICITY [*]

Source of variation	Average mean squares
Operators	$\sigma^2 + 3\sigma^2_{OXG} + 21\sigma^2_{D(O)} + 63\sigma^2_O$
Days	$\sigma^2 + 3\sigma^2_{OXG} + 21\sigma^2_{D(O)} + 147\sigma^2_D$
Grains	$\sigma^2 + 3\sigma^2_{OXG} \qquad\qquad + 36\sigma^2_G$
Day (operator)	$\sigma_2 + 3\sigma^2_{OXG} + 21\sigma^2_{D(O)}$
Day–grain interaction	$\sigma^2 \qquad\qquad + 7\sigma^2_{DXG}$
Operator–grain interaction	$\sigma^2 + 3\sigma^2_{OXG}$
Day (operator)–grain interaction	σ^2

[*] Data from Rosenfeld and Griffiths, 1953.

measurements of the raw data; Rosenfeld and Griffiths present further interesting analyses of transformed sphericity data and of the roundness data.

An analysis of the raw sphericity data is presented in tables 8.2 and 8.3. The set of average mean squares (table 8.2) is similar to but not exactly like that postulated by Rosenfeld and Griffiths, a difference that we deliberately introduce to show that different people can interpret the same data differently. Thus, unlike Rosenfeld and Griffiths, we believe that although the day-to-day variation of a single operator can be estimated, this variation is not an operator–day interaction because days are nested with respect to operators. (An interaction would be like the operator–grain interaction in this experiment, or the method-formation interaction in section 5.9.) Moreover, it is unlikely that the sphericity of the quartz grains could change from day to day, so we presume that the day–grain interaction is 0.

TABLE 8.3. ANALYSIS OF VARIANCE IN AN EXPERIMENT TO TEST VISUAL-COMPARISON TECHNIQUES FOR ESTIMATING SPHERICITY [*]

Source of variation	Sum of squares	Degrees of freedom	Mean square	F	$F_{10\%}$	Degrees of freedom used for test
Operators	865.14	6	144.19	6.94	2.33	(6, 12)
Days	42.13	2	21.07	1.01	2.81	(2, 12)
Grains	4879.55	20	243.98	24.72	1.48	(20, 120)
Day (operator)	249.51	12	20.79	2.11	1.60	(12, 120)
Day–grain interaction	203.01	40	5.08	0.91	1.28	(40, 240)
Operator–grain interaction	1184.19	120	9.87	1.76	1.21	(120, 240)
Day (operator)–grain interaction	1344.02	240	5.60			

[*] Data from Rosenfeld and Griffiths, 1953.

The analysis of variance (table 8.3) for our postulated model leads to the following conclusions:

1. There is a small day-to-day difference in the estimates of roundness, which stems mostly from the day-to-day variability of the individual operators. The additional day-to-day variability found by averaging among operators is negligible.

2. The next largest source of variability is the variation among operators; this variability is larger than the variability of a single operator from day to day.

3. The largest source of variability is that in the grains, because their sphericities are very different. Notably, had the operator variability been confused with the sphericity variability through some operators measuring only certain grains and other operators measuring only certain other grains, the operator and sphericity variabilities could not have been separated. Then it might have been difficult to identify the sphericity differences because of the larger operator variability.

Some Problems of Analytical Sensitivity

Chemical and other analytical methods are generally, if not always, suitable for certain concentration ranges of the substance being analyzed. Then observations too large or too small must be grouped as being either above or below the detection limits. When the value "zero" or "not detected" is reported for some of a group of determinations from an area where the substance is usually present, as in a set of gold assays from a mine, one believes that, if most of the geological samples contain detectable gold, the others also contain a little. Other sorts of grouped determinations are those reported as "trace" or as "1 ppm," etc. Although the problem usually arises at the low end of the frequency distribution, trouble also may occur at the high end of the range, as with spectrographic analyses. Miesch (1967b, p. B3) explains clearly the four kinds of problems, which come from distributions that are *left censored* (concentrations too small to be evaluated), *right censored* (concentrations too large to be evaluated), *left truncated* (an unknown number of values below the detection limit), and *right truncated* (an unknown number of values above the detection limit). He writes that

Every analytical method has limits of sensitivity beyond which it is ineffective for determination of concentration values. These limits may occur at both the lower and the upper bounds for a concentration range. . . . [A] spectrographic method . . . for example, may be used to determine silicon within the range from 0.002 to 10 percent. Concentrations judged to be lower or higher than this range are reported as < 0.002 percent or > 10 percent, respectively. Thus, the spectrographic data may be either left or right censored. The term "censored" is applicable here because the number of values beyond each sensitivity limit is known for any set of analyzed rock samples. In other types of problems, where the number of values beyond certain limits is unknown, the data are said to be truncated. . . .

An example of a left-truncated distribution is a size distribution of beach sand grains from which those grains below a certain size have been blown away by the wind. The geologist measuring this distribution might be certain that some grains had been blown away but not know how many. Another example of a left-truncated distribution is the weights of nuggets found in Victoria, South Australia; complete records establish the numbers and weights of large nuggets from this famous placer area, but unfortunately information on the small nuggets was never recorded. An example of a right-truncated distribution is assay values from mines where the assayer discards high values without recording their number and assays additional sample splits, on the grounds that high values are erratic.

Of the four problems left-censored distributions arise most often and therefore cause the most trouble, particularly if a large proportion of a set of observations is reported as trace or 0. The difficulty in forming ratios, because division by 0 is not allowed, is discussed in section 11.2. Another problem arises in computing standard deviations because, if many observations are the same, a "spike" is introduced into the distribution, and the standard deviation is reduced below its correct size. This problem exists even if observations of 0 or trace are replaced by a single arbitrary small value, perhaps the lowest limit of analytical sensitivity or halfway between this lowest limit and 0, as is often done, particularly if a logarithmic transformation is to be made.

Miesch (1967b) clearly explains several special statistical methods that have been devised to extract information from censored and truncated data. For three reasons these methods would seem to have limited use for most geological problems. First, if only relative sizes of observations are needed, as in exploration geochemistry and many problems of theoretical geology, censored and truncated distributions cause little trouble because the statistics, although they may be biased, will reflect relative sizes. Many of the kinds of data that are censored and truncated, such as spectroscopic analyses, are prevalent in these fields. Second, if observations are below the detection limit, even the most complicated statistical technique cannot help much because little information exists. Third, if observations are above the detection limit, a different method of chemical analysis should be used if the information is important to the investigator; money is better spent in this way than on statistical analysis.

Evaluation of Analytical Data

The reliability of analytical data can be evaluated by the statistical methods already introduced in this book, particularly by the analysis of variance. In this subsection, two representative examples are discussed.

TABLE 8.4. ANALYSIS OF VARIANCE IN GOLD DETERMINATIONS MADE BY THREE DIFFERENT METHODS *

Source of variation	Sum of squares	Degrees of freedom	Mean square	F	$F_{10\%}$
Samples	229.3	21	10.92	42.9	1.60
Analytical methods	0.456	2	0.228	0.9	2.44
Residual variability	10.69	42	0.255		

* Data from Erickson and others, 1966, p. 3.

In a study of gold mineralization in the Cortez district, Lander County, Nevada, Erickson and others (1966, p. 3) analyzed 22 gold samples by three different analytical methods—fire assay, colorimetric, and atomic absorption. In table 8.4 we compare the determinations by an analysis of variance to learn if, on the average, the three methods give the same answer. Because of the belief that gold tends to be lognormally distributed, the analysis of variance presented is the one performed on the logarithms. Since the calculated F-value of 0.9 is lower than the tabled F-value of 2.44 at the 10-percent significance level, evidently the three methods give the same results. However, if an analysis is carried out on the untransformed values, the multiple-comparison procedure (sec. 5.12) shows that the colorimetric analysis gives higher results than the other two analytical methods, which nearly agree with each other. Thus, the final conclusion must be that insufficient data were taken to draw a clear-cut conclusion about analytical variability.

As a part of the extensive study of analytical problems in analysis of the G-1 and B-1 rocks, Flanagan (1951) reported on a cooperative study of the lead content of the granite, G-1. In an interlaboratory comparison, each of six laboratories made six duplicate determinations (12 chemical analyses in all) on six samples of G-1 replicated six times. Thus there were 36 pairs of

TABLE 8.5. ANALYSIS OF VARIANCE FOR COOPERATIVE STUDY OF THE LEAD CONTENT OF GRANITE, G-1 *

Source of variation	Sum of squares	Degrees of freedom	Mean square	F	$F_{10\%}$
Samples	41.2	5	8.2	0.9	2.16
Laboratories	705.9	5	141.2	15.9	2.16
Replications	153.6	5	30.7	3.4	2.16
Residual variability	177.4	20	8.9		

* Data from Flanagan, 1951.

TABLE 8.6. MULTIPLE-COMPARISON OF MEANS FOR LEAD CONTENT ESTIMATED BY SIX LABORATORIES*

Laboratory	Type of analysis	Mean of Pb (ppm)
A	Spectrographic	55.7
B	Spectrographic	49.2
D	Spectrographic	47.8
E	Spectrographic	47.2
F	Chemical	47.1
C	Chemical	46.5

* Data from Flanagan, 1951.

analyses or 72 in all. The six samples of rock distributed to the six laboratories were randomly selected, and the experimental arrangement was that of a Latin square (Dixon and Massey, 1969, p. 310).

Table 8.5 is an analysis of variance for the results of this experiment; the numbers are those calculated by Flanagan (1951, p. 119), but their labels are changed slightly to correspond to the notation of this book. There appear to be no differences among the six samples, but there are replication differences within the laboratories, and also the laboratories find, on the average, different means. The replication variation shows that the individual laboratories cannot repeat determinations from day to day, or from week to week, as well as they can duplicate determinations at one particular time. This situation is common in chemical analysis.

In table 8.6 the means for lead estimated by the six laboratories are compared by the multiple-comparison procedure (sec. 5.12). The results show that the results from laboratory A appear to be discrepant from the rest. If the results from this laboratory are excluded, the mean value of the determinations from the other laboratories is 48 ppm, compared with the nominal value of 27 ppm for lead content of G-1 accepted before Flanagan's study. Moreover, the discrepancy between the spectrographic and chemical analyses stems from the determinations made by laboratory A. When the determinations of this laboratory are excluded, no difference between the spectrographic and colorimetric determinations is apparent.

8.6 EXPERIMENTAL DESIGNS

An experimental design is a procedure for gathering data systematically. One example is the randomized block design introduced in section 5.7. The purpose of an experimental design is to evaluate different kinds of variability.

There would be no point in defining kinds of variability if there were no way to sort them out. Experimental designs may be used to investigate the types of variability discussed in this chapter and are also pertinent for geologists who perform laboratory experiments, such as in experimental geochemistry, geophysics, and sedimentology.

In order to use an experimental design, one must have something clearly in mind that he wants to study. The variables (named *factors* in the language of experimental design) and their ranges that are to be considered must be identified. Also, the observations (often named *responses*) to be measured, whether univariate or multivariate, must be defined. The size of the experiment must be decided upon, including both the range of the factors and their number; these decisions specify the total number of observations.

Once these decisions have been made, an experimental design for the number and pattern of observations may be devised. The number of observations may be fixed initially or may be determined sequentially; the pattern may be haphazard, random, or one of many systematic types. Patterns of variables may exist with respect to space, time, or matter. Although for geology most patterns are spatial, problems may be patterned on time or on the experimental material, which often has a spatial or time pattern associated with it. Usable patterns do not include all those that might be invented; they include only those for which a statistical analysis has been devised.

A workable experimental design is achieved only through carefully specifying aims and through prescribing methods of collecting observations, which often must be done in a rigidly ordered manner. The key is planning. The reward, measured in information obtained for a certain amount of experimental effort, can be enormous compared with hit-or-miss experimentation.

These ideas may be illustrated by considering a simple fictitious experiment of sampling gold ore. Suppose that the natural variability on a scale of 100 feet is to be investigated. A decision is then made to sample at two locations 100 feet apart and to take at each location two geological samples of 5 kilograms each, as outlined in the diagram of figure 8.4. Because of local variability and sampling losses, the four samples will differ in composition. If each one is crushed, ground, pulverized, and split, eight fractions result. If

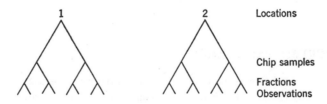

Fig. 8.4. Sampling design for a fictitious experiment of sampling gold ore.

two random assay tons (29.167 grams) are taken from each of the eight fractions and assayed, sixteen observations (assays) are obtained. Care should be taken to assay the individual assay tons in a random order and in different furnace runs (perhaps on different days and maybe even in different weeks).

From a statistical analysis of the observations, something can be learned about the variability due to assaying, to preparation, to combined local and sampling variability, and to natural variability on a scale of 100 feet. The design is an application of the nested analysis of variance (sec. 5.5), and the analysis-of-variance table has the following format:

Source of variation	Sum of squares	Degrees of freedom	Mean square	F	$F_{10\%}$
Location		1			
Geological samples		2			
Fractions		4			
Assay		8			

The previous analysis of variance sorts out and compares the four different kinds of variability.

The chain that has been outlined is perhaps the simplest that might be used to obtain an observation. In this chain the sample locations and their spacing 100 feet apart were chosen by geological judgment. The number of degrees of freedom to be obtained at each level would also require a choice based on geological and statistical judgment about the expected variability. The randomization would enter only at the sample preparation and assaying stages. The number of components of the chain to be examined separately depends on judgment about how much detail concerning variability is needed and about how many observations must be taken to learn about this variability. The randomization is a small but essential element of the whole design.

All experimental designs are basically like the one outlined. They contain arbitrary elements of judgment; they have a pattern of data collection for which a meaningful statistical analysis can be performed; and they include some places where randomization can and must be applied. The rest of this section takes up some salient points of experimental design; details about this important topic are given in an excellent book by Cochran and Cox (1957).

Pattern of Observations

The pattern of observations defines the experimental design. The best way to collect data unfortunately depends upon the functional form of the mathematical model and even upon the relationship of the coordinate system to this function. The problems can be extremely complicated, and answers are seldom available except for trivial cases. More often than not, once the data are collected, it becomes clear they they should have been collected in a different manner, one that can be specified. For instance, if one wanted to collect observations in order to investigate how a response changes with changing location, no trouble would arise if the nature of the surface to be fitted were known. Thus, if one knew that the surface was a plane, maximum information about its strike, dip, and elevation could be obtained by taking all of the observations in each of the four corners of the area of interest. In practice, one seldom if ever knows that a plane rather than some other surface is appropriate; and if a plane is inappropriate, sampling at the corners would be among the worst possible ways to gather observations.

This dilemma implies that some sort of sequential collecting process is often useful, particularly if data are expensive to collect and if only a small penalty is incurred through the delay needed to collect and analyze the data in sequence rather than all at once. The size of the penalty should be reduced as much as possible by advance planning, high-speed data processing by computer, etc. Appropriate elements of randomization must be introduced if possible into sequential sampling; otherwise, experimental effects can be confused with time trends, which may be introduced by such causes as increasing skill of the operator (sec. 7.6), changes in reagents, or seasonal changes (sec. 15.3).

Systematic Patterns

Sampling is usually done on systematic patterns because the most thorough coverage of the area of interest is obtained. In a systematic pattern, points are arranged on a grid (fig. 8.5), which is usually square; but it may be rectangular if a linear trend is suspected in the data (sec. 7.2), or it may be on polar coordinates for a radial pattern as in pyroclastic rocks dispersed outward from a volcano. Systematic patterns have been used for years in drillhole and soil sampling. Their main drawback is that unexpected periodicity in the data may wreak havoc with conclusions. If one systematically sampled traffic flow on alternate streets in midtown New York City, he might wrongly conclude that all traffic is westbound if he picked odd-numbered streets and that those few streets with eastbound traffic represented experimental error. In truth, almost all odd-numbered streets are one-way westbound, almost all even-numbered streets are one-way eastbound, and a few streets are two-way. In geology periodic effects such as folding of the Appalachian type, joint patterns, and wave patterns are well known; and other periodic patterns are doubtless still unrecognized.

If a systematic pattern is adopted, the experimental design depends on the rock body and on the physical method of sampling. If stratification is

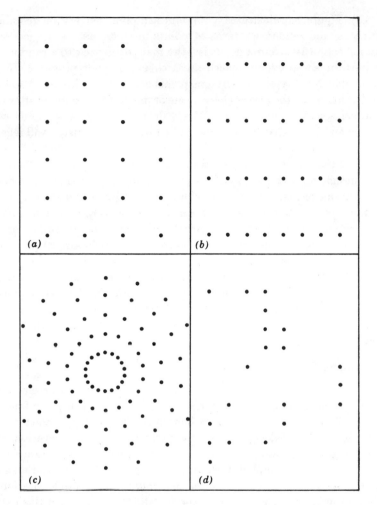

Fig. 8.5. Some arrangements of sample points. Patterns (*a*) to (*c*) are systematic; pattern (*d*) is random.

present, a systematic pattern should be used for the physical field sampling, which is concerned with natural and sampling variability. Then a random pattern should be adopted for the sample preparation and the chemical analysis.

Random Patterns

As first stated in section 3.4, and repeatedly emphasized since, a random pattern is not the same as a haphazard one. An example of a haphazard pattern that can cause trouble is the preference of chemists for certain digits (sec. 8.5). Figure 8.5 illustrates a random sampling pattern on rectangular coordinates, and random patterns of points in stereographic projection are shown in section 10.7. For a geologist, the main drawback to a random sampling pattern is that evenly spaced areal coverage is not provided. Readers who travel in Nevada can readily grasp this concept by playing the game of Keno. In this game, the player chooses one or more of 80 numbers, after which the banker selects 20 at random. Rarely do the 20 numbers chosen, which appear on an illuminated 10 by 8 grid, afford what a geologist would regard as suitable coverage for the grid area.

Thus random patterns are generally unsuitable for setting up the experimental design for the major factors of an investigation, but they become essential at one or more of the detailed working stages of data gathering, as in the sample preparation or chemical analysis. However, a random sampling pattern should be used whenever areal coverage is not required or if no stratification is present in the data, as generally occurs in sample preparation and chemical analysis.

Sometimes in geological sampling, random and systematic patterns may be used together. Thus for sampling granitic rocks in the San Bernardino Mountains (sec. 7.3) Richmond (1965) laid out a coarse grid. Then a sampler went to the grid points and at each one randomly chose the exact spot to sample.

Cluster Sampling

Cluster sampling refers to samples taken in local groups or clusters. In order to sample a large area, say, the state of California, it may be feasible to visit only 50 locations; but once one has taken the trouble to reach them, it may be desirable to take several geological samples at each location. If the local and regional variability are of about the same size, cluster sampling is a statistically efficient sampling technique as measured by information gained per dollar spent. In the extreme case one could visit a single location and obtain all the information available about California. Although this extreme seldom happens, it may be found where the stratification is vertical, as in sampling an oil field, a coal bed, a phosphate deposit, or sea water. Sampling vertically downward from a single point through all strata may be sufficient, and, indeed, samples from other locations may yield essentially identical results.

Cluster sampling has been extensively used in social science surveys, as explained in books by Stephan and McCarthy (1958) and Cochran (1963, Chaps. 9 and 10). Because of the potential gains, one should always try to apportion sampling so that an appropriate amount is done at individual locations as opposed to occupying other locations. In this book this subject is discussed with respect to drill-hole deflections (sec. 7.4), the contrast of local and regional variability in the Fresnillo mine (sec. 9.4), and geochemical sampling (sec. 15.3).

A General Rule for Choosing a Pattern of Observations

A general rule for choosing a pattern of observations is to be as systematic as possible, down to the place in the list of sources of variability where no advantage is gained by being systematic. However, it is essential to introduce some randomness at the very end of the sampling process, if not before.

NUMBER OF OBSERVATIONS. If observations are regarded as the group of values obtained at the end of a sampling process, which includes the acts of geological sampling, preparation, and chemical analysis, the appropriate number of observations to obtain may be considered. From this point of view the number of observations depends only on the following:

1. The variability associated with each of the several sources—natural, sampling, preparation, and analytical.
2. The statistic or statistics to be computed—mean, standard deviation, coefficient of variation, or other.
3. The distributional forms of the statistics—normal, lognormal, or other— and how well these forms are known.
4. The required precision.
5. The required accuracy.
6. The cost of obtaining observations, including planning, field sampling, preparation, and chemical analysis.

TYPES OF EXPERIMENTAL DESIGNS. The purposes of experimental designs may be summarized in three categories:

1. To estimate the population mean for one or more variables of interest.
2. To investigate how changes in different variables affect the population mean of a variable of interest. (The changes may be studied by varying quantitative or qualitative variables.)
3. To investigate how the population mean changes with changes in geographic location through trend-surface analysis (Chap. 9). (Although this category is a special case of the second one, it is separated because of its importance in geology.)

Several types of variable are recognized. In section 5.11 variables are classed as dependent or independent in a mathematical sense. They may be further classified as quantitative or qualitative. Quantitative variables have definite scales such as units of length and weight associated with them, whereas qualitative variables have only classifications such as species and type of basalt associated with them. Although quantitative variables may be used in a qualitative sense, as when rocks are classified into several types according to their iron content, the original ordering remains and may be important for the statistical analysis. Generally, the variables defined as dependent ones are quantitative, and the variables that affect their behavior may be either quantitative or qualitative. The variable or variables being studied are usually named the *response variable* or *variables*, and the variables that may influence their average values are named *factors*.

Although the ideal experimental design that would eliminate all sources of extraneous variability is never realized, it should be striven for, because often extraneous variability can be greatly reduced. Of the many ways to reduce experimental fluctuation, only two outstanding ones are mentioned here. First, *blocking* is the grouping together of similar material to take advantage of stratification (sec. 5.7). The term *block* has an abstract meaning; a laboratory, day, furnace run, or rock type may be a block. *Pairing* is the special name for the blocking of two items. Second, trends, which in geology are usually geographic, can be removed through trend-surface analysis to reduce the variability, a subject discussed in Chapter 9 (see in particular the Mi Vida example, sec. 9.8). Both methods take advantage of prior information about how the mean value of the response variable may be affected by some of the factors to be considered in an experiment.

Of the many types of experimental design, those for which a suitable statistical analysis exists or can be devised are of practical interest. Some experimental designs are introduced in this volume and others are discussed in the second volume. The simplest experimental design, the *completely randomized design*, is associated with the one-way analysis of variance (sec. 5.3). The *randomized-block design* takes advantage of blocking to reduce extraneous variability or experimental error and is also associated with an analysis of variance explained in detail (sec. 5.7). In order to consider more than one factor, *factorial designs* have been devised. Being relatively complex, they are not discussed in this book; the reader is referred to Cochran and Cox (1957) and Scheffé (1959). The example of operator variability in measuring quartz grains (sec. 8.5) is analyzed through a factorial design although the complexities of analysis are not explained. *Response surface designs* take advantage of the fact that, if several factors in an experiment are quantitative, other patterns than factorial designs may offer more efficient ways to gather data. One response surface design is explained in section 12.5. Experimental designs for interpretation of nonlinear models are being developed, but the work is still in its infancy, as explained by Draper and Smith (1966).

In any experimental design the replicating of observations is desirable in order to assess local variability, which by definition is independent of the trend-surface or other factors. Learning whether an observation has been replicated or merely duplicated may not be easy. If duplicate geological samples are sent to two chemists for analysis, their determinations clearly furnish replications, provided only that they do not communicate with one another. On the other hand, if duplicate geological samples are sent to the same chemist, his determinations furnish replications only if the samples are randomly numbered by the sender and if the chemist has no way, such as color, to match the duplicates. Also, a consideration of scale arises. If two diamond-drill holes are bored 10 feet apart, replicated observations undoubtedly are obtained if the scale of investigation is measured in miles. Whether a replication has been achieved or merely a duplication has resulted must always be considered. Duplicity, whether willful or not, must always be expected and guarded against because it can lead to underestimating the real variability.

REFERENCES

Antweiler, J. C., and Love, J. D., 1967, Gold-bearing sedimentary rocks in northwest Wyoming—a preliminary report: U.S. Geol. Survey Circ. 541, 12 p.

Boericke, W. F., 1939, The Jones riffle in cutting down samples: Eng. Mining Jour., v. 140, no. 6, p. 55–56.

Britten, H., 1961, Laboratory control, part 1: gold assaying: Trans. 7th Commonwealth Metall. Cong., Johannesburg, p. 1007–1021.

Cochran, W. G., and Cox, G. M., 1957, Experimental designs: New York, John Wiley & Sons, 611 p.

Cochran, W. G., 1963, Sampling techniques: New York, John Wiley & Sons, 413 p.

Coxon, C. H., and Sichel, H. S., 1959, Quality control of routine mine assaying and its influence on underground valuation: Jour. South African Inst. Mining and Metall., v. 59, no. 10, p. 489–517.

Dalrymple, G. B. and Lanphere, M. A., 1969, Potassium-argon dating: San Francisco, W. H. Freeman, 258 p.

Davis, G. R., 1963, Observations on sample preparation: Trans. Inst. Mining and Metall., v. 73, p. 255–267.

Dixon, W. J., and Massey, F. J., Jr., 1969, Introduction to statistical analysis: New York, McGraw-Hill, 638 p.

Draper, N. R., and Smith, Harry, 1966, Applied regression analysis: New York, John Wiley & Sons, 407 p.

Erickson, R. L., van Sickle, G. H., Nakagama, H. M., McCarthy, J. H., Jr., and Leong, K. W., 1966, Gold geochemical anomaly in the Cortez district, Nevada: U.S. Geol. Survey Circ. 534, 9 p.

Flanagan, F. J., 1960, The lead content of G-1, *in* Stevens, R. E., and others, Second report on a cooperative investigation of the composition of two silicate rocks: U.S. Geol. Survey Bull. 1113, p. 113–121.

Flanagan, F. J., Kellagher, R. C., and Smith, W. L., 1959, The slotted cone splitter: Jour. Sed. Petrology, v. 29, no. 1, p. 108–115.

Fleischer, Michael, and Chao, E. C. T., 1960, Some problems in the estimation of abundances of elements in the earth's crust: Internat. Geol. Cong., 21st, Copenhagen, Rept., pt. 1, p. 141–148.

Griffiths, J. C., 1967, Scientific method in analysis of sediments: New York, McGraw-Hill, 508 p.

Hawkes, H. E., and Webb, J. S., 1962, Geochemistry in mineral exploration: New York, Harper & Row, 401 p.

Hazen, S. W., Jr., 1967, Some statistical techniques for analyzing mine and mineral-deposit sample and assay data: U.S. Bur. Mines Bull. 621, 223 p.

Johannsen, Albert, 1914, Manual of petrographic methods: New York, McGraw-Hill, 649 p.

Kellagher, R. C., and Flanagan, F. J., 1956, The multiple-cone sample splitter: Jour. Sed. Petrology, v. 26, no. 3, p. 213–221.

Krumbein, W. C., and Pettijohn, F. J., 1938, Manual of sedimentary petrography: New York, Appleton-Century-Crofts, 531 p.

Lundel, G. E. F., Bright, H. A., and Hoffman, J. I., 1953, Applied inorganic analysis: New York, John Wiley & Sons, 1034 p.

McKinstry, H. E., 1948, Mining geology: Englewood Cliffs, N.J., Prentice-Hall, 680 p.

Miesch, A. T., 1967, Methods of computation for estimating geochemical abundance: U.S. Geol. Survey Prof. Paper 574-B, 14 p.

Milner, H. B., 1962, Laboratory technique: pt. 1, preparatory methods, *in* Sedimentary petrography, v. 1: New York, Macmillan, p. 80–128.

Muller, L. D., 1967, Mineral separation, *in* Zussman, J., ed., Physical methods in determinative mineralogy: New York, Academic Press, p. 1–30.

Müller, German, 1967, Methods in sedimentary petrology: New York, Hafner Publishing 283 p.

Otto, G. H., 1937, The use of statistical methods in effecting improvements on a Jones sample splitter: Jour. Sed. Petrology, v. 7, no. 3, p. 110–132.

Peele, Robert, ed., 1941, Mining engineers' handbook: New York, John Wiley & Sons, 2 vols., 2442 p.

Richmond, J. F., 1965, Chemical variation in quartz monozonite from Cactus Flat, San Bernardino Mountains, California: Am. Jour. Science, v. 263, p. 53–63.

Rosenfeld, M. A., and Griffiths, J. C., 1953, An experimental test of visual comparison technique in estimating two dimensional sphericity and roundness of quartz grains: Am. Jour. Science, v. 251, p. 553–585.

Scheffé, Henry, 1959, Analysis of variance: New York, John Wiley & Sons, 476 p.

Stephan, F. F., and McCarthy, P. J., 1958, Sampling opinions, an analysis of survey procedure: New York, John Wiley & Sons, 451 p.

Taggert, A. F., ed., 1945, Handbook of mineral dressing: New York, John Wiley & Sons. 1905 p.

Tatlock, D. B., 1966, Some alkali and titania analyses of tektites before and after G-1 precision monitoring: Geochimica et Cosmochimica Acta, v. 80, p. 123–128.

Wentworth, C. K., Wilgus, W. L., and Koch, H. L., 1934, A rotary type of sample splitter: Jour. Sed. Petrology, v. 4, no. 3, p. 127.

Wilson, E. B., Jr., 1952, An introduction to scientific research: New York, McGraw-Hill, 375 p.

APPENDIX

APPENDIX

TABLE A.1. RANDOM DIGITS*

10000	11164	36318	75061	37674	26320	75100	10431	20418	19228	91792
10001	21215	91791	76831	58678	87054	31687	93205	43685	19732	08468
10002	10438	44482	66558	37649	08882	90870	12462	41810	01806	02977
10003	36792	26236	33266	66583	60881	97395	20461	36742	02852	50564
10004	73944	04773	12032	51414	82384	38370	00249	80709	72605	67497
10005	49563	12872	14063	93104	78483	72717	68714	18048	25005	04151
10006	64208	48237	41701	73117	33242	42314	83049	21933	92813	04763
10007	51486	72875	38605	29341	80749	80151	33835	52602	79147	08868
10008	99756	26360	64516	17971	48478	09610	04638	17141	09227	10606
10009	71325	55217	13015	72907	00431	45117	33827	92873	02953	85474
10010	65285	97198	12138	53010	94601	15838	16805	61004	43516	17020
10011	17264	57327	38224	29301	31381	38109	34976	65692	98566	29550
10012	95639	99754	31199	92558	68368	04985	51092	37780	40261	14479
10013	61555	76404	86210	11808	12841	45147	97438	60022	12645	62000
10014	78137	98768	04689	87130	79225	08153	84967	64539	79493	74917
10015	62490	99215	84987	28759	19177	14733	24550	28067	68894	38490
10016	24216	63444	21283	07044	92729	37284	13211	37485	10415	36457
10017	16975	95428	33226	55903	31605	43817	22250	03918	46999	98501
10018	59138	39542	71168	57609	91510	77904	74244	50940	31553	62562
10019	29478	59652	50414	31966	87912	87154	12944	49862	96566	48825
10020	96155	95009	27429	72918	08457	78134	48407	26061	58754	05326
10021	29621	66583	62966	12468	20245	14015	04014	35713	03980	03024
10022	12639	75291	71020	17265	41598	64074	64629	63293	53307	48766
10023	14544	37134	54714	02401	63228	26831	19386	15457	17999	18306
10024	83403	88827	09834	11333	68431	31706	26652	04711	34593	22561
10025	67642	05204	30697	44806	96989	68403	85621	45556	35434	09532
10026	64041	99011	14610	40273	09482	62864	01573	82274	81446	32477
10027	17048	94523	97444	59904	16936	39384	97551	09620	63932	03091
10028	93039	89416	52795	10631	09728	68202	20963	02477	55494	39563
10029	82244	34392	96607	17220	51984	10753	76272	50985	97593	34320
10030	96990	55244	70693	25255	40029	23289	48819	07159	60172	81697
10031	09119	74803	97303	88701	51380	73143	98251	78635	27556	20712
10032	57666	41204	47589	78364	38266	94393	70713	53388	79865	92069
10033	46492	61594	26729	58272	81754	14648	77210	12923	53712	87771
10034	08433	19172	08320	20839	13715	10597	17234	39355	74816	03363
10035	10011	75004	86054	41190	10061	19660	03500	68412	57812	57929
10036	92420	65431	16530	05547	10683	88102	30176	84750	10115	69220
10037	35542	55865	07304	47010	43233	57022	52161	82976	47981	46588
10038	86595	26247	18552	29491	33712	32285	64844	69395	41387	87195
10039	72115	34985	58036	99137	47482	06204	24138	24272	16196	04393
10040	07428	58863	96023	88936	51343	70958	96768	74317	27176	29600
10041	35379	27922	28906	55013	26937	48174	04197	36074	65315	12537
10042	10982	22807	10920	26299	23593	64629	57801	10437	43965	15344
10043	90127	33341	77806	12446	15444	49244	47277	11346	15884	28131
10044	63002	12990	23510	68774	48983	20481	59815	67248	17076	78910
10045	40779	86382	48454	65269	91239	45989	45389	54847	77919	41105
10046	43216	12608	18167	84631	94058	82458	15139	76856	86019	47928
10047	96167	64375	74108	93643	09204	98855	59051	56492	11933	64958
10048	70975	62693	35684	72607	23026	37004	32989	24843	01128	74658
10049	85812	61875	23570	75754	29090	40264	80399	47254	40135	69911

TABLE A.1. (CONTINUED)

10050	40603	16152	83235	37361	98783	24838	39793	80954	76865	32713
10051	40941	53585	69958	60916	71018	90561	84505	53980	64735	85140
10052	73505	83472	55953	17957	11446	22618	34771	25777	27064	13526
10053	39412	16013	11442	89320	11307	49396	39805	12249	57656	88686
10054	57994	76748	54627	48511	78646	33287	35524	54522	08795	56273
10055	61834	59199	15469	82285	84164	91333	90954	87186	31598	25942
10056	91402	77227	79516	21007	58602	81418	87838	18443	76162	51146
10057	58299	83880	20125	10794	37780	61705	18276	99041	78135	99661
10058	40684	99948	33880	76413	63839	71371	32392	51812	48248	96419
10059	75978	64298	08074	62055	73864	01926	78374	15741	74452	49954
10060	34556	39861	88267	76068	62445	64361	78685	24246	27027	48239
10061	65990	57048	25067	77571	77974	37634	81564	98608	37224	49848
10062	16381	15069	25416	87875	90374	86203	29677	82543	37554	89179
10063	52458	88880	78352	67913	09245	47773	51272	06976	99571	33365
10064	33007	85607	92008	44897	24964	50559	79549	85658	96865	24186
10065	38712	31512	08588	61490	72294	42862	87334	05866	66269	43158
10066	58722	03678	19186	69602	34625	75958	56869	17907	81867	11535
10067	26188	69497	51351	47799	20477	71786	52560	66827	79419	70886
10068	12893	54048	07255	86149	99090	70958	50775	31768	52903	27645
10069	33186	81346	85095	37282	85536	72661	32180	40229	19209	74939
10070	79893	29448	88392	54211	61708	83452	61227	81690	42265	20310
10071	48449	15102	44126	19438	23382	14985	37538	30120	82443	11152
10072	94205	04259	68983	50561	06902	10269	22216	70210	60736	58772
10073	38648	09278	81313	77400	41126	52614	93613	27263	99381	49500
10074	04292	46028	75666	26954	34979	68381	45154	09314	81009	05114
10075	17026	49737	85875	12139	59391	81830	30185	83095	78752	40899
10076	48070	76848	02531	97737	10151	18169	31709	74842	85522	74092
10077	30159	95450	83778	46115	99178	97718	98440	15076	21199	20492
10078	12148	92231	31361	60650	54695	30035	22765	91386	70399	79270
10079	73838	77067	24863	97576	01139	54219	02959	45696	98103	78867
10080	73547	43759	95632	39555	74391	07579	69491	02647	17050	49869
10081	07277	93217	79421	21769	83572	48019	17327	99638	87035	89300
10082	65128	48334	07493	28098	52087	55519	83718	60904	48721	17522
10083	38716	61380	60212	05099	21210	22052	01780	36813	19528	07727
10084	31921	76458	73720	08657	74922	61335	41690	41967	50691	30508
10085	57238	27464	61487	52329	26150	79991	64398	91273	26824	94827
10086	24219	41090	08531	61578	08236	41140	76335	91189	66312	44000
10087	31309	49387	02330	02476	96074	33256	48554	95401	02642	29119
10088	20750	97024	72619	66628	66509	31206	55293	24249	02266	39010
10089	28537	84395	26654	37851	80590	53446	34385	86893	87713	26842
10090	97929	41220	86431	94485	28778	44997	38802	56594	61363	04206
10091	40568	33222	40486	91122	43294	94541	40988	02929	83190	74247
10092	41483	92935	17061	78252	40498	43164	68646	33023	64333	64083
10093	93040	66476	24990	41099	65135	37641	97613	87282	63693	55299
10094	76869	39300	84978	07504	36835	72748	47644	48542	25076	68626
10095	02982	57991	50765	91930	21375	35604	29963	13738	03155	59914
10096	94479	76500	39170	06629	10031	48724	49822	44021	44335	26474
10097	52291	75822	95966	90947	65031	75913	52654	63377	70664	60082
10098	03684	03600	52831	55381	97013	19993	41295	29118	18710	64851
10099	58939	28366	86765	67465	45421	74228	01095	50987	83833	37276

TABLE A.1 (CONTINUED)

10100	37100	62492	63642	47638	13925	80113	88067	42575	44078	62703
10101	53406	13855	38519	29500	62479	01036	87964	44498	07793	21599
10102	55172	81556	18856	59043	64315	38270	25677	01965	21310	28115
10103	40353	84807	47767	46890	16053	32415	60259	99788	55924	22077
10104	18899	09612	77541	57675	70153	41179	97535	82889	27214	03482
10105	68141	25340	92551	11326	60939	79355	41544	88926	09111	86431
10106	51559	91159	81310	63251	91799	41215	87412	35317	74271	11603
10107	92214	33386	73459	79359	65867	39269	57527	69551	17495	91456
10108	15089	50557	33166	87094	52425	21211	41876	42525	36625	63964
10109	96461	00604	11120	22254	16763	19206	67790	88362	01880	37911
10110	28177	44111	15705	73835	69399	33602	13660	84342	97667	80847
10111	66953	44737	81127	07493	07861	12666	85077	95972	96556	80108
10112	19712	27263	84575	49820	19837	69985	34931	67935	71903	82560
10113	68756	64757	19987	92222	11691	42502	00952	47981	97579	93408
10114	75022	65332	98606	29451	57349	39219	08585	31502	96936	96356
10115	11323	70069	90269	89266	46413	61615	66447	49751	15836	97343
10116	55208	63470	18158	25283	19335	53893	87746	72531	16826	52605
10117	11474	08786	05594	67045	13231	51186	71500	50498	59487	48677
10118	81422	86842	60997	79669	43804	78690	58358	87639	24427	66799
10119	21771	75963	23151	90274	08275	50677	99384	94022	84888	80139
10120	42278	12160	32576	14278	34231	20724	27908	02657	19023	07190
10121	17697	60114	63247	32096	32503	04923	17570	73243	76181	99343
10122	05686	30243	34124	02936	71749	03031	72259	26351	77511	00850
10123	52992	46650	89910	57395	39502	49738	87854	71066	84596	33115
10124	94518	93984	81478	67750	89354	01080	25988	84359	31088	13655
10125	00184	72186	78906	75480	71140	15199	69002	08374	22126	23555
10126	87462	63165	79816	61630	50140	95319	79205	79202	67414	60805
10127	88692	58716	12273	48176	86038	78474	76730	82931	51595	20747
10128	20094	42962	41382	16768	13261	13510	04822	96354	72001	68642
10129	60935	81504	50520	82153	27892	18029	79663	44146	72876	67843
10130	51392	85936	43898	50596	81121	98122	69196	54271	12059	62539
10131	54239	41918	79526	46274	24853	67165	12011	04923	20273	89405
10132	57892	73394	07160	90262	48731	46648	70977	58262	78359	50436
10133	02330	74736	53274	44468	53616	35794	54838	39114	68302	26855
10134	76115	29247	55342	51299	79908	36613	68361	18864	13419	34950
10135	63312	81886	29085	20101	38037	34742	78364	39356	40006	49800
10136	27632	21570	34274	56426	00330	07117	86673	46455	66866	76374
10137	06335	62111	44014	52567	79480	45886	92585	87828	17376	35254
10138	64142	87676	21358	88773	10604	62834	63971	03989	21421	76086
10139	28436	25468	75235	75370	63543	76266	27745	31714	04219	00699
10140	09522	83855	85973	15888	29554	17995	37443	11461	42909	32634
10141	93714	15414	93712	02742	34395	21929	38928	31205	01838	60000
10142	15681	53599	58185	73840	88758	10618	98725	23146	13521	47905
10143	77712	23914	08907	43768	10304	61405	53986	61116	76164	54958
10144	78453	54844	61509	01245	91199	07482	02534	08189	62978	55516
10145	24860	68284	19367	29073	93464	06714	45268	60678	58506	23700
10146	37284	06844	78887	57276	42695	03682	83240	09744	63025	60997
10147	35488	52473	37634	32569	39590	27379	23520	29714	03743	08444
10148	51595	59909	35223	44991	29830	56614	59661	83397	38421	17503
10149	90660	35171	30021	91120	78793	16827	89320	08260	09181	53622

TABLE A.1 (CONTINUED)

10150	54723	56527	53076	38235	42780	22716	36400	48028	78196	92985
10151	84828	81248	25548	34075	43459	44628	21866	90350	82264	20478
10152	65799	01914	81363	05173	23674	41774	25154	73003	87031	94368
10153	87917	38549	48213	71708	92035	92527	55484	32274	87918	22455
10154	26907	88173	71189	28377	13785	87469	35647	19695	33401	51998
10155	68052	65422	88460	06352	42379	55499	60469	76931	83430	24560
10156	42587	68149	88147	99700	56124	53239	38726	63652	36644	50876
10157	97176	55416	67642	05051	89931	19482	80720	48977	70004	03664
10158	53295	87133	38264	94708	00703	35991	76404	82249	22942	49659
10159	23011	94108	29196	65187	69974	01970	31667	54307	40032	30031
10160	75768	49549	24543	63285	32803	18301	80851	89301	02398	99891
10161	86668	70341	66460	75648	78678	27770	30245	44775	56120	44235
10162	56727	72036	50347	33521	05068	47248	67832	30960	95465	32217
10163	27936	78010	09617	04408	18954	61862	64547	52453	83213	47833
10164	31994	69072	37354	93025	38934	90219	91148	62757	51703	84040
10165	02985	95303	15182	50166	11755	56256	89546	31170	87221	63267
10166	89965	10206	95830	95406	33845	87588	70237	84360	19629	72568
10167	45587	29611	98579	42481	05359	36578	56047	68114	58583	16313
10168	01071	08530	74305	77509	16270	20889	99753	80035	55643	18291
10169	90209	68521	14293	39194	68803	32052	39413	26883	83119	69623
10170	04982	68470	27875	15480	13206	44784	83601	03172	07817	01520
10171	19740	24637	97377	32112	74283	69384	49768	64141	02024	85380
10172	50197	79869	86497	68709	42073	28498	82750	43571	77075	07123
10173	46954	67536	28968	81936	95999	04319	09932	66223	45491	69503
10174	82549	62676	31123	49899	70512	95288	15517	85352	21987	08669
10175	61798	81600	80018	84742	06103	60786	01408	75967	29948	21454
10176	57666	29055	46518	01487	30136	14349	56159	47408	78311	25896
10177	29805	64994	66872	62230	41385	58066	96600	99301	85976	84194
10178	06711	34939	19599	76247	87879	97114	74314	39599	43544	36255
10179	13934	46885	58315	88366	06138	37923	11192	90757	10831	01580
10180	28549	98327	99943	25377	17628	65468	07875	16728	22602	33892
10181	40871	61803	25767	55484	90997	86941	64027	01020	39518	34693
10182	47704	38355	71708	80117	11361	88875	22315	38048	42891	87885
10183	62611	19698	09304	29265	07636	08508	23773	56545	08015	28891
10184	03047	83981	11916	09267	67316	87952	27045	62536	32180	60936
10185	26460	50501	31731	18938	11025	18515	31747	96828	58258	97107
10186	01764	25959	69293	89875	72710	49659	66632	25314	95260	22146
10187	11762	54806	02651	52912	32770	64507	59090	01275	47624	16124
10188	31736	31695	11523	64213	91190	10145	34231	36405	65860	48771
10189	97155	48706	52239	21831	49043	18650	72246	43729	63368	53822
10190	31181	49672	17237	04024	65324	32460	01566	67342	94986	36106
10191	32115	82683	67182	89030	41370	50266	19505	57724	93358	49445
10192	07068	75947	71743	69285	30395	81818	36125	52055	20289	16911
10193	26622	74184	75166	96748	34729	61289	36908	73686	84641	45130
10194	02805	52676	22519	47848	68210	23954	63085	87729	14176	45410
10195	32301	58701	04193	30142	99779	21697	05059	26684	63516	75925
10196	26339	56909	39331	42101	01031	01947	02257	47236	19913	90371
10197	95274	09508	81012	42413	11278	19354	68661	04192	36878	84366
10198	24275	39632	09777	98800	48027	96908	08177	15364	02317	89548
10199	36116	42128	65401	94199	51058	10759	47244	99830	64255	40550

* From "A Million Random Digits" published by the Rand Corporation, Free Press Glencoe, Ill., 1955.

TABLE A-2. PERCENTAGE POINTS OF THE STANDARDIZED NORMAL DISTRIBUTION *

Percentage point	s.n.d.
99.9	− 3.0902
99.5	− 2.5758
99.0	− 2.3263
97.5	− 1.9600
95.0	− 1.6449
90.0	− 1.2816
80.0	− 0.8416
75.0	− 0.6745
70.0	− 0.5244
60.0	− 0.2533
50.0	0.0000
40.0	0.2533
30.0	0.5244
25.0	0.6745
20.0	0.8416
10.0	1.2816
5.0	1.6449
2.5	1.9600
1.0	2.3263
0.5	2.5758
0.1	3.0902

* From Table 3, Biometrika Tables for Statisticians, Volume 1, Second Edition, Cambridge University Press, 1962.

TABLE A.3. PERCENTAGE POINTS OF THE χ^2-DISTRIBUTION *

α df	99.5%	99%	97.5%	95%	90%
1	392704.10⁻¹⁰	157088.10⁻⁹	982069.10⁻⁹	393214.10⁻⁸	0·0157908
2	0·0100251	0·0201007	0·0506356	0·102587	0·210720
3	0·0717212	0·114832	0·215795	0·351846	0·584375
4	0·206990	0·297110	0·484419	0·710721	1·063623
5	0·411740	0·554300	0·831211	1·145476	1·61031
6	0·675727	0·872085	1·237347	1·63539	2·20413
7	0·989265	1·239043	1·68987	2·16735	2·83311
8	1·344419	1·646482	2·17973	2·73264	3·48954
9	1·734926	2·087912	2·70039	3·32511	4·16816
10	2·15585	2·55821	3·24697	3·94030	4·86518
11	2·60321	3·05347	3·81575	4·57481	5·57779
12	3·07382	3·57056	4·40379	5·22603	6·30380
13	3·56503	4·10691	5·00874	5·89186	7·04150
14	4·07468	4·66043	5·62872	6·57063	7·78953
15	4·60094	5·22935	6·26214	7·26094	8·54675
16	5·14224	5·81221	6·90766	7·96164	9·31223
17	5·69724	6·40776	7·56418	8·67176	10·0852
18	6·26481	7·01491	8·23075	9·39046	10·8649
19	6·84398	7·63273	8·90655	10·1170	11·6509
20	7·43386	8·26040	9·59083	10·8508	12·4426
21	8·03366	8·89720	10·28293	11·5913	13·2396
22	8·64272	9·54249	10·9823	12·3380	14·0415
23	9·26042	10·19567	11·6885	13·0905	14·8479
24	9·88623	10·8564	12·4011	13·8484	15·6587
25	10·5197	11·5240	13·1197	14·6114	16·4734
26	11·1603	12·1981	13·8439	15·3791	17·2919
27	11·8076	12·8786	14·5733	16·1513	18·1138
28	12·4613	13·5648	15·3079	16·9279	18·9392
29	13·1211	14·2565	16·0471	17·7083	19·7677
30	13·7867	14·9535	16·7908	18·4926	20·5992
40	20·7065	22·1643	24·4331	26·5093	29·0505
50	27·9907	29·7067	32·3574	34·7642	37·6886
60	35·5346	37·4848	40·4817	43·1879	46·4589
70	43·2752	45·4418	48·7576	51·7393	55·3290
80	51·1720	53·5400	57·1532	60·3915	64·2778
90	59·1963	61·7541	65·6466	69·1260	73·2912
100	67·3276	70·0648	74·2219	77·9295	82·3581

TABLE A.3 (CONTINUED)

α df	10%	5%	2.5%	1%	0.5%
1	2·70554	3·84146	5·02389	6·63490	7·87944
2	4·60517	5·99147	7·37776	9·21034	10·5966
3	6·25139	7·81473	9·34840	11·3449	12·8381
4	7·77944	9·48773	11·1433	13·2767	14·8602
5	9·23635	11·0705	12·8325	15·0863	16·7496
6	10·6446	12·5916	14·4494	16·8119	18·5476
7	12·0170	14·0671	16·0128	18·4753	20·2777
8	13·3616	15·5073	17·5346	20·0902	21·9550
9	14·6837	16·9190	19·0228	21·6660	23·5893
10	15·9871	18·3070	20·4831	23·2093	25·1882
11	17·2750	19·6751	21·9200	24·7250	26·7569
12	18·5494	21·0261	23·3367	26·2170	28·2995
13	19·8119	22·3621	24·7356	27·6883	29·8194
14	21·0642	23·6848	26·1190	29·1413	31·3193
15	22·3072	24·9958	27·4884	30·5779	32·8013
16	23·5418	26·2962	28·8454	31·9999	34·2672
17	24·7690	27·5871	30·1910	33·4087	35·7185
18	25·9894	28·8693	31·5264	34·8053	37·1564
19	27·2036	30·1435	32·8523	36·1908	38·5822
20	28·4120	31·4104	34·1696	37·5662	39·9968
21	29·6151	32·6705	35·4789	38·9321	41·4010
22	30·8133	33·9244	36·7807	40·2894	42·7956
23	32·0069	35·1725	38·0757	41·6384	44·1813
24	33·1963	36·4151	39·3641	42·9798	45·5585
25	34·3816	37·6525	40·6465	44·3141	46·9278
26	35·5631	38·8852	41·9232	45·6417	48·2899
27	36·7412	40·1133	43·1944	46·9630	49·6449
28	37·9159	41·3372	44·4607	48·2782	50·9933
29	39·0875	42·5560	45·7222	49·5879	52·3356
30	40·2560	43·7729	46·9792	50·8922	53·6720
40	51·8050	55·7585	59·3417	63·6907	66·7659
50	63·1671	67·5048	71·4202	76·1539	79·4900
60	74·3970	79·0819	83·2976	88·3794	91·9517
70	85·5271	90·5312	95·0231	100·425	104·215
80	96·5782	101·879	106·629	112·329	116·321
90	107·565	113·145	118·136	124·116	128·299
100	118·498	124·342	129·561	135·807	140·169

* From Table 8, Biometrika Tables for Statisticians, Volume 1, Second Edition, Cambridge University Press, 1962.

TABLE A.4. PERCENTAGE POINTS OF THE t_m DISTRIBUTION *

α df	10%	5%	2.5%	1%
1	3.078	6.314	12.706	31.821
2	1.886	2.920	4.303	6.965
3	1.638	2.353	3.182	4.541
4	1.533	2.132	2.776	3.747
5	1.476	2.015	2.571	3.365
6	1.440	1.943	2.447	3.143
7	1.415	1.895	2.365	2.998
8	1.397	1.860	2.306	2.896
9	1.383	1.833	2.262	2.821
10	1.372	1.812	2.228	2.764
11	1.363	1.796	2.201	2.718
12	1.356	1.782	2.179	2.681
13	1.350	1.771	2.160	2.650
14	1.345	1.761	2.145	2.624
15	1.341	1.753	2.131	2.602
16	1.337	1.746	2.120	2.583
17	1.333	1.740	2.110	2.567
18	1.330	1.734	2.101	2.552
19	1.328	1.729	2.093	2.539
20	1.325	1.725	2.086	2.528
21	1.323	1.721	2.080	2.518
22	1.321	1.717	2.074	2.508
23	1.319	1.714	2.069	2.500
24	1.318	1.711	2.064	2.492
25	1.316	1.708	2.060	2.485
26	1.315	1.706	2.056	2.479
27	1.314	1.703	2.052	2.473
28	1.313	1.701	2.048	2.467
29	1.311	1.699	2.045	2.462
30	1.310	1.697	2.042	2.457
40	1.303	1.684	2.021	2.423
60	1.296	1.671	2.000	2.390
120	1.289	1.658	1.980	2.358
∞	1.282	1.645	1.960	2.326

* From Table 12, Biometrika Tables for Statisticians, Volume 1, Second Edition, Cambridge University Press, 1962.

TABLE A.5. PERCENTAGE POINTS OF THE F-DISTRIBUTION *

$\alpha = 10\%$

df_2 \ df_1	1	2	3	4	5	6	7	8	9
1	39.864	49.500	53.593	55.833	57.241	58.204	58.906	59.439	59.858
2	8.5263	9.0000	9.1618	9.2434	9.2926	9.3255	9.3491	9.3668	9.3805
3	5.5383	5.4624	5.3908	5.3427	5.3092	5.2847	5.2662	5.2517	5.2400
4	4.5448	4.3246	4.1908	4.1073	4.0506	4.0098	3.9790	3.9549	3.9357
5	4.0604	3.7797	3.6195	3.5202	3.4530	3.4045	3.3679	3.3393	3.3163
6	3.7760	3.4633	3.2888	3.1808	3.1075	3.0546	3.0145	2.9830	2.9577
7	3.5894	3.2574	3.0741	2.9605	2.8833	2.8274	2.7849	2.7516	2.7247
8	3.4579	3.1131	2.9238	2.8064	2.7265	2.6683	2.6241	2.5893	2.5612
9	3.3603	3.0065	2.8129	2.6927	2.6106	2.5509	2.5053	2.4694	2.4403
10	3.2850	2.9245	2.7277	2.6053	2.5216	2.4606	2.4140	2.3772	2.3473
11	3.2252	2.8595	2.6602	2.5362	2.4512	2.3891	2.3416	2.3040	2.2735
12	3.1765	2.8068	2.6055	2.4801	2.3940	2.3310	2.2828	2.2446	2.2135
13	3.1362	2.7632	2.5603	2.4337	2.3467	2.2830	2.2341	2.1953	2.1638
14	3.1022	2.7265	2.5222	2.3947	2.3069	2.2426	2.1931	2.1539	2.1220
15	3.0732	2.6952	2.4898	2.3614	2.2730	2.2081	2.1582	2.1185	2.0862
16	3.0481	2.6682	2.4618	2.3327	2.2438	2.1783	2.1280	2.0880	2.0553
17	3.0262	2.6446	2.4374	2.3077	2.2183	2.1524	2.1017	2.0613	2.0284
18	3.0070	2.6239	2.4160	2.2858	2.1958	2.1296	2.0785	2.0379	2.0047
19	2.9899	2.6056	2.3970	2.2663	2.1760	2.1094	2.0580	2.0171	1.9836
20	2.9747	2.5893	2.3801	2.2489	2.1582	2.0913	2.0397	1.9985	1.9649
21	2.9609	2.5746	2.3649	2.2333	2.1423	2.0751	2.0232	1.9819	1.9480
22	2.9486	2.5613	2.3512	2.2193	2.1279	2.0605	2.0084	1.9668	1.9327
23	2.9374	2.5493	2.3387	2.2065	2.1149	2.0472	1.9949	1.9531	1.9189
24	2.9271	2.5383	2.3274	2.1949	2.1030	2.0351	1.9826	1.9407	1.9063
25	2.9177	2.5283	2.3170	2.1843	2.0922	2.0241	1.9714	1.9292	1.8947
26	2.9091	2.5191	2.3075	2.1745	2.0822	2.0139	1.9610	1.9188	1.8841
27	2.9012	2.5106	2.2987	2.1655	2.0730	2.0045	1.9515	1.9091	1.8743
28	2.8939	2.5028	2.2906	2.1571	2.0645	1.9959	1.9427	1.9001	1.8652
29	2.8871	2.4955	2.2831	2.1494	2.0566	1.9878	1.9345	1.8918	1.8560
30	2.8807	2.4887	2.2761	2.1422	2.0492	1.9803	1.9269	1.8841	1.8498
40	2.8354	2.4404	2.2261	2.0909	1.9968	1.9269	1.8725	1.8289	1.7929
60	2.7914	2.3932	2.1774	2.0410	1.9457	1.8747	1.8194	1.7748	1.7380
120	2.7478	2.3473	2.1300	1.9923	1.8959	1.8238	1.7675	1.7220	1.6843
∞	2.7055	2.3026	2.0838	1.9449	1.8473	1.7741	1.7167	1.6702	1.6315

α = 10%

10	12	15	20	24	30	40	60	120	∞
60.195	60.705	61.220	61.740	62.002	62.265	62.529	62.794	63.061	63.328
9.3916	9.4081	9.4247	9.4413	9.4496	9.4579	9.4663	9.4746	9.4829	9.4913
5.2304	5.2156	5.2003	5.1845	5.1764	5.1681	5.1597	5.1512	5.1425	5.1337
3.9199	3.8955	3.8689	3.8443	3.8310	3.8174	3.8036	3.7896	3.7753	3.7607
3.2974	3.2682	3.2380	3.2067	3.1905	3.1741	3.1573	3.1402	3.1228	3.1050
2.9369	2.9047	2.8712	2.8363	2.8183	2.8000	2.7812	2.7620	2.7423	2.7222
2.7025	2.6681	2.6322	2.5947	2.5753	2.5555	2.5351	2.5142	2.4928	2.4708
2.5380	2.5020	2.4642	2.4246	2.4041	2.3830	2.3614	2.3391	2.3162	2.2926
2.4163	2.3789	2.3396	2.2983	2.2768	2.2547	2.2320	2.2085	2.1843	2.1592
2.3226	2.2841	2.2435	2.2007	2.1784	2.1554	2.1317	2.1072	2.0818	2.0554
2.2482	2.2087	2.1671	2.1230	2.1000	2.0762	2.0516	2.0261	1.9997	1.9721
2.1878	2.1474	2.1049	2.0597	2.0360	2.0115	1.9861	1.9597	1.9323	1.9036
2.1376	2.0966	2.0532	2.0070	1.9827	1.9576	1.9315	1.9043	1.8759	1.8462
2.0954	2.0537	2.0095	1.9625	1.9377	1.9119	1.8852	1.8572	1.8280	1.7973
2.0593	2.0171	1.9722	1.9243	1.8990	1.8728	1.8454	1.8168	1.7867	1.7551
2.0281	1.9854	1.9399	1.8913	1.8656	1.8388	1.8108	1.7816	1.7507	1.7182
2.0009	1.9577	1.9117	1.8624	1.8362	1.8090	1.7805	1.7506	1.7191	1.6856
1.9770	1.9333	1.8868	1.8368	1.8103	1.7827	1.7537	1.7232	1.6910	1.6567
1.9557	1.9117	1.8647	1.8142	1.7873	1.7592	1.7298	1.6988	1.6659	1.6308
1.9367	1.8924	1.8449	1.7938	1.7667	1.7382	1.7083	1.6768	1.6433	1.6074
1.9197	1.8750	1.8272	1.7756	1.7481	1.7193	1.6890	1.6569	1.6228	1.5862
1.9043	1.8593	1.8111	1.7590	1.7312	1.7021	1.6714	1.6389	1.6042	1.5668
1.8903	1.8450	1.7964	1.7439	1.7159	1.6864	1.6554	1.6224	1.5871	1.5490
1.8775	1.8319	1.7831	1.7302	1.7019	1.6721	1.6407	1.6073	1.5715	1.5327
1.8658	1.8200	1.7708	1.7175	1.6890	1.6589	1.6272	1.5934	1.5570	1.5176
1.8550	1.8090	1.7596	1.7059	1.6771	1.6468	1.6147	1.5805	1.5437	1.5036
1.8451	1.7989	1.7492	1.6951	1.6662	1.6356	1.6032	1.5686	1.5313	1.4906
1.8359	1.7895	1.7395	1.6852	1.6560	1.6252	1.5925	1.5575	1.5198	1.4784
1.8274	1.7808	1.7306	1.6759	1.6465	1.6155	1.5825	1.5472	1.5090	1.4670
1.8195	1.7727	1.7223	1.6673	1.6377	1.6065	1.5732	1.5376	1.4989	1.4564
1.7627	1.7146	1.6624	1.6052	1.5741	1.5411	1.5056	1.4672	1.4248	1.3769
1.7070	1.6574	1.6034	1.5435	1.5107	1.4755	1.4373	1.3952	1.3476	1.2915
1.6524	1.6012	1.5450	1.4821	1.4472	1.4094	1.3676	1.3203	1.2646	1.1926
1.5987	1.5458	1.4871	1.4206	1.3832	1.3419	1.2951	1.2400	1.1686	1.0000

TABLE A.5 (CONTINUED)

$\alpha = 5\%$

df_1 df_2	1	2	3	4	5	6	7	8	9
1	161.45	199.50	215.71	224.58	230.16	233.99	236.77	238.88	240.54
2	18.513	19.000	19.164	19.247	19.296	19.330	19.353	19.371	19.385
3	10.128	9.5521	9.2766	9.1172	9.0135	8.9406	8.8868	8.8452	8.8123
4	7.7086	6.9443	6.5914	6.3883	6.2560	6.1631	6.0942	6.0410	5.9988
5	6.6079	5.7861	5.4095	5.1922	5.0503	4.9503	4.8759	4.8183	4.7725
6	5.9874	5.1433	4.7571	4.5337	4.3874	4.2839	4.2066	4.1468	4.0990
7	5.5914	4.7374	4.3468	4.1203	3.9715	3.8660	3.7870	3.7257	3.6767
8	5.3177	4.4590	4.0662	3.8378	3.6875	3.5806	3.5005	3.4381	3.3881
9	5.1174	4.2565	3.8626	3.6331	3.4817	3.3738	3.2927	3.2296	3.1789
10	4.9646	4.1028	3.7083	3.4780	3.3258	3.2172	3.1355	3.0717	3.0204
11	4.8443	3.9823	3.5874	3.3567	3.2039	3.0946	3.0123	2.9480	2.8962
12	4.7472	3.8853	3.4903	3.2592	3.1059	2.9961	2.9134	2.8486	2.7964
13	4.6672	3.8056	3.4105	3.1791	3.0254	2.9153	2.8321	2.7669	2.7144
14	4.6001	3.7389	3.3439	3.1122	2.9582	2.8477	2.7642	2.6987	2.6458
15	4.5431	3.6823	3.2874	3.0556	2.9013	2.7905	2.7066	2.6408	2.5876
16	4.4940	3.6337	3.2389	3.0069	2.8524	2.7413	2.6572	2.5911	2.5377
17	4.4513	3.5915	3.1968	2.9647	2.8100	2.6987	2.6143	2.5480	2.4943
18	4.4139	3.5546	3.1599	2.9277	2.7729	2.6613	2.5767	2.5102	2.4563
19	4.3808	3.5219	3.1274	2.8951	2.7401	2.6283	2.5435	2.4768	2.4227
20	4.3513	3.4928	3.0984	2.8661	2.7109	2.5990	2.5140	2.4471	2.3928
21	4.3248	3.4668	3.0725	2.8401	2.6848	2.5727	2.4876	2.4205	2.3661
22	4.3009	3.4434	3.0491	2.8167	2.6613	2.5491	2.4638	2.3965	2.3419
23	4.2793	3.4221	3.0280	2.7955	2.6400	2.5277	2.4422	2.3748	2.3201
24	4.2597	3.4028	3.0088	2.7763	2.6207	2.5082	2.4226	2.3551	2.3002
25	4.2417	3.3852	2.9912	2.7587	2.6030	2.4904	2.4047	2.3371	2.2821
26	4.2252	3.3690	2.9751	2.7426	2.5868	2.4741	2.3883	2.3205	2.2655
27	4.2100	3.3541	2.9604	2.7278	2.5719	2.4591	2.3732	2.3053	2.2501
28	4.1960	3.3404	2.9467	2.7141	2.5581	2.4453	2.3593	2.2913	2.2360
29	4.1830	3.3277	2.9340	2.7014	2.5454	2.4324	2.3463	2.2782	2.2229
30	4.1709	3.3158	2.9223	2.6896	2.5336	2.4205	2.3343	2.2662	2.2107
40	4.0848	3.2317	2.8387	2.6060	2.4495	2.3359	2.2490	2.1802	2.1240
60	4.0012	3.1504	2.7581	2.5252	2.3683	2.2540	2.1665	2.0970	2.0401
120	3.9201	3.0718	2.6802	2.4472	2.2900	2.1750	2.0867	2.0164	1.9588
∞	3.8415	2.9957	2.6049	2.3719	2.2141	2.0986	2.0096	1.9384	1.8799

$\alpha = 5\%$

10	12	15	20	24	30	40	60	120	∞
241.88	243.91	245.95	248.01	249.05	250.09	251.14	252.20	253.25	254.32
19.396	19.413	19.429	19.446	19.454	19.462	19.471	19.479	19.487	19.496
8.7855	8.7446	8.7029	8.6602	8.6385	8.6166	8.5944	8.5720	8.5494	8.5265
5.9644	5.9117	5.8578	5.8025	5.7744	5.7459	5.7170	5.6878	5.6581	5.6281
4.7351	4.6777	4.6188	4.5581	4.5272	4.4957	4.4638	4.4314	4.3984	4.3650
4.0600	3.9999	3.9381	3.8742	3.8415	3.8082	3.7743	3.7398	3.7047	3.6688
3.6365	3.5747	3.5108	3.4445	3.4105	3.3758	3.3404	3.3043	3.2674	3.2298
3.3472	3.2840	3.2184	3.1503	3.1152	3.0794	3.0428	3.0053	2.9669	2.9276
3.1373	3.0729	3.0061	2.9365	2.9005	2.8637	2.8259	2.7872	2.7475	2.7067
2.9782	2.9130	2.8450	2.7740	2.7372	2.6996	2.6609	2.6211	2.5801	2.5379
2.8536	2.7876	2.7186	2.6464	2.6090	2.5705	2.5309	2.4901	2.4480	2.4045
2.7534	2.6866	2.6169	2.5436	2.5055	2.4663	2.4259	2.3842	2.3410	2.2962
2.6710	2.6037	2.5331	2.4589	2.4202	2.3803	2.3392	2.2966	2.2524	2.2064
2.6021	2.5342	2.4630	2.3879	2.3487	2.3082	2.2664	2.2230	2.1778	2.1307
2.5437	2.4753	2.4035	2.3275	2.2878	2.2468	2.2043	2.1601	2.1141	2.0658
2.4935	2.4247	2.3522	2.2756	2.2354	2.1938	2.1507	2.1058	2.0589	2.0096
2.4499	2.3807	2.3077	2.2304	2.1898	2.1477	2.1040	2.0584	2.0107	1.9604
2.4117	2.3421	2.2686	2.1906	2.1497	2.1071	2.0629	2.0166	1.9681	1.9168
2.3779	2.3080	2.2341	2.1555	2.1141	2.0712	2.0264	1.9796	1.9302	1.8780
2.3479	2.2776	2.2033	2.1242	2.0825	2.0391	1.9938	1.9464	1.8963	1.8432
2.3210	2.2504	2.1757	2.0960	2.0540	2.0102	1.9645	1.9165	1.8657	1.8117
2.2967	2.2258	2.1508	2.0707	2.0283	1.9842	1.9380	1.8895	1.8380	1.7831
2.2747	2.2036	2.1282	2.0476	2.0050	1.9605	1.9139	1.8649	1.8128	1.7570
2.2547	2.1834	2.1077	2.0267	1.9838	1.9390	1.8920	1.8424	1.7897	1.7331
2.2365	2.1649	2.0889	2.0075	1.9643	1.9192	1.8718	1.8217	1.7684	1.7110
2.2197	2.1479	2.0716	1.9898	1.9464	1.9010	1.8533	1.8027	1.7488	1.6906
2.2043	2.1323	2.0558	1.9736	1.9299	1.8842	1.8361	1.7851	1.7307	1.6717
2.1900	2.1179	2.0411	1.9586	1.9147	1.8687	1.8203	1.7689	1.7138	1.6541
2.1768	2.1045	2.0275	1.9446	1.9005	1.8543	1.8055	1.7537	1.6981	1.6377
2.1646	2.0921	2.0148	1.9317	1.8874	1.8409	1.7918	1.7396	1.6835	1.6223
2.0772	2.0035	1.9245	1.8389	1.7929	1.7444	1.6928	1.6373	1.5766	1.5089
1.9926	1.9174	1.8364	1.7480	1.7001	1.6491	1.5943	1.5343	1.4673	1.3893
1.9105	1.8337	1.7505	1.6587	1.6084	1.5543	1.4952	1.4290	1.3519	1.2539
1.8307	1.7522	1.6664	1.5705	1.5173	1.4591	1.3940	1.3180	1.2214	1.0000

TABLE A.5 (CONTINUED)

$\alpha = 1\%$

df_1 / df_2	1	2	3	4	5	6	7	8	9
1	4052.2	4999.5	5403.3	5624.6	5763.7	5859.0	5928.3	5981.6	6022.5
2	98.503	99.000	99.166	99.249	99.299	99.332	99.356	99:374	99.388
3	34.116	30.817	29.457	28.710	28.237	27.911	27.672	27.489	27.345
4	21.198	18.000	16.694	15.977	15.522	15.207	14.976	14.799	14.659
5	16.258	13.274	12.060	11.392	10.967	10.672	10.456	10.289	10.158
6	13.745	10.925	9.7795	9.1483	8.7459	8.4661	8.2600	8.1016	7.9761
7	12.246	9.5466	8.4513	7.8467	7.4604	7.1914	6.9928	6.8401	6.7188
8	11.259	8.6491	7.5910	7.0060	6.6318	6.3707	6.1776	6.0289	5.9106
9	10.561	8.0215	6.9919	6.4221	6.0569	5.8018	5.6129	5.4671	5.3511
10	10.044	7.5594	6.5523	5.9943	5.6363	5.3858	5.2001	5.0567	4.9424
11	9.6460	7.2057	6.2167	5.6683	5.3160	5.0692	4.8861	4.7445	4.6315
12	9.3302	6.9266	5.9526	5.4119	5.0643	4.8206	4.6395	4.4994	4.3875
13	9.0738	6.7010	5.7394	5.2053	4.8616	4.6204	4.4410	4.3021	4.1911
14	8.8616	6.5149	5.5639	5.0354	4.6950	4.4558	4.2779	4.1399	4.0297
15	8.6831	6.3589	5.4170	4.8932	4.5556	4.3183	4.1415	4.0045	3.8948
16	8.5310	6.2262	5.2922	4.7726	4.4374	4.2016	4.0259	3.8896	3.7804
17	8.3997	6.1121	5.1850	4.6690	4.3359	4.1015	3.9267	3.7910	3.6822
18	8.2854	6.0129	5.0919	4.5790	4.2479	4.0146	3.8406	3.7054	3.5971
19	8.1850	5.9259	5.0103	4.5003	4.17C8	3.9386	3.7653	3.6305	3.5225
20	8.0960	5.8489	4.9382	4.4307	4.1027	3.8714	3.6987	3.5644	3.4567
21	8.0166	5.7804	4.8740	4.3688	4.0421	3.8117	3.6396	3.5056	3.3981
22	7.9454	5.7190	4.8166	4.3134	3.9880	3.7583	3.5867	3.4530	3.3458
23	7.8811	5.6637	4.7649	4.2635	3.9392	3.7102	3.5390	3.4057	3.2986
24	7.8229	5.6136	4.7181	4.2184	3.8951	3.6667	3.4959	3.3629	3.2560
25	7.7698	5.5680	4.6755	4.1774	3.8550	3.6272	3.4568	3.3239	3.2172
26	7.7213	5.5263	4.6366	4.1400	3.8183	3.5911	3.4210	3.2884	3.1818
27	7.6767	5.4881	4.6009	4.1056	3.7848	3.5580	3.3882	3.2558	3.1494
28	7.6356	5.4529	4.5681	4.0740	3.7539	3.5276	3.3581	3.2259	3.1195
29	7.5976	5.4205	4.5378	4.0449	3.7254	3.4995	3.3302	3.1982	3.0920
30	7.5625	5.3904	4.5097	4.0179	3.6990	3.4735	3.3045	3.1726	3.0665
40	7.3141	5.1785	4.3126	3.8283	3.5138	3.2910	3.1238	2.9930	2.8876
60	7.0771	4.9774	4.1259	3.6491	3.3389	3.1187	2.9530	2.8233	2.7185
120	6.8510	4.7865	3.9493	3.4796	3.1735	2.9559	2.7918	2.6629	2.5586
∞	6.6349	4.6052	3.7816	3.3192	3.0173	2.8020	2.6393	2.5113	2.4073

$\alpha = 1\%$

10	12	15	20	24	30	40	60	120	∞
6055.8	6106.3	6157.3	6208.7	6234.6	6260.7	6286.8	6313.0	6339.4	6366.0
99.399	99.416	99.432	99.449	99.458	99.466	99.474	99.483	99.491	99.501
27.229	27.052	26.872	26.690	26.598	26.505	26.411	26.316	26.221	26.125
14.546	14.374	14.198	14.020	13.929	13.838	13.745	13.652	13.558	13.463
10.051	9.8883	9.7222	9.5527	9.4665	9.3793	9.2912	9.2020	9.1118	9.0204
7.8741	7.7183	7.5590	7.3958	7.3127	7.2285	7.1432	7.0568	6.9690	6.8801
6.6201	6.4691	6.3143	6.1554	6.0743	5.9921	5.9084	5.8236	5.7372	5.6495
5.8143	5.6668	5.5151	5.3591	5.2793	5.1981	5.1156	5.0316	4.9460	4.8588
5.2565	5.1114	4.9621	4.8080	4.7290	4.6486	4.5667	4.4831	4.3978	4.3105
4.8492	4.7059	4.5582	4.4054	4.3269	4.2469	4.1653	4.0819	3.9965	3.9090
4.5393	4.3974	4.2509	4.0990	4.0209	3.9411	3.8596	3.7761	3.6904	3.6025
4.2961	4.1553	4.0096	3.8584	3.7805	3.7008	3.6192	3.5355	3.4494	3.3608
4.1003	3.9603	3.8154	3.6646	3.5868	3.5070	3.4253	3.3413	3.2548	3.1654
3.9394	3.8001	3.6557	3.5052	3.4274	3.3476	3.2656	3.1813	3.0942	3.0040
3.8049	3.6662	3.5222	3.3719	3.2940	3.2141	3.1319	3.0471	2.9595	2.8684
3.6909	3.5527	3.4089	3.2588	3.1808	3.1007	3.0182	2.9330	2.8447	2.7528
3.5931	3.4552	3.3117	3.1615	3.0835	3.0032	2.9205	2.8348	2.7459	2.6530
3.5082	3.3706	3.2273	3.0771	2.9990	2.9185	2.8354	2.7493	2.6597	2.5660
3.4338	3.2965	3.1533	3.0031	2.9249	2.8442	2.7608	2.6742	2.5839	2.4893
3.3682	3.2311	3.0880	2.9377	2.8594	2.7785	2.6947	2.6077	2.5168	2.4212
3.3098	3.1729	3.0299	2.8796	2.8011	2.7200	2.6359	2.5484	2.4568	2.3603
3.2576	3.1209	2.9780	2.8274	2.7488	2.6675	2.5831	2.4951	2.4029	2.3055
3.2106	3.0740	2.9311	2.7805	2.7017	2.6202	2.5355	2.4471	2.3542	2.2559
3.1681	3.0316	2.8887	2.7380	2.6591	2.5773	2.4923	2.4035	2.3099	2.2107
3.1294	2.9931	2.8502	2.6993	2.6203	2.5383	2.4530	2.3637	2.2695	2.1694
3.0941	2.9579	2.8150	2.6640	2.5848	2.5026	2.4170	2.3273	2.2325	2.1315
3.0618	2.9256	2.7827	2.6316	2.5522	2.4699	2.3840	2.2938	2.1984	2.0965
3.0320	2.8959	2.7530	2.6017	2.5223	2.4397	2.3535	2.2629	2.1670	2.0642
3.0045	2.8685	2.7256	2.5742	2.4946	2.4118	2.3253	2.2344	2.1378	2.0342
2.9791	2.8431	2.7002	2.5487	2.4689	2.3860	2.2992	2.2079	2.1107	2.0062
2.8005	2.6648	2.5216	2.3689	2.2880	2.2034	2.1142	2.0194	1.9172	1.8047
2.6318	2.4961	2.3523	2.1978	2.1154	2.0285	1.9360	1.8363	1.7263	1.6006
2.4721	2.3363	2.1915	2.0346	1.9500	1.8600	1.7628	1.6557	1.5330	1.3805
2.3209	2.1848	2.0385	1.8783	1.7908	1.6964	1.5923	1.4730	1.3246	1.0000

* From Table 18, Biometrika Tables for Statisticians, Volume 1, Second Edition, Cambridge University Press, 1962.

TABLE A.6. PERCENTILES OF THE STUDENTIZED RANGE *

$$\alpha = 10\%$$

df \ k	2	3	4	5	6	7	8	9	10
1	8.93	13.44	16.36	18.49	20.15	21.51	22.64	23.62	24.48
2	4.13	5.73	6.77	7.54	8.14	8.63	9.05	9.41	9.72
3	3.33	4.47	5.20	5.74	6.16	6.51	6.81	7.06	7.29
4	3.01	3.98	4.59	5.03	5.39	5.68	5.93	6.14	6.33
5	2.85	3.72	4.26	4.66	4.98	5.24	5.46	5.65	5.82
6	2.75	3.56	4.07	4.44	4.73	4.97	5.17	5.34	5.50
7	2.68	3.45	3.93	4.28	4.55	4.78	4.97	5.14	5.28
8	2.63	3.37	3.83	4.17	4.43	4.65	4.83	4.99	5.13
9	2.59	3.32	3.76	4.08	4.34	4.54	4.72	4.87	5.01
10	2.56	3.27	3.70	4.02	4.26	4.47	4.64	4.78	4.91
11	2.54	3.23	3.66	3.96	4.20	4.40	4.57	4.71	4.84
12	2.52	3.20	3.62	3.92	4.16	4.35	4.51	4.65	4.78
13	2.50	3.18	3.59	3.88	4.12	4.30	4.46	4.60	4.72
14	2.49	3.16	3.56	3.85	4.08	4.27	4.42	4.56	4.68
15	2.48	3.14	3.54	3.83	4.05	4.23	4.39	4.52	4.64
16	2.47	3.12	3.52	3.80	4.03	4.21	4.36	4.49	4.61
17	2.46	3.11	3.50	3.78	4.00	4.18	4.33	4.46	4.58
18	2.45	3.10	3.49	3.77	3.98	4.16	4.31	4.44	4.55
19	2.45	3.09	3.47	3.75	3.97	4.14	4.29	4.42	4.53
20	2.44	3.08	3.46	3.74	3.95	4.12	4.27	4.40	4.51
24	2.42	3.05	3.42	3.69	3.90	4.07	4.21	4.34	4.44
30	2.40	3.02	3.39	3.65	3.85	4.02	4.16	4.28	4.38
40	2.38	2.99	3.35	3.60	3.80	3.96	4.10	4.21	4.32
60	2.36	2.96	3.31	3.56	3.75	3.91	4.04	4.16	4.25
120	2.34	2.93	3.28	3.52	3.71	3.86	3.99	4.10	4.19
∞	2.33	2.90	3.24	3.48	3.66	3.81	3.93	4.04	4.13

TABLE A.6 (CONTINUED)

df \ k	11	12	13	14	15	16	17	18	19	20
1	25.24	25.92	26.54	27.10	27.62	28.10	28.54	28.96	29.35	29.71
2	10.01	10.26	10.49	10.70	10.89	11.07	11.24	11.39	11.54	11.68
3	7.49	7.67	7.83	7.98	8.12	8.25	8.37	8.48	8.58	8.68
4	6.49	6.65	6.78	6.91	7.02	7.13	7.23	7.33	7.41	7.50
5	5.97	6.10	6.22	6.34	6.44	6.54	6.63	6.71	6.79	6.86
6	5.64	5.76	5.87	5.98	6.07	6.16	6.25	6.32	6.40	6.47
7	5.41	5.53	5.64	5.74	5.83	5.91	5.99	6.06	6.13	6.19
8	5.25	5.36	5.46	5.56	5.64	5.72	5.80	5.87	5.93	6.00
9	5.13	5.23	5.33	5.42	5.51	5.58	5.66	5.72	5.79	5.85
10	5.03	5.13	5.23	5.32	5.40	5.47	5.54	5.61	5.67	5.73
11	4.95	5.05	5.15	5.23	5.31	5.38	5.45	5.51	5.57	5.63
12	4.89	4.99	5.08	5.16	5.24	5.31	5.37	5.44	5.49	5.55
13	4.83	4.93	5.02	5.10	5.18	5.25	5.31	5.37	5.43	5.48
14	4.79	4.88	4.97	5.05	5.12	5.19	5.26	5.32	5.37	5.43
15	4.75	4.84	4.93	5.01	5.08	5.15	5.21	5.27	5.32	5.38
16	4.71	4.81	4.89	4.97	5.04	5.11	5.17	5.23	5.28	5.33
17	4.68	4.77	4.86	4.93	5.01	5.07	5.13	5.19	5.24	5.30
18	4.65	4.75	4.83	4.90	4.98	5.04	5.10	5.16	5.21	5.26
19	4.63	4.72	4.80	4.88	4.95	5.01	5.07	5.13	5.18	5.23
20	4.61	4.70	4.78	4.85	4.92	4.99	5.05	5.10	5.16	5.20
24	4.54	4.63	4.71	4.78	4.85	4.91	4.97	5.02	5.07	5.12
30	4.47	4.56	4.64	4.71	4.77	4.83	4.89	4.94	4.99	5.03
40	4.41	4.49	4.56	4.63	4.69	4.75	4.81	4.86	4.90	4.95
60	4.34	4.42	4.49	4.56	4.62	4.67	4.73	4.78	4.82	4.86
120	4.28	4.35	4.42	4.48	4.54	4.60	4.65	4.69	4.74	4.78
∞	4.21	4.28	4.35	4.41	4.47	4.52	4.57	4.61	4.65	4.69

TABLE A.6 (CONTINUED)

$\alpha = 5\%$

df \ k	2	3	4	5	6	7	8	9	10
1	17.97	26.98	32.82	37.08	40.41	43.12	45.40	47.36	49.07
2	6.08	8.33	9.80	10.88	11.74	12.44	13.03	13.54	13.99
3	4.50	5.91	6.82	7.50	8.04	8.48	8.85	9.18	9.46
4	3.93	5.04	5.76	6.29	6.71	7.05	7.35	7.60	7.83
5	3.64	4.60	5.22	5.67	6.03	6.33	6.58	6.80	6.99
6	3.46	4.34	4.90	5.30	5.63	5.90	6.12	6.32	6.49
7	3.34	4.16	4.68	5.06	5.36	5.61	5.82	6.00	6.16
8	3.26	4.04	4.53	4.89	5.17	5.40	5.60	5.77	5.92
9	3.20	3.95	4.41	4.76	5.02	5.24	5.43	5.59	5.74
10	3.15	3.88	4.33	4.65	4.91	5.12	5.30	5.46	5.60
11	3.11	3.82	4.26	4.57	4.82	5.03	5.20	5.35	5.49
12	3.08	3.77	4.20	4.51	4.75	4.95	5.12	5.27	5.39
13	3.06	3.73	4.15	4.45	4.69	4.88	5.05	5.19	5.32
14	3.03	3.70	4.11	4.41	4.64	4.83	4.99	5.13	5.25
15	3.01	3.67	4.08	4.37	4.59	4.78	4.94	5.08	5.20
16	3.00	3.65	4.05	4.33	4.56	4.74	4.90	5.03	5.15
17	2.98	3.63	4.02	4.30	4.52	4.70	4.86	4.99	5.11
18	2.97	3.61	4.00	4.28	4.49	4.67	4.82	4.96	5.07
19	2.96	3.59	3.98	4.25	4.47	4.65	4.79	4.92	5.04
20	2.95	3.58	3.96	4.23	4.45	4.62	4.77	4.90	5.01
24	2.92	3.53	3.90	4.17	4.37	4.54	4.68	4.81	4.92
30	2.89	3.49	3.85	4.10	4.30	4.46	4.60	4.72	4.82
40	2.86	3.44	3.79	4.04	4.23	4.39	4.52	4.63	4.73
60	2.83	3.40	3.74	3.98	4.16	4.31	4.44	4.55	4.65
120	2.80	3.36	3.68	3.92	4.10	4.24	4.36	4.47	4.56
∞	2.77	3.31	3.63	3.86	4.03	4.17	4.29	4.39	4.47

TABLE A.6 (CONTINUED)

df \ k	11	12	13	14	15	16	17	18	19	20
1	50.59	51.96	53.20	54.33	55.36	56.32	57.22	58.04	58.83	59.56
2	14.39	14.75	15.08	15.38	15.65	15.91	16.14	16.37	16.57	16.77
3	9.72	9.95	10.15	10.35	10.52	10.69	10.84	10.98	11.11	11.24
4	8.03	8.21	8.37	8.52	8.66	8.79	8.91	9.03	9.13	9.23
5	7.17	7.32	7.47	7.60	7.72	7.83	7.93	8.03	8.12	8.21
6	6.65	6.79	6.92	7.03	7.14	7.24	7.34	7.43	7.51	7.59
7	6.30	6.43	6.55	6.66	6.76	6.85	6.94	7.02	7.10	7.17
8	6.05	6.18	6.29	6.39	6.48	6.57	6.65	6.73	6.80	6.87
9	5.87	5.98	6.09	6.19	6.28	6.36	6.44	6.51	6.58	6.64
10	5.72	5.83	5.93	6.03	6.11	6.19	6.27	6.34	6.40	6.47
11	5.61	5.71	5.81	5.90	5.98	6.06	6.13	6.20	6.27	6.33
12	5.51	5.61	5.71	5.80	5.88	5.95	6.02	6.09	6.15	6.21
13	5.43	5.53	5.63	5.71	5.79	5.86	5.93	5.99	6.05	6.11
14	5.36	5.46	5.55	5.64	5.71	5.79	5.85	5.91	5.97	6.03
15	5.31	5.40	5.49	5.57	5.65	5.72	5.78	5.85	5.90	5.96
16	5.26	5.35	5.44	5.52	5.59	5.66	5.73	5.79	5.84	5.90
17	5.21	5.31	5.39	5.47	5.54	5.61	5.67	5.73	5.79	5.84
18	5.17	5.27	5.35	5.43	5.50	5.57	5.63	5.69	5.74	5.79
19	5.14	5.23	5.31	5.39	5.46	5.53	5.59	5.65	5.70	5.75
20	5.11	5.20	5.28	5.36	5.43	5.49	5.55	5.61	5.66	5.71
24	5.01	5.10	5.18	5.25	5.32	5.38	5.44	5.49	5.55	5.59
30	4.92	5.00	5.08	5.15	5.21	5.27	5.33	5.38	5.43	5.47
40	4.82	4.90	4.98	5.04	5.11	5.16	5.22	5.27	5.31	5.36
60	4.73	4.81	4.88	4.94	5.00	5.06	5.11	5.15	5.20	5.24
120	4.64	4.71	4.78	4.84	4.90	4.95	5.00	5.04	5.09	5.13
∞	4.55	4.62	4.68	4.74	4.80	4.85	4.89	4.93	4.97	5.01

TABLE A.6 (CONTINUED)

$$\alpha = 1\%$$

k df	2	3	4	5	6	7	8	9	10
1	90.03	135.0	164.3	185.6	202.2	215.8	227.2	237.0	245.6
2	14.04	19.02	22.29	24.72	26.63	28.20	29.53	30.68	31.69
3	8.26	10.62	12.17	13.33	14.24	15.00	15.64	16.20	16.69
4	6.51	8.12	9.17	9.96	10.58	11.10	11.55	11.93	12.27
5	5.70	6.98	7.80	8.42	8.91	9.32	9.67	9.97	10.24
6	5.24	6.33	7.03	7.56	7.97	8.32	8.61	8.87	9.10
7	4.95	5.92	6.54	7.01	7.37	7.68	7.94	8.17	8.37
8	4.75	5.64	6.20	6.62	6.96	7.24	7.47	7.68	7.86
9	4.60	5.43	5.96	6.35	6.66	6.91	7.13	7.33	7.49
10	4.48	5.27	5.77	6.14	6.43	6.67	6.87	7.05	7.21
11	4.39	5.15	5.62	5.97	6.25	6.48	6.67	6.84	6.99
12	4.32	5.05	5.50	5.84	6.10	6.32	6.51	6.67	6.81
13	4.26	4.96	5.40	5.73	5.98	6.19	6.37	6.53	6.67
14	4.21	4.89	5.32	5.63	5.88	6.08	6.26	6.41	6.54
15	4.17	4.84	5.25	5.56	5.80	5.99	6.16	6.31	6.44
16	4.13	4.79	5.19	5.49	5.72	5.92	6.08	6.22	6.35
17	4.10	4.74	5.14	5.43	5.66	5.85	6.01	6.15	6.27
18	4.07	4.70	5.09	5.38	5.60	5.79	5.94	6.08	6.20
19	4.05	4.67	5.05	5.33	5.55	5.73	5.89	6.02	6.14
20	4.02	4.64	5.02	5.29	5.51	5.69	5.84	5.97	6.09
24	3.96	4.55	4.91	5.17	5.37	5.54	5.69	5.81	5.92
30	3.89	4.45	4.80	5.05	5.24	5.40	5.54	5.65	5.76
40	3.82	4.37	4.70	4.93	5.11	5.26	5.39	5.50	5.60
60	3.76	4.28	4.59	4.82	4.99	5.13	5.25	5.36	5.45
120	3.70	4.20	4.50	4.71	4.87	5.01	5.12	5.21	5.30
∞	3.64	4.12	4.40	4.60	4.76	4.88	4.99	5.08	5.16

TABLE A.6 (CONTINUED)

df \ k	11	12	13	14	15	16	17	18	19	20
1	253.2	260.0	266.2	271.8	277.0	281.8	286.3	290.4	294.3	298.0
2	32.59	33.40	34.13	34.81	35.43	36.00	36.53	37.03	37.50	37.95
3	17.13	17.53	17.89	18.22	18.52	18.81	19.07	19.32	19.55	19.77
4	12.57	12.84	13.09	13.32	13.53	13.73	13.91	14.08	14.24	14.40
5	10.48	10.70	10.89	11.08	11.24	11.40	11.55	11.68	11.81	11.93
6	9.30	9.48	9.65	9.81	9.95	10.08	10.21	10.32	10.43	10.54
7	8.55	8.71	8.86	9.00	9.12	9.24	9.35	9.46	9.55	9.65
8	8.03	8.18	8.31	8.44	8.55	8.66	8.76	8.85	8.94	9.03
9	7.65	7.78	7.91	8.03	8.13	8.23	8.33	8.41	8.49	8.57
10	7.36	7.49	7.60	7.71	7.81	7.91	7.99	8.08	8.15	8.23
11	7.13	7.25	7.36	7.46	7.56	7.65	7.73	7.81	7.88	7.95
12	6.94	7.06	7.17	7.26	7.36	7.44	7.52	7.59	7.66	7.73
13	6.79	6.90	7.01	7.10	7.19	7.27	7.35	7.42	7.48	7.55
14	6.66	6.77	6.87	6.96	7.05	7.13	7.20	7.27	7.33	7.39
15	6.55	6.66	6.76	6.84	6.93	7.00	7.07	7.14	7.20	7.26
16	6.46	6.56	6.66	6.74	6.82	6.90	6.97	7.03	7.09	7.15
17	6.38	6.48	6.57	6.66	6.73	6.81	6.87	6.94	7.00	7.05
18	6.31	6.41	6.50	6.58	6.65	6.73	6.79	6.85	6.91	6.97
19	6.25	6.34	6.43	6.51	6.58	6.65	6.72	6.78	6.84	6.89
20	6.19	6.28	6.37	6.45	6.52	6.59	6.65	6.71	6.77	6.82
24	6.02	6.11	6.19	6.26	6.33	6.39	6.45	6.51	6.56	6.61
30	5.85	5.93	6.01	6.08	6.14	6.20	6.26	6.31	6.36	6.41
40	5.69	5.76	5.83	5.90	5.96	6.02	6.07	6.12	6.16	6.21
60	5.53	5.60	5.67	5.73	5.78	5.84	5.89	5.93	5.97	6.01
120	5.37	5.44	5.50	5.56	5.61	5.66	5.71	5.75	5.79	5.83
∞	5.23	5.29	5.35	5.40	5.45	5.49	5.54	5.57	5.61	5.65

* From Table of Upper Percentage Points of the Studentized Range, James Pachares, Biometrika, Vol 46, parts 3–4, 1959.

TABLE A.7. MULTIPLYING FACTOR FOR THE GEOMETRIC MEAN *

				Sample Size				
T	2	5	8	10	13	15	25	50
0.05	1.025	1.041	1.045	1.046	1.047	1.048	1.049	1.050
0.10	1.050	1.082	1.091	1.093	1.096	1.097	1.100	1.103
0.15	1.076	1.125	1.138	1.143	1.147	1.149	1.154	1.158
0.20	1.102	1.169	1.187	1.194	1.200	1.203	1.210	1.216
0.25	1.128	1.214	1.238	1.247	1.255	1.259	1.268	1.276
0.30	1.154	1.260	1.291	1.302	1.312	1.317	1.330	1.340
0.35	1.180	1.307	1.345	1.359	1.372	1.378	1.393	1.406
0.40	1.207	1.356	1.401	1.418	1.433	1.441	1.460	1.476
0.45	1.234	1.406	1.459	1.479	1.498	1.506	1.530	1.548
0.50	1.261	1.457	1.519	1.542	1.564	1.574	1.602	1.625
0.55	1.288	1.509	1.581	1.608	1.633	1.645	1.678	1.705
0.60	1.315	1.563	1.645	1.675	1.705	1.719	1.757	1.789
0.65	1.343	1.618	1.711	1.746	1.780	1.796	1.840	1.876
0.70	1.371	1.675	1.779	1.818	1.857	1.876	1.926	1.968
0.75	1.399	1.733	1.849	1.894	1.938	1.958	2.016	2.064
0.80	1.427	1.792	1.922	1.971	2.021	2.045	2.110	2.165
0.85	1.456	1.853	1.996	2.052	2.108	2.134	2.208	2.270
0.90	1.485	1.915	2.074	2.135	2.197	2.227	2.310	2.381
0.95	1.514	1.979	2.153	2.221	2.291	2.323	2.417	2.496
1.00	1.543	2.044	2.235	2.310	2.387	2.424	2.528	2.617
1.05	1.573	2.111	2.320	2.403	2.487	2.528	2.644	2.744
1.10	1.602	2.180	2.407	2.498	2.591	2.636	2.765	2.876
1.15	1.632	2.250	2.497	2.596	2.698	2.748	2.891	3.014
1.20	1.662	2.321	2.589	2.698	2.810	2.864	3.022	3.159
1.25	1.693	2.395	2.685	2.803	2.926	2.985	3.159	3.311
1.30	1.724	2.470	2.783	2.911	3.045	3.111	3.301	3.470
1.35	1.754	2.547	2.884	3.023	3.169	3.241	3.450	3.636
1.40	1.786	2.626	2.988	3.139	3.298	3.376	3.604	3.809
1.45	1.817	2.706	3.096	3.259	3.431	3.515	3.766	3.991
1.50	1.849	2.788	3.206	3.382	3.569	3.661	3.933	4.181
1.55	1.880	2.873	3.320	3.510	3.711	3.811	4.108	4.379
1.60	1.913	2.959	3.437	3.642	3.859	3.967	4.291	4.587
1.65	1.945	3.047	3.558	3.777	4.012	4.129	4.480	4.804
1.70	1.977	3.137	3.682	3.918	4.171	4.297	4.678	5.031
1.75	2.010	3.229	3.810	4.062	4.334	4.471	4.883	5.269
1.80	2.043	3.323	3.942	4.212	4.504	4.651	5.097	5.517
1.85	2.077	3.420	4.077	4.366	4.680	4.838	5.320	5.776
1.90	2.110	3.518	4.216	4.525	4.861	5.031	5.552	6.048
1.95	2.144	3.619	4.359	4.688	5.049	5.232	5.794	6.331
2.00	2.178	3.721	4.506	4.857	5.243	5.439	6.045	6.628

* From The Lognormal Frequency Distribution in Relation to Gold Assay Data, U.S. Bureau of Mines Report of Investigations, in press

TABLE A.8. PERCENTILES OF THE DISTRIBUTION OF r WHEN $\rho = 0$

N	$r_{.95}$	$r_{.975}$	$r_{.99}$	$r_{.995}$	$r_{.9995}$	N	$r_{.95}$	$r_{.975}$	$r_{.99}$	$r_{.995}$	$r_{.9995}$
5	.805	.878	.934	.959	.991	20	.378	.444	.516	.561	.679
6	.729	.811	.882	.917	.974	22	.360	.423	.492	.537	.652
7	.669	.754	.833	.875	.951	24	.344	.404	.472	.515	.629
8	.621	.707	.789	.834	.925	26	.330	.388	.453	.496	.607
9	.582	.666	.750	.798	.898	28	.317	.374	.437	.479	.588
10	.549	.632	.715	.765	.872	30	.306	.361	.423	.463	.570
11	.521	.602	.685	.735	.847	40	.264	.312	.366	.402	.501
12	.497	.576	.658	.708	.823	50	.235	.279	.328	.361	.451
13	.476	.553	.634	.684	.801	60	.214	.254	.300	.330	.414
14	.457	.532	.612	.661	.780	80	.185	.220	.260	.286	.361
15	.441	.514	.592	.641	.760	100	.165	.196	.232	.256	.324
16	.426	.497	.574	.623	.742	250	.104	.124	.147	.163	.207
17	.412	.482	.558	.606	.725	500	.074	.088	.104	.115	.147
18	.400	.468	.543	.590	.708	1000	.052	.062	.074	.081	.104
19	.389	.456	.529	.575	.693	∞	0	0	0	0	0
N	$-r_{.05}$	$-r_{.025}$	$-r_{.01}$	$-r_{.005}$	$-r_{.0005}$	N	$-r_{.05}$	$-r_{.025}$	$-r_{.01}$	$-r_{.005}$	$-r_{.0005}$

Greek Alphabet of Capital and Lower-Case Letters

alpha A α a	nu N ν	
beta B β	xi Ξ ξ	
gamma Γ γ	omicron O o	
delta Δ δ ∂	pi Π π	
epsilon E ϵ ε	rho P ρ	
zeta Z ζ	sigma Σ σ s	
eta H η	tau T τ	
theta Θ θ ϑ	upsilon Υ υ	
iota I ι	phi Φ ϕ φ	
kappa K κ	chi X χ	
lambda Λ λ	psi Ψ ψ	
mu M μ	omega Ω ω	

Partial List of Symbols

Q^2	single-degree-of-freedom statistic	169
r	correlation coefficient	279
s	sample standard deviation	51
s^2	sample variance	51
s_p^2	pooled sample variance	95
s_u^2	sample variance of logarithms	217
s.n.d.	standardized normal distribution	85
SS	sum of squares	51
t	student's t-statistic	87
T	sum total of observations in a sample	140
T	theoretical frequency of observations	212
T_r^2	sum of squares of the replication totals	159
T_t^2	sum of squares of the treatment totals	159
u	a transformed observation	52
u	logarithm of an observation	217
$\bar{u}$	sample mean of logarithms	217
V^2	a statistic to estimate the variance of lognormally distributed observations	217
w	observation	23
$\bar{w}$	sample mean	49

Greek Letters

α	percentage risk of type I error	107
α	mean of the lognormal distribution	217
β	percentage risk of type II error	107
β^2	variance of the lognormal distribution	217
γ	population coefficient of variation	54
δ	bias	80

Index

367